Information Science and Statistics

Series Editors

M. Jordan
R. Nowak
B. Schölkopf

Information Science and Statistics

object-oriented probabilistic network, 95–101
  instance tree, 100
  object-orientation
    classes, 96
    definition, 95
    inheritance, 96
    objects, 96
observation
  essential, 134
  requisite, 134
Occam's Razor, 184
Oil Wildcatter, 92
OOBN, *see* object-oriented probabilistic
    network
overfitting, 259, 281

P
pa, *see* parent
parameters
  elicitation of, 176
parent, 18
parent divorcing, 192
partial conflict, 295
partial order, 227
path, 19
  blocked in DAG, 33
  blocked in undirected graph, 19
  directed, 19
pattern, 241
PC algorithm, 246
PDAG, 241
perfect map, 61
perfect recall, *see* no-forgetting
poly-tree, 114
posterior probability distribution, *see*
    probability distribution, posterior
potential calculus, 50
prediction, 103
Prim's algorithm, 267
Probabilistic graphical model
  qualitative component, 69
  quantitative component, 69
  two-phase construction process, 69
probabilistic network
  characteristics, 147
  when to use, 146
probability
  conditional, 41
probability axioms, 40
probability calculus, 50
  chain rule of, 62
  distributive law of, 51
  fundamental rule of, 42, 54

probability distribution
  decomposition, 39
  for variables, 43
  graphical representation of conditional, 46
  joint, 39, 44
  marginal, 46
  posterior, 49
  recursive factorization of, 39
probability of future decisions, 138
probability potential, 46
  combination, 50
  division, 50
  marginalization, *see* marginalization
  normalization of, 47, 49
  vacuous, 48
process
  underlying distribution, 238
projection, *see* marginalization
propagation, *see* junction tree, propagation
proportional scaling, 312

R
reasoning
  abductive, 18, 30
  causal, 18
  deductive, 18, 30
  diagnostic, 18
  inter-causal, *see* explaining away
recursive factorization, *see* probability
    distribution, recursive factorization
    of
regularity constraint, 81
relevant network, 115
relevant variable, 135
retrieval of experience, 284
rule of total probability, 44
rule-based system, 5

S
score function
  decomposable, 261
selector variable, 197
sensitivity function, 312
sensitivity value, 315
serial connection, *see* causal network, serial
    connection in
Single Policy Updating, 137
skeleton, *see* graph, skeleton
smoothing, 103
soluble, *see* influence diagram, limited
    memory, soluble
Sore Throat, 219

sparse DAG, 273
SPU, *see* Single Policy Updating
stability assumption, 238
straw model, 292
structure
    elicitation of, 154
    uncertainty, 197
    verification, *see* model verification
structure learning
    greedy hill climbing, 257, 275
subclass, 101
subtype, *see* variable, subtype

T
tail, 118
temporal links, 102
temporal order, 196
test decision, 224
time-sliced Bayesian network, 103
topological ordering, 63
tree, 20
Tree-augmented Naive Bayes model, 269
Type I error, 245
Type II error, 245

U
undirected dependence relation, 201
undirected graph, *see* graph, undirected
unfolded network, 100
utility function, 79

V
v-structure, *see* collider
v-structures
    equivalent, 241
vacuous potential, *see* probability potential,
        vacuous
value function, *see* utility function, 329
value nodes, 80
value of information, 327
variable, 20–23

barren, 53, 88, 115
basic, 96
binding link, 97
bound, 97
category, 22
chance, 21
conditioned, 71
conditioning, 71
decision, 21
decision future, 84
decision history, 84
decision past, 84
designated state, 217
deterministic, 21
domain, 21
elimination, 115, 119
equivalence, 98
extremal decision, 139
hidden,latent, 239
identification of, 149
kind, 22
no impact state, 217
nuisance, 115
policy, 84
qualified name, 97
random, 21
scope, 97
set by intervention, 215
simple name, 97
strong marginal, 126
strong type checking, 98
subtype, 22, 180
synergy, 220
target, 115
taxonomy, 22
type, 152
vs. vertex, 21
weak marginal, 126
well-defined, 149
vertex
    symbols, 23
    vs. variable, 21
virtual cases, 281
virtual evidence, *see* evidence, soft

Uffe B. Kjærulff • Anders L. Madsen

Bayesian Networks and Influence Diagrams: A Guide to Construction and Analysis

Second Edition

Springer

Uffe B. Kjærulff
Department of Computer Science
Aalborg University
Aalborg, Denmark

Anders L. Madsen
HUGIN EXPERT A/S
Aalborg, Denmark

Department of Computer Science
Aalborg University
Aalborg, Denmark

ISSN 1613-9011
ISBN 978-1-4939-0029-9   ISBN 978-1-4614-5104-4 (eBook)
DOI 10.1007/978-1-4614-5104-4
Springer New York Heidelberg Dordrecht London

Printed on acid-free paper

Springer is part of Springer Science+Business Media (www.springer.com)

To our wives and children

Preface

This book is a monograph on practical aspects of probabilistic networks (a.k.a. probabilistic graphical models) and is intended to provide a comprehensive guide for practitioners who wish to understand, construct, and analyze decision support systems based on probabilistic networks, including a number of different variants of Bayesian networks and influence diagrams. This book consists of three parts:

- *Part I: Fundamentals of probabilistic networks*, including Chaps. 1–5, covering a brief introduction to probabilistic graphical models, the basic graph-theoretic terminology, the basic (Bayesian) probability theory, the key concepts of (conditional) dependence and independence, the different varieties of probabilistic networks, and the methods for making inference in these kinds of models. This part can be skipped by readers with fundamental knowledge about probabilistic networks.
- *Part II: Model construction*, including Chaps. 6–8, covering methods and techniques for elicitation of model structure and parameters, a large number of useful techniques and tricks to solve commonly recurring modeling problems, and methods for constructing probabilistic networks automatically from data, possibly through fusion of data and expert knowledge. Chapters 6 and 7 offer concrete advice and techniques on issues related to model construction, and Chap. 8 explains the theory and methods behind learning of Bayesian networks from data.
- *Part III: Model analysis*, including Chaps. 9–11, covering conflict analysis for detecting conflicting pieces of evidence (observations) or evidence that conflicts with the model, sensitivity analysis of a model with respect to both variations of evidence and model parameters, and value of information analysis. This part explains the theory and methods underlying the three different kinds of analyses.

Probabilistic networks have become an increasingly popular paradigm for probabilistic inference, addressing such tasks as diagnosis, prediction, decision making, classification, and data mining. From its infancy in the mid-1980s till today there has been a rapid development of algorithms for construction, inference, learning, and analysis of probabilistic networks, and since the turn of the millennium there has been a steep increase in the number of new applications of probabilistic networks. Its popularity stems from a number of factors:

- The graphical-based language for probabilistic networks is a powerful tool for expressing causal interactions while at the same time expressing dependence and independence relations among entities of a problem domain. Being graphical and compact the language furthermore provides an excellent intuitive means of communicating ideas among knowledge engineers and problem-domain experts.
- Although inference in complex probabilistic networks can be quite demanding (or even intractable), inference can often be performed efficiently in models of hundreds or even thousands of variables.
- Inference in probabilistic networks is based on a well-established theoretical foundation of probability calculus and decision theory and hence provides mathematically coherent methods for deriving conclusions under uncertainty where multiple sources of information and complex interaction patterns are involved.
- There exist efficient algorithms for learning and adaptation of probabilistic networks from data and it is possible to fuse data and expert knowledge.
- Probabilistic networks are "white boxes" in the sense that the model components (variables, links, probability and utility parameters) are open to interpretation, which makes it possible to perform a whole range of different analyses of the networks (e.g., conflict analysis, (in)dependence analyses, sensitivity analysis, and value of information analysis).
- There exist a number of powerful software tools that make it easy to construct and make inference in probabilistic networks.

As mentioned above, this book takes a practical point of departure and is intended primarily for those who wish to construct and analyze probabilistic networks without necessarily having a deep understanding neither of their underlying theory and methods nor of alternative paradigms supporting belief updating under uncertainty. Hence, the scope of this book is narrow, focusing almost exclusively on issues relevant for understanding, constructing, and analyzing the different variants of Bayesian networks and influence diagrams. Other methods for inference and support for decision making under uncertainty, therefore, get limited attention.

The intended audience of this book is practitioners as well as students of artificial intelligence, however, with a primary focus on the former. The theory behind probabilistic networks is explained to quite some depth, enabling the dedicated practitioner as well as the student to gain a solid theoretical understanding to a level sufficient to further develop and implement methods for model construction, inference, and analysis. Therefore, to support this understanding, exercises have been included in all chapters (except Chap. 1) for the reader to check his/her level of understanding. Answers to selected exercises and more can be found at hugin.com/developer/publications/bnid.

For a quick overview, the different kinds of probabilistic network models considered in this book can be characterized very briefly as follows:

- *Discrete Bayesian networks* represent factorizations of joint probability distributions over finite sets of discrete random variables. The variables are represented by the nodes of the network, and the links of the network represent the properties of (conditional) dependences and independences among the variables. A set of local probability distributions conditional on the configuration of the conditioning (parent) variables is specified for each variable.
- *Conditional linear Gaussian (CLG) Bayesian networks* represent factorizations of joint probability distributions over finite sets of random variables where some are discrete and some continuous. Each continuous variable is assumed to follow a linear Gaussian distribution conditional on the configuration of its discrete parent variables.
- *Discrete influence diagrams* are (discrete) Bayesian networks augmented with (discrete) decision variables and (discrete) utility functions. An influence diagram is capable of computing expected utilities of various decision options given the information known at the time of the decision.
- *Conditional linear-quadratic Gaussian (CLQG) influence diagrams* combine CLG Bayesian networks, discrete influence diagrams, and quadratic utility functions into a single framework supporting decision making under uncertainty with both continuous and discrete variables.
- *Limited-memory influence diagrams (LIMIDs)* relax two fundamental assumptions of influence diagrams: the no-forgetting assumption implying perfect recall of past observations and decisions and the assumption of a total order on the decisions. LIMIDs allow us to model more types of decision problems than the ordinary influence diagrams.
- *Object-oriented probabilistic networks* are hierarchically specified probabilistic networks (i.e., one of the above), allowing the knowledge engineer (model builder) to work on different levels of abstraction, as well as exploiting the usual concepts of encapsulation and inheritance known from object-oriented programming paradigms.

The book provides numerous examples, hopefully helping the reader to gain a good understanding of the various concepts, some of which are known to be hard to understand at a first encounter.

Even though probabilistic networks provide an intuitive language for constructing knowledge-based models for probabilistic inference, knowledge engineers can often benefit from a deeper understanding of the principles underlying these models. For example, knowing the rules for reading statements of dependence and independence encoded in the structure of a network may prove very valuable in evaluating whether the network correctly models the dependence and independence properties of the target problem. This, in turn, may be crucial to achieving, for example, correct posterior probability distributions from the model. Also, having a basic understanding of the relations between the structure of a network and the complexity of inference may prove useful in the model construction phase, avoiding structures that are likely to result in problems of poor performance of the final decision support system.

We present such basic concepts, principles, and methods underlying probabilistic models that practitioners need to acquaint themselves with.

In Chap. 1, we provide a bit of background and contextual introduction to Bayesian networks and influence diagrams. To give the reader a first understanding of probabilistic networks, we present a very simple Bayesian network and show how it can be augmented with explicit representation of decision options and a utility function, turning it into an influence diagram. Also, we discuss briefly the notions of causality, construction of probabilistic networks, and applicability (i.e., when to use probabilistic networks).

In Chap. 2, we describe the fundamental concepts of the graphical language used to construct probabilistic networks as well as the rules for reading statements of (conditional) dependence and independence encoded in network structures. We present two equivalent criteria for reading these statements, namely Pearl's d-separation criterion (Pearl 1988) and the criterion of directed Markov property by Lauritzen, Dawid, Larsen & Leimer (1990a).

In Chap. 3, we present the uncertainty calculus used in probabilistic networks to represent the numerical counterpart of the graphical structure, namely classical (Bayesian) probability calculus. We shall see how a basic axiom of probability calculus leads to recursive factorizations of joint probability distributions into products of conditional probability distributions and how such factorizations along with local statements of conditional independence can be expressed naturally in graphical terms.

In Chap. 4, we see how putting the basic notions of Chaps. 2 and 3 together we get the notion of discrete Bayesian networks. Also, we present a range of derived types of network models, including conditional Gaussian models where discrete and continuous variables coexist, influence diagrams that are Bayesian networks augmented with decision variables and utility functions, limited-memory influence diagrams that allow the knowledge engineer to reduce model complexity through assumptions about limited memory of past events, object-oriented models that allow the knowledge engineer to construct hierarchical models consisting of reusable submodels, and dynamic Bayesian networks that provide a framework for modeling phenomena evolving over time.

In Chap. 5, we explain the principles underlying inference in these different kinds of probabilistic networks.

In Chap. 6, we discuss the art of constructing a probabilistic network and the characteristics of problem domains that can be successfully modeled by probabilistic networks. The different phases of model construction are discussed, including design (how to identify the right set of variables, how to elicit the structure of a probabilistic network, and how to verify a network structure), implementation (elicitation of probability and utility parameters), test, and analysis (i.e., troubleshooting the model).

In Chap. 7, we present a large number of techniques and tricks for solving commonly occurring modeling problems in probabilistic networks. The set of techniques and tricks includes various structure-related techniques (parent divorcing, temporal transformation of causal relations, modeling of structural and functional uncertainty, modeling of undirected dependence relations, bidirectional relations, and naive Bayes models), probability distribution-related techniques (modeling of measurement error, (different) expert opinions, node absorption, value set by intervention, independence of causal influence, and mixture of Gaussian distributions), and decision-related techniques (modeling of test decisions, missing informational links, missing observations, hypothesis of highest probability, and constraints on decisions).

In Chap. 8, we describe how probabilistic networks can be constructed automatically from data or from a combination of data and problem domain expertise. The underlying theory of structure learning is explained, and different constraint-based learning algorithms, search and score-based algortihms, and structure restricted algorithms are presented. The expectation-maximization (EM) algorithm is described for learning the values of probability parameters from data as well as from data and problem domain expertise (penalized EM). Finally, we describe how the values of the probability parameters of a probabilistic network can be learned sequentially (adaptation).

In Chap. 9, we describe a method for performing conflict analysis in a probabilistic network, which aims at detecting pieces of evidence that might be in conflict with one another (i.e., pointing in different directions with respect to output from the network) or in conflict with the network model. Also, the topics of tracing and resolution of conflicts are discussed.

In Chap. 10, we describe how to analyze the sensitivity of the output of a probabilistic network (e.g., diagnosis and classification) to changes in the values of observed variables (evidence) as well as probability parameters. Also, we describe how parameters can be adjusted to make a probabilistic network produce required posterior probabilities.

Finally, in Chap. 11, we describe methods for performing value-of-information analysis in Bayesian networks and influence diagrams.

Compared to the first edition, this second edition of the book contains additional material: Sections 6.3.3 and 6.3.4 elaborate on model construction, Section 8.3 describes search and score-based structure learning, Section 8.4 presents a worked example on structure learning, Section 10.3 describes two-way parameter sensitivity analysis, Section 10.4 describes parameter tuning, and an appendix provides a quick reference to model construction with lists of recommendations and pitfalls for the model builder. Finally, lists of examples, figures, and tables have been added.

Uffe B. Kjærulff
Anders L. Madsen

Contents

Part I   Fundamentals

1   Introduction 3
    1.1   Expert Systems 3
        1.1.1   Representation of Uncertainty 4
        1.1.2   Normative Expert Systems 5
    1.2   Rule-Based Systems 5
        1.2.1   Causality 6
        1.2.2   Uncertainty in Rule-Based Systems 7
        1.2.3   Explaining Away 8
    1.3   Bayesian Networks 8
        1.3.1   Inference in Bayesian Networks 9
        1.3.2   Construction of Bayesian Networks 10
        1.3.3   An Example 11
    1.4   Bayesian Decision Problems 13
    1.5   When to Use Probabilistic Nets 14
    1.6   Concluding Remarks 15

2   Networks 17
    2.1   Graphs 18
    2.2   Graphical Models 20
        2.2.1   Variables 20
        2.2.2   Vertices Vs. Variables 21
        2.2.3   Taxonomy of Vertices/Variables 22
        2.2.4   Vertex Symbols 23
        2.2.5   Summary of Notation 23
    2.3   Evidence 24
    2.4   Causality 24
    2.5   Flow of Information in Causal Networks 25
        2.5.1   Serial Connections 26
        2.5.2   Diverging Connections 28

	2.5.3	Converging Connections	29
	2.5.4	Intercausal Inference (Explaining Away)	30
	2.5.5	The Importance of Correct Modeling of Causality	31
2.6	Two Equivalent Irrelevance Criteria		32
	2.6.1	d-Separation Criterion	33
	2.6.2	Directed Global Markov Criterion	35
2.7	Summary		36

3   Probabilities 39
- 3.1   Basics 40
  - 3.1.1   Events 40
  - 3.1.2   Axioms 40
  - 3.1.3   Conditional Probability 41
- 3.2   Probability Distributions for Variables 43
  - 3.2.1   Rule of Total Probability 44
  - 3.2.2   Graphical Representation 46
- 3.3   Probability Potentials 46
  - 3.3.1   Normalization 47
  - 3.3.2   Evidence Potentials 49
  - 3.3.3   Potential Calculus 50
  - 3.3.4   Barren Variables 53
- 3.4   Fundamental Rule and Bayes' Rule 54
  - 3.4.1   Interpretation of Bayes' Rule 55
- 3.5   Bayes' Factor 58
- 3.6   Independence 59
  - 3.6.1   Independence and DAGs 60
- 3.7   Chain Rule 62
- 3.8   Summary 64

4   Probabilistic Networks 69
- 4.1   Belief Update 70
  - 4.1.1   Discrete Bayesian Networks 71
  - 4.1.2   Conditional Linear Gaussian Bayesian Networks 75
- 4.2   Decision Making Under Uncertainty 79
  - 4.2.1   Discrete Influence Diagrams 80
  - 4.2.2   Conditional LQG Influence Diagrams 89
  - 4.2.3   Limited Memory Influence Diagrams 93
- 4.3   Object-Oriented Probabilistic Networks 95
  - 4.3.1   Chain Rule 100
  - 4.3.2   Unfolded OOPNs 100
  - 4.3.3   Instance Trees 100
  - 4.3.4   Inheritance 101
- 4.4   Dynamic Models 102
  - 4.4.1   Time-Sliced Networks Represented as OOPNs 104
- 4.5   Summary 105

5   Solving Probabilistic Networks 111
    5.1   Probabilistic Inference 112
         5.1.1   Inference in Discrete Bayesian Networks 112
         5.1.2   Inference in CLG Bayesian Networks 125
    5.2   Solving Decision Models 128
         5.2.1   Solving Discrete Influence Diagrams...................... 128
         5.2.2   Solving CLQG Influence Diagrams 132
         5.2.3   Relevance Reasoning 134
         5.2.4   Solving LIMIDs 136
    5.3   Solving OOPNs 140
    5.4   Summary... 140

Part II   Model Construction

6   Eliciting the Model 145
    6.1   When to Use Probabilistic Networks 146
         6.1.1   Characteristics of Probabilistic Networks 147
         6.1.2   Some Criteria for Using Probabilistic Networks 148
    6.2   Identifying the Variables of a Model 149
         6.2.1   Well-Defined Variables 149
         6.2.2   Types of Variables 152
    6.3   Eliciting the Structure 154
         6.3.1   A Basic Approach 154
         6.3.2   Idioms 156
         6.3.3   An Example: Extended Chest Clinic Model 163
         6.3.4   The Generic Structure of Probabilistic Networks 173
    6.4   Model Verification.. 174
    6.5   Eliciting the Numbers 176
         6.5.1   Eliciting Subjective Conditional Probabilities 177
         6.5.2   Eliciting Subjective Utilities 179
         6.5.3   Specifying CPTs and UTs Through Expressions 180
    6.6   Concluding Remarks 183
    6.7   Summary... 185

7   Modeling Techniques 191
    7.1   Structure-Related Techniques.. 191
         7.1.1   Parent Divorcing 192
         7.1.2   Temporal Transformation 196
         7.1.3   Structural and Functional Uncertainty 197
         7.1.4   Undirected Dependence Relations 201
         7.1.5   Bidirectional Relations 204
         7.1.6   Naive Bayes Model 206
    7.2   Probability Distribution-Related Techniques 208
         7.2.1   Measurement Uncertainty 209
         7.2.2   Expert Opinions 211

7.2.3   Node Absorption......................................   213
7.2.4   Set Value by Intervention   214
7.2.5   Independence of Causal Influence   216
7.2.6   Mixture of Gaussian Distributions   221
7.3   Decision-Related Techniques   224
7.3.1   Test Decisions   224
7.3.2   Missing Informational Links   227
7.3.3   Missing Observations   229
7.3.4   Hypothesis of Highest Probability   231
7.3.5   Constraints on Decisions   233
7.4   Summary...   235

8   **Data-Driven Modeling**   237
8.1   The Task and Basic Assumptions   238
8.1.1   Basic Assumptions   240
8.1.2   Equivalent Models   240
8.2   Constraint-Based Structure Learning.....................   242
8.2.1   Statistical Hypothesis Tests   242
8.2.2   Structure Constraints   245
8.2.3   PC Algorithm   246
8.2.4   PC* Algorithm......................................   251
8.2.5   NPC Algorithm   251
8.3   Search and Score-Based Structure Learning   256
8.3.1   Space of Structures   256
8.3.2   Search Procedures   257
8.3.3   Score Functions   258
8.3.4   Learning Structure Restricted Models   265
8.4   Worked Example on Structure Learning   271
8.4.1   PC Algorithm   272
8.4.2   NPC Algorithm   273
8.4.3   Search and Score-Based Algorithm...................   275
8.4.4   Chow–Liu Tree   276
8.4.5   Comparison   277
8.5   Batch Parameter Learning   278
8.5.1   Expectation–Maximization Algorithm   279
8.5.2   Penalized EM Algorithm   281
8.6   Sequential Parameter Learning   283
8.7   Summary...   285

Part III   Model Analysis

9   **Conflict Analysis**   291
9.1   Evidence-Driven Conflict Analysis.......................   292
9.1.1   Conflict Measure...................................   292

9.1.2 Tracing Conflicts 294
9.1.3 Conflict Resolution 295
9.2 Hypothesis-Driven Conflict Analysis 297
9.2.1 Cost-of-Omission Measure............................... 297
9.2.2 Evidence with Conflict Impact........................... 297
9.3 Summary.. 299

10 Sensitivity Analysis 303
10.1 Evidence Sensitivity Analysis 304
10.1.1 Distance and Cost-of-Omission Measures 305
10.1.2 Identify Minimum and Maximum Beliefs................. 306
10.1.3 Impact of Evidence Subsets 307
10.1.4 Discrimination of Competing Hypotheses 308
10.1.5 What-If Analysis... 309
10.1.6 Impact of Findings....................................... 309
10.2 Parameter Sensitivity Analysis 311
10.2.1 Sensitivity Function 312
10.2.2 Sensitivity Value 314
10.2.3 Admissible Deviation 316
10.3 Two-Way Parameter Sensitivity Analysis 317
10.3.1 Sensitivity Function 317
10.4 Parameter Tuning... 320
10.5 Summary.. 323

11 Value of Information Analysis.. 327
11.1 VOI Analysis in Bayesian Networks 328
11.1.1 Entropy and Mutual Information 328
11.1.2 Hypothesis-Driven Value of Information Analysis 329
11.2 VOI Analysis in Influence Diagrams 333
11.3 Summary.. 336

Quick Reference to Model Construction 341

List of Examples.. 351

List of Figures 355

List of Tables.. 365

List of Symbols 369

Reference.. 371

Index.. 377

Part I
Fundamentals

Chapter 1
Introduction

The desire to have computers perform intellectually challenging tasks has existed ever since the invention of the general-purpose computer that could be programmed to execute an arbitrary set of manipulations on numbers and symbols. Solving an intellectually challenging task can be characterized as a process of deriving conclusions (new pieces of knowledge) by manipulating a (large) body of knowledge, typically including definitions of entities (objects, concepts, events, phenomena, etc.), relations among them, and observations of states (values) of some of the entities.

As a prototypical example of a decision problem, imagine a physician who is consulted by a patient complaining about stomach pain. The physician then conducts an interview of the patient and possibly makes some investigations to localize the origin of the pain, to find other symptoms of the disorder, etc. Based on her knowledge about pathophysiological cause–effect mechanisms involving stomach pain as well as on the information revealed from the patient's medical records, from the interview, from the symptoms observed, etc., the physician makes a diagnosis and a treatment plan.

By formulating the physician's knowledge in an appropriate formal (computer) language for which there exist methods for making inferences to manipulate pieces of knowledge formulated in this language, the reasoning conducted by the physician can be automated and carried out by a computer. Probabilistic network is an example of such a language that has gained a lot of popularity over the last couple of decades.

This chapter provides brief accounts on the context of probabilistic networks, what they are, and when to use them.

1.1 Expert Systems

A system that is able to perform tasks that are supposed to be intellectually demanding is often said to exhibit artificial intelligence (AI) or to be an expert system if the system's problem-solving ability is restricted to a particular area of expertise.

U.B. Kjærulff and A.L. Madsen, *Bayesian Networks and Influence Diagrams: A Guide to Construction and Analysis*, ISS 22, DOI 10.1007/978-1-4614-5104-4_1,
© Springer Science+Business Media New York 2013

Many definitions of artificial intelligence have been proposed. In this book, we shall consider techniques that enable us to construct devices and services that are able to:

- Perform probabilistic inference to support belief updating and decision making under uncertainty
- Acquire knowledge from data/experience
- Solve problems efficiently and respond to new situations

We shall refrain from discussing whether or not this makes the devices or services exhibit AI and leave this decision to the reader.

In any case, a probabilistic network is always constructed to solve a particular problem within a given problem domain (area of expertise). Therefore, the label "expert system" can often be attached to systems that perform reasoning and decision making by means of probabilistic networks.

The motivation for constructing an expert system is typically to automate some recurring task involving belief updating and decision making under uncertainty, possibly involving extraction of information/knowledge from data.

Several other expert system paradigms have been suggested. This book is not intended to provide accounts on such competing paradigms, but for the sake of historical context of probabilistic networks, we shall briefly mention some of the important and well-known alternatives to probabilistic networks.

1.1.1   Representation of Uncertainty

Randomness and uncertain judgment is inherent in most real-world decision problems. We therefore need a method (paradigm) that supports representation of quantitative measures of uncertain statements as well as evidence about states of "the world" and a method for combining the statements and the evidence in such a way that consistent posterior probabilities and expected utilities can be provided to support belief updating and decision making under uncertainty.

Probability theory is the prevailing method for dealing with uncertainty, and it is the one in focus in this book. However, other methods have been proposed, as some researchers find probability theory inappropriate for presenting some forms of quantitative uncertainty. Probability theory deals with uncertainty of well-defined occurrences; that is, the source of ambiguity is occurrence. Situations involving ambiguously and/or vaguely *defined* occurrences might be better represented by other methods. Let us very briefly mention the two most prominent alternative methods for dealing with uncertainty. Readers interested in more detailed accounts are referred to the literature.

Dempster and Shafer (Dempster 1968, Shafer 1976) developed belief theory to be able to assign measures of uncertainty to sets of events without necessarily having to assign or assess uncertainty for single events. For example, if you receive a message that your golf partner is free for a match of golf "next Sunday," you might be willing to assign a measure of uncertainty to the pair of (mutually exclusive) events that you and your partner will be playing this Sunday or Sunday next week, but unwilling to assign measures to the two individual possibilities. Rather than focusing on events, belief theory focuses on evidence (on sets of events).[1]

Fuzzy methods (Zadeh 1965, Zadeh & Kacprzyk 1992) address situations where the ambiguity lies in the nature of events rather than in their occurrence. Typical examples of ambiguous concepts include everyday concepts like beauty, intelligence, size, and speed. For example, both the statement "Paul is tall" (S) and the statement "Paul is not tall" ($\neg S$) might be plausible, and we therefore wish to assign some degree of plausibility to $S \wedge \neg S$, which is in contrast with ordinary logic where $S \wedge \neg S$ is always false. Expert systems based on fuzzy logic have achieved some popularity, maybe especially so in applications involving control loops (fuzzy control).

1.1.2 Normative Expert Systems

The objective in some early attempts to construct expert systems was to create a model of the decision making performed by some (human) expert and let a system containing such a model perform tasks that previously needed human expertise. Today, a more realistic approach is normally taken where a model of the problem domain is created rather than a model of the expert such that systems containing such a model support experts in performing their tasks rather than substituting them.

Systems containing models of problem domains that use classical probability calculus and decision theory as their basis for supporting belief updating and decision making under uncertainty are often referred to as *normative expert systems*, as their behavior is governed by a set of fundamental rules (or axioms).

In some sense, Bayesian networks can be seen as an extension of one of the earliest methods for knowledge representation and manipulation, namely, logical rules. Let us therefore dwell a little on rule-based systems.

1.2 Rule-Based Systems

One of the earliest methods for knowledge representation and manipulation was logical rules of the form

$$R_1: \text{if } s_1 \text{ then } s_2,$$

where statement s_2 (the consequence) can be concluded with certainty whenever statement s_1 (the condition) is observed to hold. If another rule states that

$$R_2: \text{if } s_2 \text{ then } s_3,$$

---

[1] Adapted from example by Bender (1996).

then s_3 can be concluded through *forward chaining* involving rules R_1 and R_2 once s_1 is known to hold.

Notice that such rules are asymmetric in the sense that the condition and consequence statements are not interchangeable; observing the consequence statement does not allow us to conclude that the condition statement holds.

1.2.1   Causality

Assume that the occurrence of some event c is known to cause the effect e and that the relationship between c and e is known to be deterministic (logical). Then, obviously, observing c, we can conclude e. Observing e, on the other hand, does not make us able to conclude c, unless c is known to be the only cause of e. Thus, in formulating the causal relationship between c and e as a rule, we would obviously want to formulate it as "if c then e" rather than "if e then c."

From this insight, we conclude that rules like R_1 and R_2 express causal relationships, where s_1, say, plays the role of the cause and s_2 the role of the effect, the only possible exception being if s_1, as an effect, only can be caused by s_2.

A rule-based system, like any other knowledge representation scheme, represents a certain part of the world (the problem domain) only up to some precision. This implies that certain (causal) mechanisms might be ignored as being unimportant for the precision (or level of detail) at which conclusions need to be drawn. For example, in a medical expert system, a disorder causing some symptom, s, might be ignored if it only appears in, say, less than one out of a million cases. If ignoring such a rare explanation for s leaves only one possible cause (disorder), say d, for the symptom, it might at a first consideration seem reasonable to state a rule like "if s then d."

Violating the "causal direction" in formulating rules is, however, not advisable. For example, in a medical expert system, consider the causal chain

$$\text{Smoking} \rightarrow \text{Bronchitis} \rightarrow \text{Dyspnoea},$$

denoting the concatenation of rules

$$R_3: \text{if Smoking then Bronchitis,}$$

and

$$R_4: \text{if Bronchitis then Dyspnoea,}$$

Here Bronchitis is a disorder, Dyspnoea (a medical term for shortness of breath) a symptom of Bronchitis, and Smoking a cause of Bronchitis, sometimes referred to as a piece of background information. Assume that instead of R_4, we formulated the rule

$$R_4': \text{if Dyspnoea then Bronchitis,}$$

which would make Smoking and Dyspnoea be competing explanations for Bronchitis. Then upon observing that the patient smokes, we would be able to conclude only that the patient might suffer from bronchitis, but would not be able to conclude anything about the patient's breathing characteristics. In effect, rules R_3 and R_4' collectively express independence between Smoking and Dyspnoea, which is obviously wrong.

1.2.2   Uncertainty in Rule-Based Systems

As clearly demonstrated by rules R_3 and R_4, crisp logic is inappropriate for representing the nature of the causal relations among Smoking, Bronchitis, and Dyspnoea. Only a certain proportion of the smoking patients entering a chest clinic suffer from bronchitis. Similarly, dyspnoea appears as a symptom only for some of the patients suffering from bronchitis. In terms of having uncertainty associated with the (cause–effect) rules, these examples are by no means exceptional. The vast majority of cause–effect mechanisms of interest in our attempts to model parts of the world in expert (or AI) systems are uncertain.

In order to make up for this fact, a method for rule-based systems with uncertainty was developed in the 1970s by the team behind the medical expert system MYCIN (Shortliffe & Buchanan 1975). Associated with each rule in MYCIN is a numerical value in the interval $[-1, +1]$ called a *certainty factor* (CF).

This factor indicates the strength of the conclusion of the rule whenever its condition is satisfied. In particular, given the evidence available,

$$CF = \begin{cases} +1 & \text{when the conclusion is certainly true} \\ -1 & \text{when the conclusion is certainly false} \\ 0 & \text{when no information about the conclusion can be derived} \end{cases}$$

Certainty factors are, however, nothing but an ad hoc device for dealing with uncertainty. Heckerman (1986) proved that certainty factors cannot be defined consistently if the domains of the variables have more than two elements. More precisely, certain factors can be proved to be consistent only for binary variables, where the rules induce a singly connected tree in which there is exactly one variable with no parents.[2]

---

[2]The variables in the condition of a rule are often referred to as the "parents" of the consequence variable, which is often referred to as the "child" variable. For example, variables temperature and humidity in rule "if temperature $=$ high and humidity $=$ high then comfort $=$ low" are parents of comfort. A parent–child relation is depicted by two nodes in a graph (or tree) interconnected by a directed link from the parent to the child.

Fig. 1.1 Graphical
representation of rules "if C_1
then E_1" and "if C_1 and C_2
then E_2"

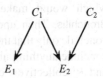

1.2.3   Explaining Away

Consider the small rule-based system depicted in Fig. 1.1, where C_1 can cause E_1
and E_2 and C_2 can cause E_2. The CF method provides a formula for combining
evidence from E_1 and E_2 and applying it to C_1. Unfortunately, however, the
CF method provides no mechanism for applying E_1 to C_2, which is needed to
implement the "explaining-away" mechanism, where evidence on E_1 makes C_1
more probable, in turn, making the competing explanation C_2 for E_2 less probable.

1.3   Bayesian Networks

Having realized that rule-based systems with certainty factors have serious limitations as a method for knowledge representation and support for reasoning under
uncertainty, researchers turned their attention towards a probabilistic interpretation
of certainty factors, leading to the definition of Bayesian networks (Kim &
Pearl 1983, Pearl 1988). A Bayesian network can be described briefly as an
acyclic directed graph (DAG) which defines a factorization of a joint probability
distribution over the variables that are represented by the nodes of the DAG, where
the factorization is given by the directed links of the DAG. More precisely, for a
DAG, $\mathcal{G} = (V, E)$, where V denotes a set of nodes (or vertices) and E a set of
directed links (or edges) between pairs of the nodes, a joint probability distribution,
$P(X_V)$, over the set of (typically discrete) variables X_V indexed by V can be
factorized as

$$P(X_V) = \prod_{v \in V} P(X_v \mid X_{\text{pa}(v)}), \tag{1.1}$$

where $X_{\text{pa}(v)}$ denotes the (preferably small) set of parent variables of variable X_v
for each node $v \in V$. The factorization in (1.1) expresses a set of independence
assumptions, which are represented by the DAG in terms of pairs of nodes that are
not directly connected to one another by a directed link. It is the existence of such
independence assumptions and the small set of parents for each node that makes it
possible to specify the conditional probabilities and to perform inference efficiently
in a Bayesian network.

Each conditional probability distribution, $P(X_v | X_{pa(v)})$, represents a set of "rules," where each "rule" (conditional probability) takes the form

$$R_5: \text{ if } X_{pa(v)} = x_{pa(v)} \text{ then } X_v = x_v \text{ with probability } z,$$

where x_v and $x_{pa(v)}$ denote, respectively, a value assigned to X_v and a vector of values assigned to the parent variables of X_v. For example, if one of five possible values can be assigned to X_v and it has four parents each of which can be assigned one of three possible values, then $P(X_v | X_{pa(v)})$ represents a collection of $5 \times 3^4 = 405$ rules of the kind shown in rule R_5.

Actually, the notion of rules is only implicitly apparent in Bayesian networks. The explicit notion is that of conditional probability distributions, $P(X_v | X_{pa(v)})$, where, rather than as in rule R_5, each term is formulated as a conditional probability (parameter) of the form

$$P(X_v = x_v | X_{pa(v)} = x_{pa(v)}) = z$$

or even simpler as

$$P(x_v | x_{pa(v)}) = z.$$

1.3.1 Inference in Bayesian Networks

Contrary to rule-based systems with certainty factors, inference in Bayesian networks is always consistent, and the ability to handle the explaining-away problem is embedded naturally in the way in which inference is performed in Bayesian networks. However, in general, it is an NP-hard task to solve the inference problem in Bayesian networks (Cooper 1990); even approximate inference is NP-hard (Dagum & Luby 1993). Fortunately, efficient inference algorithms have been developed such that inference in Bayesian networks can be done in fractions of a second even for large networks containing hundreds or even thousands of variables (Lauritzen & Spiegelhalter 1988, Jensen, Lauritzen & Olesen 1990). Efficiency of inference, however, is highly dependent on the structure of the DAG, so networks with a relatively small number of variables sometimes resist exact inference, in which case approximate methods must be applied.

As Bayesian networks most often represent causal statements of the kind $X \rightarrow Y$, where X is a cause of Y and where Y often takes the role of an observable effect of X, which typically cannot be observed itself, we need to derive the posterior probability distribution $P(X | Y = y)$ given the observation $Y = y$ using the prior distribution $P(X)$ and the conditional probability distribution $P(Y | X)$ specified in the model. Reverend Thomas Bayes (1702–1761) provided the famous Bayes' rule for performing this calculation:

$$P(X | Y = y) = \frac{P(Y = y | X) P(X)}{P(Y = y)},$$

where $P(Y = y) = \sum_x P(Y = y \mid X = x) P(X = x)$. This rule (or theorem) plays a central role in statistical inference because the probability of a cause can be inferred when its effect has been observed. Olmsted (1983) and Shachter (1986) developed a method for inference in Bayesian networks, which involved multiple applications of Bayes' rule. Lauritzen & Spiegelhalter (1988) and Jensen et al. (1990) developed inference methods for Bayesian networks based on message passing in a tree structure (junction tree) derived from the structure of the Bayesian network. The latter approach is the prevailing inference method used in modern software packages for inference in probabilistic networks.

1.3.2   Construction of Bayesian Networks

As described above, a Bayesian network can be described in terms of a qualitative component, consisting of a DAG, and a quantitative component, consisting of a joint probability distribution that factorizes into a set of conditional probability distributions governed by the structure of the DAG.

The construction of a Bayesian network thus runs in two phases. First, given the problem at hand, one identifies the relevant variables and the (causal) relations among them. The resulting DAG specifies a set of dependence and independence assumptions that will be enforced on the joint probability distribution, which is next to be specified in terms of a set of conditional probability distributions, $P(X_v \mid X_{\text{pa}(v)})$, one for each "family," $\{v\} \cup \text{pa}(v)$, of the DAG.

A Bayesian network can be constructed manually, (semi-)automatically from data, or through a combination of a manual and a data-driven process, where partial knowledge about structure, as well as parameters (i.e., conditional probabilities), blends with statistical information extracted from databases of cases (i.e., previous joint observations of values of the variables).

Manual construction of a Bayesian network can be a labor-intensive task, requiring a great deal of skill and creativity as well as close communication with problem-domain experts. Extensive guidance on how to manually construct a probabilistic network is the core of this book. This includes methods and hints on how to elicit the network structure (with emphasis on the importance of maintaining a causal perspective), methods for eliciting and specifying the parameter values of the network, and numerous tricks that can be applied for solving prototypical modeling problems.

Once constructed (be it manually or automatically), the parameters of a Bayesian network may be continuously updated as new information arrives. Thus, a model for which rough guesses on the parameter values are provided initially will gradually improve itself as it gets presented with more and more cases.

1.3.3 An Example

As a simple example, let us consider a problem concerning reasoning about starting problems for a car. Assume for simplicity that we only consider two competing causes for starting problems, namely, no fuel and dirty spark plugs. Also assume that, apart from starting problems, the only observation we can make is reading the fuel gauge. Now, if the car will not start and the fuel gauge reads "empty," then we conclude that "no fuel" is probably the cause of the problem, and we strongly reduce our suspicion that dirty spark plugs might be causing the problem.

Let us see how to automate that reasoning process in a Bayesian network. First, we identify four variables and the possible values (states) that they may attain (in this case, no more than two states for each variable is necessary). The variables and their possible states are shown in Table 1.1.

Figure 1.2 shows the structure of the Bayesian network for this simple problem, where Fuel? and Spark_plugs have causal influences on Start? and Fuel? has a causal influence on Fuel_gauge.

A (conditional) probability table needs to be specified for each variable. Assume that when knowing nothing about the states of the other variables, we would expect that there is fuel on the car (i.e., Fuel? = yes) in 999 out of 1,000 cases. Therefore, respecting the order of states in Table 1.1, we specify the probability distribution for Fuel? as

$$P(\text{Fuel?}) = (0.001, 0.999).$$

Similarly, expecting that the spark plugs are clean in 95 out of 100 cases, we specify

$$P(\text{Spark\_plugs}) = (0.05, 0.95).$$

For Fuel_gauge, we need to specify two conditional probability distributions, one for each possible state of Fuel?. For Start?, we need to specify four conditional probability distributions, one for each combination of possible states of Fuel?

Table 1.1 The four variables and their possible states for the "car would not start" problem

Variable	Possible states
Start?	{no, yes}
Spark_plugs	{dirty, clean}
Fuel?	{no, yes}
Fuel_gauge	{empty, not_empty}

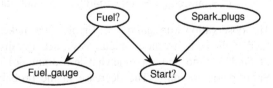

Fig. 1.2 Bayesian network for the "car would not start" problem

Table 1.2 Conditional
probability distributions for
Fuel_gauge given Fuel?,
$P(\text{Fuel\_gauge} \,|\, \text{Fuel?})$

	Fuel?	
Fuel_gauge	no	yes
empty	0.995	0.001
not_empty	0.005	0.999

Table 1.3 Conditional
probability distributions for
Start? given Fuel? and
Spark_plugs,
$P(\text{Start?} \,|\, \text{Fuel?}, \text{Spark\_plugs})$

		Start?	
Fuel?	Spark_plugs	no	yes
no	dirty	1	0
no	clean	1	0
yes	dirty	0.1	0.9
yes	clean	0.01	0.99

and Spark_plugs. These probability distributions appear in the conditional probability tables shown in Tables 1.2 and 1.3, respectively, where we expect the fuel
gauge to read empty with probability 0.995 if there is no fuel, the car to start
with probability 0.99 when there is fuel on the car and the spark plugs are clean,
etc. In Table 1.3, the probability of 0.01 for Start? = no when Fuel? = yes and
Spark_plugs = clean captures other causes not explicitly considered in our simple
model.

From the probabilities specified, we can compute that

$$P(\text{Start?} = \text{no}) = 0.016,$$

that is, we expect the car to start in 984 out of 1,000 cases (or with probability
0.984). Now, if we fix the value of Start? to no, then, using Bayes' rule, we get

$$P(\text{Fuel?} = \text{no} \,|\, \text{Start?} = \text{no}) = 0.065$$

and

$$P(\text{Spark\_plugs} = \text{dirty} \,|\, \text{Start?} = \text{no}) = 0.326.$$

Thus, our best guess is that dirty spark plugs are causing the problem, although the
probability of dirty spark plugs might not be high enough and the probability of "no
fuel" not low enough to settle with the conclusion that dirty spark plugs are causing
our problem. Making the observation that Fuel_gauge = empty and repeating the
computations, we find that

$$P(\text{Fuel?} = \text{no} \,|\, \text{Start?} = \text{no}, \text{Fuel\_gauge} = \text{empty}) = 0.986$$

and

$$P(\text{Spark\_plugs} = \text{dirty} \,|\, \text{Start?} = \text{no}, \text{Fuel\_gauge} = \text{empty}) = 0.054.$$

The observation Fuel_gauge = Empty thus makes us strongly believe that "no
fuel" is the cause of the problem, as we see a dramatic increase in the probability
of Fuel? = no and a (somewhat less dramatic) decrease in the probability of
Spark_plugs = dirty. The decrease in the probability of Spark_plugs = dirty
illustrates the explaining-away effect.

1.4 Bayesian Decision Problems

Most often, the outputs of interest of a Bayesian network are the posterior probabilities of the variables representing the problem that we wish to reason about (e.g., possible diagnoses). These probabilities are often combined with costs and benefits (utilities) of performing one or more actions to solve the problem. That is, from the posterior probabilities and the utilities, we compute expected utilities for each possible decision option (e.g., different treatment alternatives). The decision option with the highest expected utility should then be selected. Based on a number of studies, Tversky & Kahneman (1981) have shown that people usually do *not* make decisions that maximize their expected utility, so supporting human decisions by recommendations from decision support systems can often improve the quality of decisions.

A Bayesian network can be augmented with decision variables, representing decision options, and utility functions, representing preferences, that may depend on both random (or chance) variables and decision variables. Networks so augmented are called influence diagrams and can be used to compute expected utilities for the various decision options given the observations (and decisions) made.

Assume that we wish to augment our Bayesian network in Fig. 1.2 with a decision variable, say Action, with states {no_action, add_fuel, clean_spark_plugs} and a utility function, say U, that depends on the states of chance variables Fuel? and Spark_plugs and on our decision variable. Figure 1.3 shows the structure of the augmented network, where the links from Fuel_gauge and Start? to Action indicate that the states of Fuel_gauge and Start? are known prior to making the decision on which action to perform. Table 1.4 shows our utility function, where we assign a utility value of 1 to combinations of states of Action, Fuel?, and Spark_plugs where the action is supposed to solve a problem; otherwise, we assign a value of 0 to the utility function.

With the evidence that Start? = no and Fuel_gauge = empty, we find that

$$EU(\text{Action}) = (0.009, 0.986, 0.054);$$

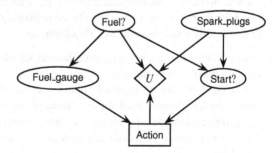

Fig. 1.3 Influence diagram for the "car would not start" problem

Table 1.4 Utility function for the "car would not start" decision problem

		Action		
Fuel?	Spark_plugs	no_action	add_fuel	clean_spark_plugs
no	dirty	0	1	1
no	clean	0	1	0
yes	dirty	0	0	1
yes	clean	1	0	0

that is, the expected utilities (EU) of decision options no_action, add_fuel, and clean_spark_plugs are 0.009, 0.986, and 0.054, respectively. Since EU(add_fuel) is greater than both EU(no_action) and EU(clean_spark_plugs), we select decision option add_fuel.

Note that, coincidentally, $P(\text{Fuel?} = \text{no}) = \text{EU}(\text{add\_fuel})$ because of the way in which we have defined the utility function. In general, the domain of a utility function is the set of real numbers. If one defines the utility values on, say, a monetary scale, the expected utilities of one's decision options can be interpreted directly as expected gains or losses on the chosen scale, say dollars.

1.5   When to Use Probabilistic Nets

There are many good reasons to choose probabilistic networks as the framework for supporting belief updating and decision problems under uncertainty. As indicated above, these include (among others):

- Coherent and mathematically sound handling of uncertainty
- Normative decision making
- Automated construction and adaptation of models based on data
- Intuitive and compact representation of cause–effect relations and (conditional) dependence and independence relations
- Efficient solution of queries given evidence

There are, however, some requirements to the nature of the problem that should be fulfilled for probabilistic networks to be an appropriate choice of method. Here, we shall just briefly mention some key requirements:

- The variables and events (i.e., possible values of the variables) of the problem domain must be well-defined.
- Knowledge should be available about the (causal) relations among the variables, the conditional probabilities quantifying the relations, and the utilities (preferences) associated with the various decision options.
- Uncertainty should be associated with at least some of the relations among the variables.
- The problem at hand should most probably contain an element of decision making involving a desire to maximize the expected utility of a decision.

Data are often available in the form of joint observations of a subset of the variables pertaining to the problem domain. Each set of joint observations pertains to a particular instance (case) of the problem domain. For example, data can be extracted from a database of customers, where a lot of features (e.g., gender, age, marital status, and income) are recorded for each customer (case), and analyzed statistically to derive both structure and parameters of a probabilistic network. Such automatically generated models can reveal a lot of information about dependence and independence relations (and sometimes even causal mechanisms) among the variables and thus provide new knowledge about the problem domain. Sometimes, however, the available data do not originate from variables with clearly understood semantics, or the patterns of interactions among variables are complex. In such cases, a neural network model might be better suited, where the model consists of a function that attempts to match each input case with a desired output by iteratively tweaking a large number of coefficients (weights) until convergence (i.e., until the distance between the desired and the actual outputs is sufficiently small).

1.6   Concluding Remarks

In this brief introduction, we have only touched superficially upon a few key characteristics of probabilistic networks. These and many more will be presented in much greater detail in the chapters to come.

Careful introductions to the fundamental concepts, theories, and methods underlying probabilistic networks as well as definitions of Bayesian networks, influence diagrams, and their variants are provided in the remaining chapters of Part I, that is, Chaps. 2–5. These chapters can be skipped if you already know enough about the basics of probabilistic networks and wish to dive directly into Part II covering topics on model elicitation, modeling techniques, and learning models from data or Part III covering topics on model analysis.

Chapter 2
Networks

Probabilistic networks are graphical models of (causal) interactions among a set of variables, where the variables are represented as vertices (nodes) of a graph and the interactions (direct dependences) as directed edges (links or arcs) between the vertices. Any pair of unconnected vertices of such a graph indicates (conditional) independence between the variables represented by these vertices under particular circumstances that can easily be read from the graph. Hence, probabilistic networks capture a set of (conditional) dependence and independence properties associated with the variables represented in the network.

Graphs have proven themselves an intuitive language for representing such dependence and independence statements and thus provide an excellent language for communicating and discussing dependence and independence relations among problem-domain variables. A large and important class of assumptions about dependence and independence relations expressed in factorized representations of joint probability distributions can be represented compactly in a class of graphs known as acyclic, directed graphs (DAGs).

Chain graphs are a generalization of DAGs capable of representing a broader class of dependence and independence assumptions (Frydenberg 1989, Wermuth & Lauritzen 1990). The added expressive power comes, however, with the cost of a significant increase in the semantic complexity, making specification of joint probability factors much less intuitive. Thus, despite their expressive power, chain graph models have gained little popularity as practical models for decision support systems, and we shall therefore focus exclusively on models that factorize according to DAGs.

As indicated above, probabilistic network is a class of probabilistic models that have gotten their name from the fact that the joint probability distributions represented by these models can be naturally described in graphical terms, where the vertices of a graph (or network) represent variables over which a joint probability distribution is defined and the presence and absence of edges represent dependence and independence properties among the variables.

U.B. Kjærulff and A.L. Madsen, *Bayesian Networks and Influence Diagrams: A Guide to Construction and Analysis*, ISS 22, DOI 10.1007/978-1-4614-5104-4_2,

Probabilistic networks can be seen as compact representations of "fuzzy" cause–effect rules that, contrary to ordinary (logical) rule-based systems, is capable of performing deductive and abductive reasoning as well as intercausal reasoning. Deductive reasoning (sometimes referred to as *causal reasoning*) follows the direction of the causal links between variables of a model; for example, knowing that a patient suffers from angina, we can conclude (with high probability) the patient has fever and a sore throat. Abductive reasoning (sometimes referred to as *diagnostic reasoning*) goes against the direction of the causal links; for example, observing that a patient has a sore throat provides supporting evidence for angina being the correct diagnosis.

The property, however, that sets inference in probabilistic networks apart from other automatic reasoning paradigms is its ability to make *intercausal* reasoning: Getting evidence that supports solely a single hypothesis (or a subset of hypotheses) automatically leads to decreasing belief in the unsupported, competing hypotheses. This property is often referred to as the *explaining away* effect. For example, there are a large number of possible causes that a car will not start, one being lack of fuel. Observing that the fuel gauge indicates no fuel provides strong evidence that lack of fuel is the cause of the problem, while the beliefs in other possible causes decrease substantially (i.e., they are explained away by the observation; see Sect. 1.3.3 on page 11). The ability of probabilistic networks to automatically perform such intercausal inference is a key contribution to their reasoning power.

Often the graphical aspect of a probabilistic network is referred to as its *qualitative* aspect and the probabilistic, numerical part as its *quantitative* aspect. This chapter is devoted to the qualitative aspect of probabilistic networks. In Sect. 2.1, we introduce some basic graph notation that will be used throughout the book. Section 2.2 discusses the notion of variables, which is the key entity of probabilistic networks. Another key concept is that of "evidence," which we shall touch upon in Sect. 2.3. Maintaining a causal perspective in the model, construction process can prove valuable, as mentioned briefly in Sect. 2.4. Sections 2.5 and 2.6 are devoted to an in-depth treatment on the principles and rules for flow of information in DAGs. We carefully explain the properties of the three basic types of connections in a DAG (i.e., serial, diverging, and converging connections) through examples and show how these combine directly into the d-separation criterion and how they support intercausal (explaining away) reasoning. We also present an alternative to the d-separation criterion known as the directed global Markov criterion, which in many cases proves to be a more efficient method for reading off dependence and independence statements of a DAG.

2.1   Graphs

A graph is a pair $\mathcal{G} = (V, E)$, where V is a finite set of distinct vertices and $E \subseteq V \times V$ is a set of edges. An ordered pair $(u, v) \in E$ denotes a *directed edge* from vertex u to vertex v, and u is said to be a *parent* of v and v a *child* of u. The set of parents and children of a vertex v shall be denoted by $\mathrm{pa}(v)$ and $\mathrm{ch}(v)$, respectively.

Fig. 2.1 (a) An acyclic,
directed graph (DAG). (b)
Moralized graph

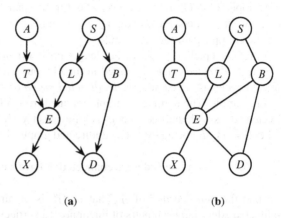

(a)   (b)

As we shall see later, depending on what they represent, vertices are displayed as labeled circles, ovals, or polygons, directed edges as arrows, and undirected edges as lines. Figure 2.1a shows a graph with eight vertices and eight edges (all directed), where, for example, the vertex labeled E has two parents labeled T and L. The labels of the vertices are referring to (1) the names of the vertices, (2) the names of the variables represented by the vertices, or (3) descriptive labels associated with the variables represented by the vertices.[1]

We often use the intuitive notation $u \overset{\mathcal{G}}{\to} v$ to denote $(u,v) \in E$ (or just $u \to v$ if $\mathcal{G}$ is understood). If $(u,v) \in E$ and $(v,u) \in E$, the edge between u and v is an *undirected edge* denoted by $\{u,v\} \in E$ or $u \overset{\mathcal{G}}{-} v$ (or just $u - v$). We shall use the notation $u \sim v$ to denote that $u \to v$, $v \to u$, or $u - v$. Vertices u and v are said to be *connected* in $\mathcal{G}$ if $u \overset{\mathcal{G}}{\sim} v$. If $u \to v$ and $w \to v$, then these edges are said to meet *head-to-head* at v.

If E does not contain undirected edges, then $\mathcal{G}$ is a *directed graph*, and if E does not contain directed edges, then it is an *undirected graph*. As mentioned above, we shall not deal with mixed cases of both directed and undirected edges.

A *path* $\langle v_1, \ldots, v_n \rangle$ is a sequence of distinct vertices such that $v_i \sim v_{i+1}$ for each $i = 1, \ldots, n-1$; the *length* of the path is $n-1$. The path is a *directed path* if $v_i \to v_{i+1}$ for each $i = 1, \ldots, n-1$; v_i is then an *ancestor* of v_j and v_j a *descendant* of v_i for each $j > i$. The set of ancestors and descendants of v are denoted an(v) and de(v), respectively. The set nd$(v) = V \setminus$ de$(v) \cup \{v\}$ are called the *non-descendants* of v. The *ancestral set* An$(U) \subseteq V$ of a set $U \subseteq V$ of a graph $\mathcal{G} = (V, E)$ is the set of vertices $U \cup \bigcup_{u \in U}$ an(u).

A path $\langle v_1, \ldots, v_n \rangle$ from v_1 to v_n of an undirected graph, $\mathcal{G} = (V, E)$, is *blocked* by a set $S \subseteq V$ if $\{v_2, \ldots, v_{n-1}\} \cap S \neq \emptyset$. There is a similar concept for paths of acyclic, directed graphs (see below), but the definition is somewhat more complicated (see Proposition 2.4 on page 33).

---

[1] See Sect. 2.2 for the naming conventions used for vertices and variables.

A graph $\mathcal{G} = (V, E)$ is *connected* if for any pair $\{u, v\} \subseteq V$, there is a path $\langle u, \ldots, v \rangle$ in $\mathcal{G}$. A connected graph $\mathcal{G} = (V, E)$ is a *tree* if for any pair $\{u, v\} \subseteq V$, there is a unique path $\langle u, \ldots, v \rangle$ in $\mathcal{G}$.

A *cycle* is a path, $\langle v_1, \ldots, v_n \rangle$, of length greater than two with the exception that $v_1 = v_n$; a *directed cycle* is defined in the obvious way. A directed graph with no directed cycles is called an *acyclic, directed graph* or simply a *DAG*; see Fig. 2.1a for an example. The undirected graph obtained from a DAG, $\mathcal{G}$, by replacing all its directed edges with undirected ones is known as the *skeleton* of $\mathcal{G}$.

Let $\mathcal{G} = (V, E)$ be a DAG. The undirected graph, $\mathcal{G}^m = (V, E^m)$, where

$$E^m = \{\{u, v\} \mid u \text{ and } v \text{ are connected or have a common child in } \mathcal{G}\},$$

is called the *moral graph* of $\mathcal{G}$. That is, $\mathcal{G}^m$ is obtained from $\mathcal{G}$ by first adding undirected edges between pairs of unconnected vertices that share a common child and then replacing all directed edges with undirected edges; see Fig. 2.1b for an example.

The set of vertices connected to a vertex $v \in V$ in the moral graph of a DAG $\mathcal{G} = (V, E)$ (i.e., the set of vertices $\mathrm{pa}(v) \cup \mathrm{ch}(v) \cup \bigcup_{u \in \mathrm{ch}(v)} \mathrm{pa}(u)$) is denoted the *Markov blanket* of v.

2.2   Graphical Models

On a structural (or qualitative) level, probabilistic network models are graphs with the vertices representing variables and utility functions and the edges representing different kinds of relations among the variables and utility functions.

2.2.1   Variables

A *chance variable* represents an exhaustive set of mutually exclusive events, referred to as the *domain* of the variable. These events are also often called states, levels, values, choices, options, etc. The domain of a variable can be discrete or continuous; discrete domains are always finite.

Example 2.1 (Sample Variable Domains).   Some sample variable domains can be

$$\{\mathsf{false}, \mathsf{true}\}$$
$$\{\mathsf{red}, \mathsf{green}, \mathsf{blue}\}$$
$$\{1, 3, 5, 7\}$$
$$\{-1.7, 0, 2.32, 5\}$$
$$\{< 0, 0 - 5, > 5\}$$
$$] - \infty; \infty[$$
$$\{] - \infty; 0[, [0; 5[, [5; 10[\}$$

The penultimate domain in the above list represents a domain for a continuous variable; the remaining ones represent domains for discrete variables. □

Throughout this book, we shall use capital letters (possibly indexed) to denote variables or sets of variables and lowercase letters (possibly indexed) to denote particular values of variables. Thus, $X = x$ may either denote the fact that variable X attains the value x or the fact that the set of variables $X = (X_1, \ldots, X_n)$ attains the (vector) of values $x = (x_1, \ldots, x_n)$. By $\mathrm{dom}(X) = (x_1, \ldots, x_{\|X\|})$, we shall denote the domain of X, where $\|X\| = |\mathrm{dom}(X)|$ is the number of possible distinct values of X. If $X = (X_1, \ldots, X_n)$, then $\mathrm{dom}(X)$ is the Cartesian product (or product space) over the domains of the variables in X. Formally,

$$\mathrm{dom}(X) = \mathrm{dom}(X_1) \times \cdots \times \mathrm{dom}(X_n),$$

and thus, $\|X\| = \prod_i \|X_i\|$. For two (sets of) variables X and Y, we shall write either $\mathrm{dom}(X \cup Y)$ or $\mathrm{dom}(X, Y)$ to denote $\mathrm{dom}(X) \times \mathrm{dom}(Y)$. If $z \in \mathrm{dom}(Z)$, then by z_X, we shall denote the projection of z to $\mathrm{dom}(X)$, where $X \cap Z \neq \emptyset$.

Example 2.2 (Cartesian Product and Projecting Variable Domains). Assume that $\mathrm{dom}(X) = $ (false, true) and $\mathrm{dom}(Y) = $ (red, green, blue). Then $\mathrm{dom}(X, Y) = $ ((false, red), (false, green), (false, blue), (true, red), (true, green), (true, blue)}. For $z = $ (true, blue), we get $z_X = $ true and $z_Y = $ blue. □

Chance Variables and Decision Variables

There are basically two categories of variables, namely, variables representing random events and variables representing choices under the control of some, typically human, agent. Consequently, the first category of variables is often referred to as *chance variables* (or *random variables*) and the second category as *decision variables*. Note that a random variable can depend functionally on other variables in which case it is sometimes referred to as a *deterministic (random) variable*. Sometimes, it is important to distinguish between truly random variables and deterministic variables, but unless this distinction is important, we shall treat them uniformly and refer to them simply as "random variables," or just "variables."

The problem of identifying those entities of a domain that qualify as variables is not necessarily trivial. Also, identifying the "right" set of variables can be nontrivial. These questions, however, will not be further touched upon in this chapter but will be discussed in detail in Chap. 6.

2.2.2 Vertices Vs. Variables

The notions of variables and vertices (or nodes) are often used interchangeably for models containing no decision variables and utility functions (e.g., Bayesian networks). For models that contain decision variables and utility Functions, it is

Table 2.1 The taxonomy for variables/vertices

Category	Kind	Subtype
Chance	Discrete	Labeled
Decision	Continuous	Boolean
Utility		Numbered
		Interval

Note that the subtype dimension only applies for discrete chance and decision variables

convenient to distinguish between variables and vertices, as a vertex does not necessarily represent a variable. In this book, we shall therefore maintain that distinction.

As indicated above, we shall use lowercase letters like u, v, w (or sometimes α, β, γ, etc.) to denote vertices and uppercase letters like U, V, W to denote sets of vertices. Vertex names will sometimes be used in the subscripts of variable names to identify the variables corresponding to vertices. For example, if v is a vertex representing a variable, then we denote that variable by X_v. If v represents a utility function, then $X_{pa(v)}$ denotes the domain of the function, which is a set of chance and/or decision variables.

2.2.3 Taxonomy of Vertices/Variables

For convenience, we shall use the following terminology for classifying variables and/or vertices of probabilistic networks.

First, as discussed above, there are three main classes of vertices in probabilistic networks, namely, vertices representing chance variables, vertices representing decision variables, and vertices representing utility functions. We define the *category* of a vertex to represent this dimension of the taxonomy.

Second, chance and decision variables as well as utility functions can be discrete or continuous. This dimension of the taxonomy will be characterized by the *kind* of the variable or vertex.

Finally, for discrete chance and decision variables, we shall distinguish between labeled, Boolean, numbered, and interval variables. For example, referring to Example 2.1 on page 20, the first domain is the domain of a Boolean variable, the second and the fifth are domains of labeled variables, the third and the fourth are domains of numbered variables, and the last is the domain of an interval variable. This dimension of the taxonomy is referred to by the *subtype* of discrete variables and is useful for providing mathematical expressions of specifications of conditional probability tables and utility tables, as discussed in Chap. 6.

Table 2.1 summarizes the variable/vertex taxonomy.

Table 2.2 Vertex symbols

Category	Kind	Symbol
Chance	Discrete	⬭
	Continuous	⬭
Decision	Discrete	▢
	Continuous	▣
Utility	Discrete	◇
	Continuous	◈

Table 2.3 Notation used for vertices, variables, and utility functions

S, U, V, W	Sets of vertices
V	Set of vertices of a model
V_Δ	The subset of V that represent discrete variables
V_Γ	The subset of V that represent continuous variables
$u, v, w, \ldots$	Vertices
$\alpha, \beta, \gamma, \ldots$	Vertices
X, Y_i, Z_k	Variables or sets of variables
X_W	Subset of variables corresponding to set of vertices W
$\mathcal{X}$	The set of variables of a model; note that $\mathcal{X} = \mathcal{X}_V$
$\mathcal{X}_W$	Subset of $\mathcal{X}$, where $W \subseteq V$
X_u, X_α	Variables corresponding to vertices u and α, respectively
x, y_i, z_k	Configurations/states of (sets of) variables
x_Y	Projection of configuration x to dom(Y), $X \cap Y \neq \emptyset$
$\mathcal{X}_C$	The set of chance variables of a model
$\mathcal{X}_D$	The set of decision variables of a model
$\mathcal{X}_\Delta$	The subset of discrete variables of $\mathcal{X}$
$\mathcal{X}_\Gamma$	The subset of continuous variables of $\mathcal{X}$
$\mathcal{U}$	The set of utility functions of a model
V_U	The subset of V representing utility functions
$u(X)$	Utility function $u \in \mathcal{U}$ with the set of variables X as domain

2.2.4 Vertex Symbols

Throughout this book, we shall be using ovals to indicate discrete chance variables, rectangles to indicate discrete decision variables, and diamonds to indicate discrete utility functions. Continuous variables and utility functions are indicated with double borders. See Table 2.2 for an overview.

2.2.5 Summary of Notation

Table 2.3 summarizes the notation used for vertices (upper part), variables (middle part), and utility functions (lower part).

Fig. 2.2 (a) Hard evidence
on X. (b) Soft (or hard)
evidence on X

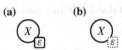

2.3 Evidence

A key inference task with a probabilistic network is computation of posterior
probabilities of the form $P(x \mid \varepsilon)$, where, in general, ε is *evidence* (i.e., information)
received from external sources in the form of a *likelihood distribution* over the states
of a set of variables, X, of the network, also often called an *evidence function*
(or *potential*[2]) for X. An evidence function, $\mathcal{E}_X$, for a set, X, of variables is a
function $\mathcal{E}_X : \mathrm{dom}(X) \to \mathbb{R}^+$.

Example 2.3 (Evidence Function).   If $\mathrm{dom}(X) = (x_1, x_2, x_3)$, then $\mathcal{E}_X = (1, 0, 0)$
is an evidence function indicating that $X = x_1$ with certainty. If $\mathcal{E}_X = (1, 2, 0)$,
then with certainty $X \neq x_3$ and $X = x_2$ is twice as likely as $X = x_1$.   □

An evidence function that assigns a zero probability to all but one state is often
said to provide *hard evidence*; otherwise, it is said to provide *soft evidence*.[3] We
shall often leave out the "hard" or "soft" qualifier and simply talk about evidence if
the distinction is immaterial. Hard evidence on a variable X is also often referred to
as *instantiation* of X, or we say that X has been *observed*.

We shall attach the label $\boxed{\varepsilon}$ to vertices representing variables with hard evidence
and the label $\vdots\overline{\varepsilon}\vdots$ to vertices representing variables with either soft or hard evidence.
For example, hard evidence on variable X (like $\mathcal{E}_X = (1, 0, 0)$ in Example 2.3)
is indicated as shown in Fig. 2.2a, and soft evidence (like $\mathcal{E}_X = (1, 2, 0)$ in
Example 2.3) is indicated as shown in Fig. 2.2b.

2.4 Causality

Causality plays an important role in the process of constructing probabilistic
network models. There are a number of reasons why proper modeling of causal
relations is important or helpful, although, in a Bayesian network model, it is not
strictly necessary to have the directed links of the model follow a causal interpretation. In models with explicit representation of decisions (influence diagrams), the
directed links into chance variables must represent causal relations. We shall only

---

[2]See Sect. 3.3 on page 46.

[3]In the literature, soft evidence is often called *virtual evidence*.

briefly touch upon the issue of causality and stress a few important points about causal modeling. The reader is referred to Pearl's work for an in-depth treatment of the subject (Pearl 2000).

A variable X is said to be a direct cause of Y if setting the value of X by force, the value of Y may change and there is no other variable Z that is a direct cause of Y such that X is a direct cause of Z; see Pearl (2000) for details.

As an example, consider the variables Flu and Fever. Common sense tells us that flu is a cause of fever, not the other way around. This fact can be verified from the thought experiment of forcefully setting the states of Flu and Fever: Killing fever with an aspirin or by taking a cold shower will have no effect on the state of Flu, whereas eliminating a flu would make the body temperature go back to normal (assuming flu is the only effective cause of fever).

To correctly represent the dependence and independence relations that exist among a set of variables of a problem domain, it is useful to have the causal relations among the variables represented in terms of directed links from causes to effects. That is, if X is a direct cause of Y, we should make sure to add a directed link from X to Y. If done the other way around (i.e., $Y \rightarrow X$), we may end up with a model that does not properly represent the dependence and independence relations of the problem domain. In subsequent sections, we shall see several examples of the importance of respecting the causal relations in this sense.

That said, however, in a Bayesian network, one does not *have* to construct a model where the links can be interpreted as causal relations; it just makes the model much more intuitive, eases the process of getting the dependence and independence relations right, and significantly eases the process of eliciting the conditional probabilities of the model. In Sect. 2.5.5 on page 31, we shall briefly return to the issue of the importance of correctly modeling the causal relationships in probabilistic networks.

2.5 Flow of Information in Causal Networks

As mentioned above, the DAG of a probabilistic network model is a graphical representation of the dependence and independence properties of the joint probability distribution of the model. In this section, we shall see how to read these properties from a DAG. In doing this, it is convenient to consider each possible basic kind of connection that can exist in a DAG.

To illustrate the different kinds of connections, consider the example in Fig. 2.3 on the following page, which shows the structure of a probabilistic network for the following small fictitious example, where each variable has two possible states, no and yes.

Example 2.4 (Burglary or Earthquake (Pearl 1988)). Mr. Holmes is working in his office when he receives a *phone call* from his neighbor Dr. Watson, who tells him that Holmes' *burglar alarm* has gone off. Convinced that a *burglar* has broken into

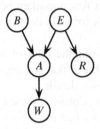

- W: Phone call from Watson
- A: Burglary alarm
- B: Burglary
- R: Radio news
- E: Earthquake

Fig. 2.3 Structure of a probabilistic network model for the "Burglary or Earthquake" story of Example 2.4 on the previous page.

his house, Holmes rushes to his car and heads for home. On his way, he listens to the radio, and in the *news*, it is reported that there has been a small *earthquake* in the area. Knowing that earthquakes have a tendency to make burglar alarms go off, he returns to his work.   □

Notice that all of the links in the network of Fig. 2.3 are causal: Burglary or earthquake can cause the alarm to go off, earthquake can cause a report on earthquake in the radio news, and the alarm can cause Dr. Watson to call Mr. Holmes.

We see three different kinds of connections in the network of Fig. 2.3:

- Two *serial connections* $B \to A \to W$ and $E \to A \to W$
- One *diverging connection* $A \leftarrow E \to R$
- One *converging connection* $B \to A \leftarrow E$

In the following subsections, we discuss each of these three possible kinds of connections in terms of their ability to transmit information given evidence and given no evidence on the middle variable, and we shall see that it is the converging connection that provides the ability of probabilistic networks to perform intercausal reasoning (explaining away).

2.5.1   *Serial Connections*

Let us consider the serial connection (causal chain) depicted in Fig. 2.4 on the facing page, referring to Example 2.4 on the previous page.

We need to consider two cases, namely, with and without hard evidence (see Sect. 2.3 on page 24) on the middle variable (Alarm).

First, assume we do not have definite knowledge about the state of Alarm. Then evidence about Burglary will make us update our belief about the state of Alarm, which in turn will make us update our belief about the state of Watson_calls. The opposite is also true: If we receive information about the state of Watson_calls, that will influence our belief about the state of Alarm, which in turn will influence our belief about Burglary.

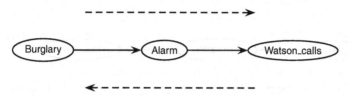

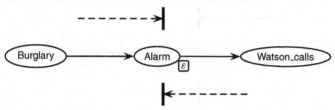

Fig. 2.4 Serial connection (causal chain) with no hard evidence on Alarm. Evidence on Burglary will affect our belief about the state of Watson_calls and vice versa

Fig. 2.5 Serial connection (causal chain) with hard evidence on Alarm. Evidence on Burglary will have no affect on our belief about the state of Watson_calls and vice versa

So, in conclusion, as long as we do not know the state of Alarm for sure, information about either Burglary or Watson_calls will influence our belief on the state of the other variable. This is illustrated in Fig. 2.4 by the two dashed arrows, signifying that evidence may be transmitted through a serial connection as long as we do not have definite knowledge about the state of the middle variable.

Now, assume we do have definite knowledge about the state of Alarm (see Fig. 2.5). Then a change in our belief about the state of Burglary (Watson_calls) given a change in our belief about Watson_calls (Burglary) can only happen through a change in our belief on Alarm which then in turn will make us change our belief about Burglary (Watson_calls). But, since our belief on the state of Alarm cannot be changed, as we have observed its state with certainty, any information about the state of Watson_calls (Burglary) will not make us change our belief about Burglary (Watson_calls).

In conclusion, when the state of the middle variable of a serial connection is known for sure (i.e., we have hard evidence on it), then transmission of evidence between the other two variables cannot take place through this connection. This is illustrated in Fig. 2.5 by the two dashed arrows ending at the observed variable, indicating that transmission of evidence is blocked.

The general rule for transmission of evidence in serial connections can thus be stated as follows:

Proposition 2.1 (Serial Connection). *Information may be transmitted through a serial connection $X \rightarrow Y \rightarrow Z$ unless the state of Y is known.*

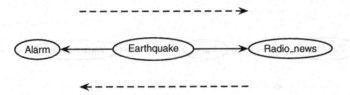

Fig. 2.6 Diverging connection with no evidence on Earthquake. Evidence on Alarm will affect our belief about the state of Radio_news and vice versa

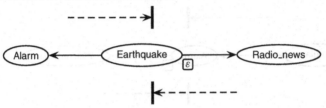

Fig. 2.7 Diverging connection with hard evidence on Earthquake. Evidence on Alarm will not affect our belief about the state of Radio_news and vice versa

2.5.2   Diverging Connections

Consider the diverging connection depicted in Fig. 2.6, referring to Example 2.4 on page 25.

Again, we consider the cases with and without hard evidence on the middle variable (Earthquake).

First, assume we do not know the state of Earthquake for sure. Then receiving information about Alarm will of course influence our belief about Earthquake, as earthquake is a possible explanation for alarm. The updated belief about the state of Earthquake will in turn make us update our belief about the state of Radio_news. The opposite case (i.e., receiving information about Radio_news) will, of course, lead to a similar conclusion. So, we get a result that is similar to the result for serial connections, namely, that evidence can be transmitted through a diverging connection if we do not have definite knowledge about the state of the middle variable. This result is illustrated in Fig. 2.6.

Next, assume the state of Earthquake is known for sure (i.e., we have received hard evidence on that variable). Now, if information is received about the state of either Alarm or Radio_news, then this information is not going to change our belief about the state of Earthquake, and consequently, we are not going to update our belief about the other, yet unobserved, variable. Again, this result is similar to the case for serial connections and is illustrated in Fig. 2.7.

The general rule for transmission of evidence in diverging connections can be stated as follows:

Proposition 2.2 (Diverging Connection). *Information may be transmitted through a diverging connection $X \leftarrow Y \rightarrow Z$ unless the state of Y is known.*

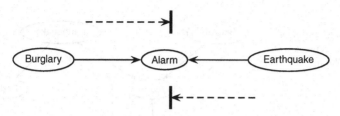

Fig. 2.8 Converging connection with no evidence on Alarm or any of its descendants. Information about Burglary will not affect our belief about the state of Earthquake and vice versa

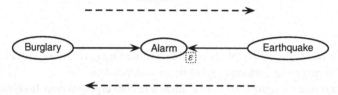

Fig. 2.9 Converging connection with (possibly soft) evidence on Alarm or any of its descendants. Information about Burglary will affect our belief about the state of Earthquake and vice versa

2.5.3 Converging Connections

Consider the converging connection depicted in Fig. 2.8, referring to Example 2.4 on page 25.

First, if no evidence is available about the state of Alarm, then information about the state of Burglary will not provide any derived information about the state of Earthquake. In other words, burglary is not an indicator of earthquake and vice versa (again, of course, assuming correctness of the model). Thus, contrary to serial and diverging connections, a converging connection will not transmit information if no evidence is available for the middle variable. This fact is illustrated in Fig. 2.8.

Second, if evidence is available on Alarm, then information about the state of Burglary will provide an explanation for the evidence that was received about the state of Alarm and thus either confirm or disconfirm Earthquake as the cause of the evidence received for Alarm. The opposite, of course, also holds true. Again, contrary to serial and diverging connections, converging connections allow transmission of information whenever evidence about the middle variable is available. This fact is illustrated in Fig. 2.9.

The rule illustrated in Fig. 2.8 tells us that if nothing is known about a common effect of two (or more) causes, then the causes are independent; that is, receiving information about one of them will have no impact on the belief about the other(s). However, as soon as some evidence is available on a common effect, the causes become dependent. If, for example, Mr. Holmes receives a phone call from Dr. Watson, telling him that his burglar alarm has gone off, burglary and earthquake become competing explanations for this effect, and receiving information about the possible state of one of them obviously either confirms or disconfirms the other one as the cause of the (possible) alarm. Note that even if the information received

Fig. 2.10 Flow of
information in the "Burglary
or Earthquake" network with
evidence on Alarm

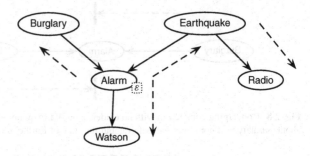

from Dr. Watson might not be totally reliable (amounting to receiving soft evidence
on Alarm), Burglary and Earthquake still become dependent.

The general rule for transmission of evidence in converging connections can then
be stated as follows:

Proposition 2.3 (Converging Connection). *Information may only be transmitted through a* converging connection $X \to Y \leftarrow Z$ *if evidence on Y or one of
its descendants is available.*

2.5.4 Intercausal Inference (Explaining Away)

Before we conclude this section by discussing the importance of correct modeling
of causality, let us dwell a bit on what we have learned so far concerning flow of
information in DAGs.

Example 2.5 (The Power of DAGs).   Consider the "Burglary or Earthquake" model
in Example 2.4 on page 25 and assume that the burglar alarm goes off (i.e., we have
evidence on variable Alarm); see Fig. 2.10.

Using Propositions 2.1 and 2.2, we find that information flows to all the
remaining variables of the network, indicated by the dashed arrows in Fig. 2.10.
The information flowing opposite the direction of the causal links (indicated by
the dashed arrows pointing upwards) represent abductive (or diagnostic) reasoning,
whereas information flowing in the direction of the causal links (indicated by the
dashed arrows pointing downwards) represent deductive (or causal) reasoning.

Now, assume that we receive an additional piece of evidence, namely, a radio
report on a recent earthquake, which provides strong evidence that the earthquake
is responsible for the burglar alarm. Thus, we have now also evidence on variable
Radio; see Fig. 2.11 on the facing page.

Fig. 2.11 Flow of
information in the "Burglary
or Earthquake" network with
additional evidence on Radio
after evidence on Alarm

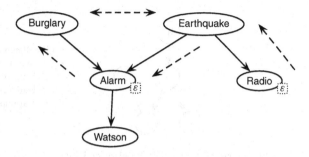

Using Propositions 2.2 and 2.3, we find that information flows to variables
Earthquake and Burglary, indicated by the dashed arrows in Fig. 2.11. The (hard)
evidence on variable Alarm prevents information from flowing down to variable Watson.

The property of converging connections, $X \to Y \leftarrow Z$, that information about
the state of X (Z) provides an explanation for an observed effect on Y, and hence
confirms or disconfirms Z (X) as the cause of the effect, is often referred to as the
explaining-away effect or as *intercausal inference*.

The flow of information from Radio to Burglary is an example of the explaining-
away effect. The bidirected dashed arrow in Fig. 2.11 indicates the intercausal
inference taking place between Earthquake and Burglary, in this case explaining
away burglary as the cause of the alarm.   □

As shown by the example, representing causal interactions between domain
variables by way of a DAG provides a compact yet powerful means of performing
deductive, abductive, and intercausal reasoning.

The ability to perform intercausal inference is unique for graphical models
and is one of the key differences between automatic reasoning systems based on
probabilistic networks and systems based on, for example, production rules. In a
rule-based system, we would need dedicated rules for taking care of intercausal
reasoning.

2.5.5 The Importance of Correct Modeling of Causality

It is a common modeling mistake to let arrows point from effect to cause, leading to
faulty statements of (conditional) dependence and independence and, consequently,
faulty inference. For example, in the "Burglary or Earthquake" example on page 25
one might put a directed link from W (Watson_calls) to A (Alarm) because the
fact that Dr. Watson makes a phone call to Mr. Holmes "points to" the fact that

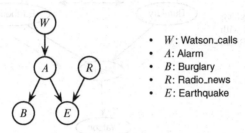

- W: Watson_calls
- A: Alarm
- B: Burglary
- R: Radio_news
- E: Earthquake

Fig. 2.12 Wrong model for the "Burglary or Earthquake" story of Example 2.4 on page 25, where the links are directed from effects to causes, leading to faulty statements of (conditional) dependence and independence

Mr. Holmes' alarm has gone off. Experience shows that this kind of reasoning is common when people are building their first probabilistic networks and is probably due to a mental flow of information model, where evidence acts as the "input" and the derived conclusions as the "output."

Using this faulty modeling approach, the "Burglary or Earthquake" model in Fig. 2.3 on page 26 would have all its links reversed (see Fig. 2.12). Using Proposition 2.2 on page 28 on the model in Fig. 2.12, we find that B and E are dependent when nothing is known about A, and, using Proposition 2.3 on page 30, we find that A and R are dependent whenever evidence about E is available. Neither of these conclusions are, of course, true and will make the model make wrong inferences.

Having the causal relations among domain variables be mapped to directed links $X \rightarrow Y$, where X is a cause of Y, is thus (at least) helpful, if not crucial, to having the model correctly represent the dependence and independence properties of the problem domain.

Another reason why respecting a causal modeling approach is important stems from the potential difficulties in specifying the conditional probability of $X = x$ given that $Y = y$ when $Y \rightarrow X$ does not reflect a causal relationship. For example, it might be difficult for Mr. Holmes to specify the probability that a burglar has broken into his house given that he knows the alarm has gone off, as the alarm might have gone off for other reasons. Thus, specifying the probability that the alarm goes off given its possible causes might be easier and more natural, providing a sort of complete description of a local phenomenon. We shall leave the discussion of this important issue for now and resume in Chaps. 4 and 6.

2.6 Two Equivalent Irrelevance Criteria

Propositions 2.1–2.3 comprise the components needed to formulate a general rule for reading off the statements of relevance and irrelevance relations for two (sets of) variables, possibly given a third variable (or set of variables). This general rule is known as the d-separation criterion (where d stands for directional) and is due to Pearl (1988).

In Chap. 3, we show that for any joint probability distribution that factorizes according to a DAG, $\mathcal{G}$ (see Chap. 3 for a definition), independence statements involving variables X_u and X_v (again, see Chap. 3 for a definition) are equivalent to similar statements about d-separation of vertices u and v in $\mathcal{G}$.

Thus, the d-separation criterion may be used to answer queries of the kind "are X and Y independent given Z" (in a probabilistic sense) or, more generally, queries of the kind "is information about X irrelevant for our belief about the state of Y given information about Z," where X and Y are individual variables and Z is either the empty set of variables or an individual variable.

The d-separation criterion may also be used with sets of variables, although this may be cumbersome. On the other hand, answering such queries is efficient using the directed global Markov criterion (Lauritzen, Dawid, Larsen & Leimer 1990b), which is a criterion that is equivalent to the d-separation criterion.

As statements of (conditional) d-separation play a key role in probabilistic networks, some shorthand notation is convenient. We shall use the standard notation $u \perp_\mathcal{G} v$ to denote that vertices u and v are d-separated in DAG $\mathcal{G}$, or simply $u \perp v$ if $\mathcal{G}$ is obvious from the context. By $u \perp v \,|\, w$, we denote the statement that u and v are d-separated given (hard) evidence on w. By $U \perp V$, we denote the fact that $u \perp v$ for each $u \in U$ and each $v \in V$. If vertices u and v are not d-separated, we shall often say that they are d-connected. We shall use $\not\perp$ to denote d-connection.

Example 2.6 (Burglary or Earthquake, page 25). Some of the d-separation/ d-connection properties observed in the "Burglary or Earthquake" example are:

(1) Burglary $\perp$ Earthquake
(2) Burglary $\not\perp$ Earthquake | Alarm
(3) Burglary $\perp$ Radio_report
(4) Burglary $\perp$ Watson_calls | Alarm

$\square$

Also, notice that d-separation and d-connection depend on the information available; that is, it depends on what you know (and do not know). Also, note that d-separation and d-connection relations are always symmetric; that is, $u \perp v \equiv v \perp u$.

2.6.1 d-Separation Criterion

Propositions 2.1–2.3 can be summarized into a rule known as *d-separation* (Pearl 1988):

Proposition 2.4 (d-Separation). *A path $\pi = \langle u, \ldots, v \rangle$ in a DAG, $\mathcal{G} = (V, E)$, is said to be blocked by $S \subseteq V$ if π contains a vertex w such that either*

- $w \in S$ *and the edges of π do not meet head-to-head at w, or*

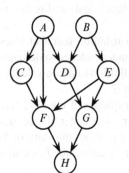

(1) C and G are d-connected

(2) C and E are d-separated

(3) C and E are d-connected given evidence on G

(4) A and G are d-separated given evidence on D and E

(5) A and G are d-connected given evidence on D

Fig. 2.13 Sample DAG with a few sample dependence (d-connected) and independence (d-separated) statements

- $w \notin S$, de$(w) \cap S = \emptyset$, and the edges of π meet head-to-head at w.

For three disjoint subsets A, B, S of V, A and B are said to be d-separated if all paths between A and B are blocked by S.

Proposition 2.4 on the previous page says, for example, that two vertices u and v are d-separated if for each path between u and v, there is a vertex w such that the edges of the path meet head-to-head at w.

Example 2.7 (d-Separation). We may use Proposition 2.4 to determine if, for example, variables C and G are d-separated in the DAG in Fig. 2.13; that is, are C and G independent when no evidence about any of the variables is available? First, we find that there is a diverging connection $C \leftarrow A \rightarrow D$ allowing transmission of information from C to D via A. Second, there is a serial connection $A \rightarrow D \rightarrow G$ allowing transmission of information from A to G via D. So, information can thus be transmitted from C to G via A and D, meaning that C and G are not d-separated (i.e., they are d-connected).

C and E, on the other hand, are d-separated since each path from C to E contains a converging connection, and since no evidence is available, each such connection will not allow transmission of information. Given evidence on one or more of the variables in the set $\{D, F, G, H\}$, C and E will, however, become d-connected. For example, evidence on H will allow the converging connection $D \rightarrow G \leftarrow E$ to transmit information from D to E via G, as H is a child of G. Then information may be transmitted from C to E via the diverging connection $C \leftarrow A \rightarrow D$ and the converging connection $D \rightarrow G \leftarrow E$.   □

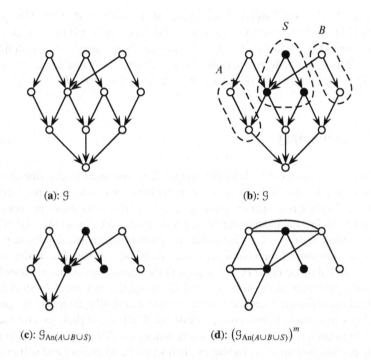

Fig. 2.14 (a) A DAG, $\mathcal{G}$. (b) $\mathcal{G}$ with subsets A, B, and S indicated, where the variables in S are assumed to be observed. (c) The subgraph induced by the ancestral set of $A \cup B \cup S$. (d) The moral graph of the DAG of part (c)

2.6.2 Directed Global Markov Criterion

The directed global Markov criterion (Lauritzen et al. 1990b) provides a criterion that is equivalent to that of the d-separation criterion but which in some cases may prove more efficient in terms of requiring less inspections of possible paths between the involved vertices of the graphs.

Proposition 2.5 (Directed Global Markov Criterion). *Let $\mathcal{G} = (V, E)$ be a DAG and A, B, S be disjoint sets of V. Then each pair of vertices $(\alpha \in A, \beta \in B)$ are d-separated by S whenever each path from α to β is blocked by vertices in S in the graph*

$$\left(\mathcal{G}_{\mathrm{An}(A \cup B \cup S)}\right)^m.$$

Although the criterion might look somewhat complicated at a first glance, it is actually quite easy to apply. The criterion says that $A \perp\!\!\!\perp_\mathcal{g} B \,|\, S$ if all paths from A to B passes at least one vertex in S in the moral graph of the sub-DAG induced by the ancestral set of $A \cup B \cup S$.

Example 2.8 (Directed Global Markov Criterion). Let us consider the DAG, $\mathcal{g} = (V, E)$, in Fig. 2.14a, and let the subsets $A, B, S \subseteq V$ be given as shown in Fig. 2.14b. That is, we ask if $A \perp\!\!\!\perp_\mathcal{g} B \,|\, S$. Using Proposition 2.5, we first remove each vertex not belonging to the ancestral set An($A \cup B \cup S$). This gives us the DAG in Fig. 2.14c. Second, we moralize the resulting sub-DAG, which gives us the undirected graph in Fig. 2.14d. Then, to answer the query, we consult this graph to see if there is a path from a vertex in A to a vertex in B that do not contain a vertex in S. Since this is indeed the case, we conclude that $A \not\perp_g B \mid S$.   □

2.7 Summary

In this chapter, we first defined some key concepts used to describe the qualitative part (i.e., the graphical structure) of probabilistic networks that are given by DAGs. We defined the notion of the moral graph of a DAG, which plays an important role in understanding the independence properties represented by a DAG and in generating a junction tree for making inference in a probabilistic network (cf. Chap. 5).

We introduced the taxonomy of variables and vertices (the nodes of the DAG of a probabilistic network that represent the chance and decision variables and the utility functions of the network) and discussed the notions of product spaces over the domains of variables and projections down to smaller-dimensional spaces, which play a crucial role in inference processes (in Chap. 3, we shall see how the operation of projection is defined for probability functions). Also, the notion of evidence was briefly touched upon, including the distinction between hard and soft evidence.

To understand the notion of (conditional) independence in probabilistic networks, we discussed the three fundamental constructs (serial, diverging, and converging connections) in terms of which any path $X \sim Y \sim Z$ of a DAG can be described. When the directed links of a DAG can be interpreted as causal links, the three fundamental constructs have clear relevance properties, which can be described collectively by means of the d-separation criterion that can be used to determine if information about one set of variables is relevant to another set of variables, possibly given information about the states of a third set of variables (i.e., can information flow from the first set to the second set given the third set). An equivalent criterion (the directed, global Markov criterion), which uses the notion of moral graphs, was also presented.

In Chap. 3, we shall see that if a joint probability distribution factorizes according to the structure of a DAG, then the DAG is a graphical representation of the independence properties of the distribution. We briefly discussed the importance of having the directed edges of a DAG of a probabilistic network represent causal relations (i.e., the directed links should point from cause to effect). Otherwise, problems might arise both in terms of getting the right (conditional) independence properties of the network and in terms of ease of specification of the conditional probabilities (parameters) of the network. This important issue will be further discussed in Chaps. 4 and 6.

Exercises

Exercise 2.1. For the discrete variables X, Y, and Z assume that $\text{dom}(X) = (0, 1)$, $\text{dom}(Y) = (\text{good}, \text{bad})$, and $\text{dom}(Z) = (\text{low}, \text{average}, \text{high})$.

(a) What is $\|(X, Y, Z)\|$?
(b) Specify $\text{dom}(X, Y = \text{good}, Z)$.
(c) Let $w = (1, \text{good}, \text{high})$. Specify $w_{\{X,Z\}}$ and w_Y.
(d) Specify evidence functions for Z that indicate:

    (i) $Z = \text{low}$ with certainty
    (ii) $Z \neq \text{low}$ and $Z = \text{high}$ is three times as likely as $Z \neq \text{average}$

Exercise 2.2. Direct the links below such that the parent nodes represent the causes and the child nodes represent the effects.

(a) Flu — Fever
(b) State_of_battery — Age_of_battery
(c) Living_standard — Education
(d) Age — Number_of_children
(e) Occupation — Education
(f) Fake_die — Number_of_6s

Exercise 2.3. Consider the DAG in Fig. 2.13 on page 34. Use the d-separation criterion to test which of the following statements are true.

(a) $A \perp B$
(b) $A \perp B \mid C$
(c) $A \perp B \mid \{C, D\}$
(d) $B \not\perp F$
(e) $B \perp F \mid E$
(f) $B \perp F \mid \{D, E\}$
(g) $F \perp G$
(h) $F \perp G \mid E$
(i) $F \perp G \mid \{A, E\}$

Exercise 2.4. Draw the graphs $\left(\mathcal{G}_{\text{An}(A \cup B \cup S)}\right)^m$ with $\mathcal{G}$ given by the DAG in Fig. 2.13 on page 34 and A, B, S given in Exercise 2.3(a)–(i) and verify your answers in Exercise 2.3 using the directed global Markov criterion.

Chapter 3
Probabilities

As mentioned in Chap. 2, probabilistic networks have a qualitative aspect and a corresponding quantitative aspect, where the qualitative aspect is given by a graphical structure in the form of a DAG that represents the (conditional) dependence and independence properties of a joint probability distribution defined over a set of variables that are indexed by the vertices of the DAG.

The fact that the structure of a probabilistic network can be characterized as a DAG derives from basic axioms of probability calculus leading to recursive factorization of a joint probability distribution into a product of lower-dimensional conditional probability distributions. First, any joint probability distribution can be decomposed (or factorized) into a product of conditional distributions of different dimensionality, where the dimensionality of the largest distribution is identical to the dimensionality of the joint distribution.[1] This gives rise to a densely connected DAG. Second, statements of local conditional independences manifest themselves as reductions of dimensionalities of some of the conditional probability distributions. Most often, these independence statements give rise to dramatic reductions of complexity of the DAG such that the resulting DAG appears to be quite sparse.

In fact, a joint probability distribution, P, can be decomposed recursively in this fashion if and only if there is a DAG that correctly represents the (conditional) dependence and independence properties of P. This means that a set of conditional probability distributions specified according to a DAG, $\mathcal{G} = (V, E)$, (i.e., a distribution $P(A \mid \mathrm{pa}(A))$ for each $A \in V$) define a joint probability distribution.

Therefore, a probabilistic network model can be specified either through direct specification of a joint probability distribution or through a DAG (typically) expressing cause–effect relations and a set of corresponding conditional probability distributions. Obviously, a model is almost always specified in the latter fashion.

---

[1] The dimensionality of a function is defined as the number of variables over which the function is defined.

U.B. Kjærulff and A.L. Madsen, *Bayesian Networks and Influence Diagrams: A Guide to Construction and Analysis*, ISS 22, DOI 10.1007/978-1-4614-5104-4_3,
© Springer Science+Business Media New York 2013

This chapter presents some basic axioms of probability calculus from which the famous Bayes' rule follows as well as the chain rule for decomposing a joint probability distribution into a product of conditional distributions. We shall also present the fundamental operations needed to perform inference in probabilistic networks.

3.1 Basics

This section introduces some axioms of probability theory and the fundamental concept of conditional probability, which provides the basis for probability calculus in discrete Bayesian networks.

3.1.1 Events

The language of probability consists of statements (propositions) about probabilities of events. The probability of an event a is denoted $P(a)$. An *event* can be considered as an outcome of an experiment (e.g., a coin flip), a particular observation of a value of a variable (or set of variables), an assignment of a value to a variable (or set of variables), etc. As a probabilistic network define a probability distribution over a set of variables, V, in our context, an event is a configuration, $x \in \text{dom}(X)$, (i.e., a vector of values) of a subset of variables $X \subseteq V$. $\text{dom}(X)$ is sometimes called an *event space*.

Example 3.1 (Burglary or Earthquake, page 25). Assume we observe $W = $ yes and $R = $ yes. This evidence is given by the event $\varepsilon = (W = \text{yes}, R = \text{yes}) \in \text{dom}(W) \times \text{dom}(R)$, where $\text{dom}(W) \times \text{dom}(R)$ is the event space. The probability $P(\varepsilon)$ denotes the probability of this event, namely, the probability that both $W = $ yes and $R = $ yes have been observed. $\square$

Let S denote an event space and let $A, B \subseteq S$. Even though A and B are subsets of S (i.e., subsets of events), we might sometimes refer to them simply as events. Elementary events are denoted by lowercase letters; for example, $a \in A \subseteq S$. By $\neg A$, we denote $S \setminus A$.

3.1.2 Axioms

The following three axioms, known as Kolmogorov's axioms or the axioms of probability, provide the basis for (Bayesian) probability calculus.

Axiom 3.1. $P(A) \in \mathbb{R} \land P(A) \geq 0$ *for any event* $A \subseteq S$.

Axiom 3.2. $P(S) = 1$.

Axiom 3.3. *For any two mutually exclusive events A and B, the probability that either A or B occur is*

$$P(A \ or \ B) \equiv P(A \cup B) = P(A) + P(B).$$

In general, if events $A_1, \ldots, A_n$ are pairwise mutually exclusive, then

$$P\left(\bigcup_i^n A_i\right) = P(A_1) + \cdots + P(A_n) = \sum_i^n P(A_i).$$

Axiom 3.1 simply says that a probability is a nonnegative real number.

Axiom 3.2 says that the probability that some elementary event in the event space will occur is 1. For example, the probability that the flip of a coin will result in either heads or tails is $P(\{\text{heads}, \text{tails}\}) = 1$ (assuming that $\{\text{heads}, \text{tails}\}$ is an exhaustive set of possible states of a coin).

Axiom 3.3 says that if two events cannot co-occur (i.e., they are mutually exclusive), then the probability that either one of them occurs equals the sum of the probabilities of their individual occurrences.

Example 3.2 (Probability of Mutually Exclusive Events). Consider the events "The roll of the die yields a 1" and "The roll of the die yields a 6." Obviously, these events are mutually exclusive, and the probability that one of them is true equals the sum of the probabilities that the first event is true and that the second event is true. Thus, intuitively, Axiom 3.3 makes sense.   □

Note that if a set of events, $\{a_1, \ldots, a_n\}$, is an exhaustive set of outcomes of some experiment (e.g., rolling a die), then $\sum_i P(a_i) = 1$.

Some consequences of the axioms are that $P(\neg A) = 1 - P(A)$ and that $P(A) \leq P(B)$ if $A \subseteq B$. The latter is called the monotonicity property, from which it follows that $P(A) \leq 1$ for all $A \subseteq S$. In the general case with A and B not necessarily being mutually exclusive, it furthermore follows that $P(A \cup B) = P(A) + P(B) - P(A \cap B)$. For proofs, see, for example, Grinstead & Snell (1997).

3.1.3   Conditional Probability

The basic concept in the Bayesian treatment of uncertainty is that of *conditional probability*: Given event B, the conditional probability of event A is x, written as

$$P(A \mid B) = x.$$

This means that if B has occurred and everything else known is irrelevant for A, then the probability that A occurs is x.

Example 3.3 (Burglary or Earthquake, page 25). Let us assume that the alarm sounds in eight of every ten cases when there is an earthquake but no burglary. This fact would then be expressed as the conditional probability $P(A = \text{yes} | B = \text{no}, E = \text{yes}) = 0.8$. $\qquad\qquad\qquad\qquad\qquad\qquad\qquad\qquad\qquad\qquad\quad$ $\square$

Definition 3.1 (Fundamental Rule). For any two events A and B, the probability that both A and B occur is

$$P(A \text{ and } B) \equiv P(A, B) = P(B|A)P(A) = P(A|B)P(B).$$

$P(A, B)$ is called the *joint probability* of the events A and B.

Definition 3.1 says that the probability of the co-occurrence of two events, A and B, can be computed as the product of the probability of event A (B) occurring conditional on the fact that event B (A) has already occurred and the probability of event B (A) occurring.

Example 3.4 (Balls in An Urn). Assume we have an urn with 2 red, 3 green, and 5 blue balls. The probabilities of picking a red, a green, or a blue ball are

$$P(\text{red}) = \frac{2}{10} = 0.2, \quad P(\text{green}) = \frac{3}{10} = 0.3, \quad P(\text{blue}) = \frac{5}{10} = 0.5.$$

By Axiom 3.3 on the previous page, we get the probability of picking either a green or a blue ball:

$$P(\text{green or blue}) = P(\text{green}) + P(\text{blue}) = 0.8.$$

Similarly, the probability of picking either a red, a green, or a blue is 1. Without replacement, the color of the second ball depends on the color of the first ball. If we first pick a red ball (and keep it), then the probabilities of picking a red, a green, or a blue ball as the next one are, respectively,

$$P(\text{2nd-is-red} | \text{1st-was-red}) = \frac{2-1}{10-1} = \frac{1}{9},$$

$$P(\text{2nd-is-green} | \text{1st-was-red}) = \frac{3}{10-1} = \frac{3}{9},$$

$$P(\text{2nd-is-blue} | \text{1st-was-red}) = \frac{5}{10-1} = \frac{5}{9}.$$

By Definition 3.1, we get the probability that the 1st ball is red and the 2nd is red:

$$P(\text{1st-was-red}, \text{2nd-is-red}) = P(\text{2nd-is-red} | \text{1st-was-red}) P(\text{1st-was-red})$$

$$= \frac{1}{9} \cdot \frac{1}{5} = \frac{1}{45}$$

Similarly, the probabilities that the 1st ball is red and the 2nd is green/blue are

$$P(\text{1st-was-red, 2nd-is-green}) = P(\text{2nd-is-green} \mid \text{1st-was-red}) P(\text{1st-was-red})$$

$$= \frac{1}{3} \cdot \frac{1}{5} = \frac{1}{15},$$

$$P(\text{1st-was-red, 2nd-is-blue}) = P(\text{2nd-is-blue} \mid \text{1st-was-red}) P(\text{1st-was-red})$$

$$= \frac{5}{9} \cdot \frac{1}{5} = \frac{1}{9},$$

respectively. $\qquad\square$

3.2 Probability Distributions for Variables

Discrete probabilistic networks are defined over a (finite) set of variables, each of which represents a finite set of exhaustive and mutually exclusive states (or events); see Sect. 2.2 on page 20. Thus, (conditional) probability distributions for variables (i.e., over exhaustive sets of mutually exclusive events) play a very central role in probabilistic networks.

If X is a (random) variable with domain $\mathrm{dom}(X) = (x_1, \ldots, x_{\|X\|})$, then $P(X)$ denotes a probability distribution (i.e., a vector of probabilities summing to 1), where

$$P(X) = \big(P(X = x_1), \ldots, P(X = x_{\|X\|}) \big).$$

If no confusion is possible, we shall often use $P(x)$ as short for $P(X = x)$, $P(y)$ as short for $P(Y = y)$, etc. If the probability distribution for a variable Y is given conditional on a variable (or set of variables) X, then we shall use the notation $P(Y \mid X)$. That is, for each possible value (state), $x \in \mathrm{dom}(X)$, we have a probability distribution $P(Y \mid X = x)$; again, if no confusion is possible, we shall often write $P(Y \mid x)$.

Example 3.5 (Balls in An Urn, page 42). Let X_1 represent the following exhaustive set of mutually exclusive events:

$$\mathrm{dom}(X_1) = (\text{"1st ball is red," "1st ball is green," "1st ball is blue"}).$$

If we define X_1 to denote the random variable "The color of the 1st ball drawn from the urn," then we may define $\mathrm{dom}(X_1) = (\mathsf{red}, \mathsf{green}, \mathsf{blue})$. Similarly, if we define X_2 to denote the random variable "The color of the 2nd ball drawn from the urn," then $\mathrm{dom}(X_2) = \mathrm{dom}(X_1)$. From Example 3.4 on the preceding page, we get

$$P(X_1) = \left(\frac{2}{10}, \frac{3}{10}, \frac{5}{10} \right),$$

$$P(X_2 | X_1 = \text{red}) = \left(\frac{1}{9}, \frac{3}{9}, \frac{5}{9} \right).$$

$P(X_2 | X_1)$ can be described in terms of a table of three (conditional) distributions:

$$P(X_2 | X_1) =$$

	$X_1 = \text{red}$	$X_1 = \text{green}$	$X_1 = \text{blue}$
$X_2 = \text{red}$	$\frac{1}{9}$	$\frac{2}{9}$	$\frac{2}{9}$
$X_2 = \text{green}$	$\frac{3}{9}$	$\frac{2}{9}$	$\frac{3}{9}$
$X_2 = \text{blue}$	$\frac{5}{9}$	$\frac{5}{9}$	$\frac{4}{9}$

Note that the probabilities in each column sum to 1.   □

3.2.1   Rule of Total Probability

Let $P(X, Y)$ be a joint probability distribution for two variables X and Y with $\text{dom}(X) = (x_1, \ldots, x_m)$ and $\text{dom}(Y) = (y_1, \ldots, y_n)$. Using the fact that $\text{dom}(X)$ and $\text{dom}(Y)$ are exhaustive sets of mutually exclusive states of X and Y, respectively, Axiom 3.3 on page 41 gives us the *rule of total probability*:

$$\forall i : P(x_i) = P(x_i, y_1) + \cdots + P(x_i, y_n) = \sum_{j=1}^{n} P(x_i, y_j). \tag{3.1}$$

Using (3.1), we can calculate $P(X)$ from $P(X, Y)$:

$$P(X) = \left(\sum_{j=1}^{n} P(x_1, y_j), \ldots, \sum_{j=1}^{n} P(x_m, y_j) \right).$$

In a more compact notation, we may write $P(X)$ as

$$P(X) = \sum_{j=1}^{n} P(X, y_j),$$

or even shorter as

$$P(X) = \sum_{Y} P(X, Y), \tag{3.2}$$

denoting the fact that we sum over all indices of Y. We shall henceforth refer to the operation in (3.2) as *marginalization* or *projection*.[2] Also, we sometimes refer to this operation as "marginalizing out Y" of $P(X, Y)$.

Example 3.6 (Balls in An Urn, page 42). Using Definition 3.1 on page 42 for each combination of states of X_1 and X_2 of Example 3.5 on page 43, we can compute

$$P(X_1 = \text{red}, X_2 = \text{red}) = P(X_1 = \text{red}) P(X_2 = \text{red} \mid X_1 = \text{red})$$

$$= \frac{2}{10} \cdot \frac{1}{9}$$

$$= \frac{1}{45},$$

etc. That is, we get $P(X_1, X_2)$ by combining $P(X_1)$ and $P(X_2 \mid X_1)$:

$$P(X_1, X_2) = \begin{array}{c|ccc} & X_1 = \text{red} & X_1 = \text{green} & X_1 = \text{blue} \\ \hline X_2 = \text{red} & \dfrac{1}{45} & \dfrac{1}{15} & \dfrac{1}{9} \\ X_2 = \text{green} & \dfrac{1}{15} & \dfrac{1}{15} & \dfrac{1}{6} \\ X_2 = \text{blue} & \dfrac{1}{9} & \dfrac{1}{6} & \dfrac{2}{9} \end{array}$$

(Note that the numbers in the table sum to 1.) Now, through marginalization, we get

$$P(X_2) = P(X_1 = \text{red}, X_2) + P(X_1 = \text{green}, X_2) + P(X_1 = \text{blue}, X_2)$$

$$= \begin{pmatrix} \dfrac{1}{45} \\ \dfrac{1}{15} \\ \dfrac{1}{9} \end{pmatrix} + \begin{pmatrix} \dfrac{1}{15} \\ \dfrac{1}{15} \\ \dfrac{1}{6} \end{pmatrix} + \begin{pmatrix} \dfrac{1}{9} \\ \dfrac{1}{6} \\ \dfrac{2}{9} \end{pmatrix} = \begin{pmatrix} \dfrac{1}{5} \\ \dfrac{3}{10} \\ \dfrac{1}{2} \end{pmatrix}.$$

That is, the probabilities of getting a red, a green, and a blue ball in the second draw are, respectively, 0.2, 0.3, and 0.5, given that we know nothing about the color of the first ball. That is, $P(X_2) = P(X_1) = (0.2, 0.3, 0.5)$, whereas, for example, $P(X_2 \mid X_1 = \text{red}) = (0.1111, 0.3333, 0.5556)$; that is, once the color of the first ball is known, our belief about the color of the second changes.   □

---

[2]See Sect. 3.3.3 on page 50 for more on marginalization.

Fig. 3.1 Graphical
representation of
$P(X | Y_1, \ldots, Y_n)$

3.2.2   Graphical Representation

The conditional probability distributions of probabilistic networks are of the form

$$P(X | Y),$$

where X is a single variable and Y is a (possibly empty) set of variables. X and Y are sometimes called the *head* and the *tail*, respectively, of $P(X | Y)$. If $Y = \emptyset$ (i.e., the empty set), $P(X | Y)$ is often called a *marginal probability distribution* and is then written as $P(X)$. This relation between X and $Y = \{Y_1, \ldots, Y_n\}$ can be represented graphically as the DAG illustrated in Fig. 3.1, where the child vertex is labeled X and the parent vertices are labeled Y_1, Y_2, etc.

Example 3.7 (Burglary or Earthquake, page 25). Consider variables B (Burglary), E (Earthquake), and A (Alarm), where B and E are possible causes of A. A natural way of specifying the probabilistic relations between these three variables would be through a conditional probability distribution for A given B and E. Thus, for each combination of outcomes (states) of B and E, we need to specify a probability distribution over the states of A:

$$P(A | B, E) = \quad$$

A	$B = $ no $E = $ no	$B = $ no $E = $ yes	$B = $ yes $E = $ no	$B = $ yes $E = $ yes
no	0.99	0.1	0.1	0.01
yes	0.01	0.9	0.9	0.99

This conditional probability table expresses the probability (whether obtained as an objective frequency or a subjective belief) of an alarm if either a burglary or an earthquake has taken place (but not both) to be 0.9.   □

3.3   Probability Potentials

In working with probabilistic networks, the notion of "potentials" often appears to be convenient. Formally, a *probability potential* is a nonnegative function defined over the product space over the domains of a set of variables. We shall use Greek letters (ϕ, ψ, etc.) to denote potentials and sometimes use subscripts to identify their domain (e.g., ϕ_X denotes a potential defined on $\text{dom}(X)$) or to indicate that a potential ϕ_X is a marginal potential of ϕ.[3]

3.3.1 Normalization

A (probability) potential, ϕ_X defined on $\text{dom}(X)$, is turned into a probability distribution, $P(X)$, through the operation known as *normalization*. We shall use the notation $\eta(\phi_X)$ to denote normalization of ϕ_X, where $\eta(\phi_X)$ is defined as

$$\eta(\phi_X) \triangleq \frac{\phi_X}{\sum_X \phi_X} \tag{3.3}$$

Hence, $P(X) = \eta(\phi_X)$. The conditional probability distribution $P(Y \mid X)$ can be obtained from the joint distribution $P(X, Y)$ through *conditional normalization* with respect to X:

$$\eta_X(P(X, Y)) \triangleq \frac{P(X, Y)}{\sum_Y P(X, Y)} = P(Y \mid X).$$

In general,

$$\eta_{X'}(P(X)) \triangleq \frac{P(X)}{\sum_{X \setminus X'} P(X)} = P(X \setminus X' \mid X'), \tag{3.4}$$

where X' is a subset of the set of variables X. In particular,

$$\eta(P(X)) = \eta_\emptyset(P(X)) = P(X),$$

whenever $P(X)$ is a probability distribution over X. This also holds true for conditional distributions:

$$\eta_Y(P(X \mid Y)) = P(X \mid Y),$$

since

$$\sum_X P(X \mid Y) = 1_Y, \tag{3.5}$$

---

[3]Note that the two interpretations are consistent. See Sect. 3.3.3 on page 50 for details on marginalization.

where 1_Y denotes a vector of 1s over dom(Y). A uniform potential, for example, 1_Y, is called a *vacuous potential*. Intuitively, a vacuous potential can be thought of as a non-informative (or superfluous) potential.

We shall be using the notion of potentials extensively in Chaps. 4 and 5, but for now, we will just give a couple of simple examples illustrating the usefulness of this notion.

Example 3.8 (Normalization). Let $P(A, B) = P(A)P(B|A)$ be a factorization of the joint distribution for the Boolean variables A and B and assume that $P(A)$ and $P(B|A)$ are given by the potentials ϕ and ψ, respectively, where

$$\phi = (1, 2) \quad \text{and} \quad \psi = \begin{array}{c|cc} B & A = \text{false} & A = \text{true} \\ \hline \text{false} & 5 & 7 \\ \text{true} & 3 & 1 \end{array}$$

Then

$$P(A) = \eta(\phi) = \left(\frac{1}{3}, \frac{2}{3}\right)$$

and

$$P(B|A) = \eta_A(\psi) = \frac{\psi}{\sum_B \psi} = \begin{array}{c|cc} B & A = \text{false} & A = \text{true} \\ \hline \text{false} & \frac{5}{8} & \frac{7}{8} \\ \text{true} & \frac{3}{8} & \frac{1}{8} \end{array}$$

Coincidentally, because the normalization vector $\sum_B \psi = (8, 8)$ is uniform, we get[4]

$$P(A, B) = \eta(\phi * \psi) = \begin{array}{c|cc} B & A = \text{false} & A = \text{true} \\ \hline \text{false} & \frac{5}{24} & \frac{14}{24} \\ \text{true} & \frac{3}{24} & \frac{2}{24} \end{array}$$

In general, with a nonuniform vector (or matrix) of normalization constants $\sum_B \psi$ in $\eta_A(\psi) = \psi / \sum_B \psi$, we have $P(A, B) \neq \eta(\phi * \psi)$. If, for example, had ψ instead been defined as

$$\psi = \begin{array}{c|cc} B & A = \text{false} & A = \text{true} \\ \hline \text{false} & 5 & 7 \\ \text{true} & 3 & 2 \end{array}$$

---

[4]See Sect. 3.3.3 for a definition of combination of potentials.

we would have $\sum_B \psi = (8, 9)$ and

$$P(A, B) = \begin{array}{c|cc} B & A = \text{false} & A = \text{true} \\ \hline \text{false} & \frac{5}{24} & \frac{14}{27} \\ \text{true} & \frac{3}{24} & \frac{4}{27} \end{array}$$

whereas

$$\eta(\phi * \psi) = \begin{array}{c|cc} B & A = \text{false} & A = \text{true} \\ \hline \text{false} & \frac{5}{26} & \frac{14}{26} \\ \text{true} & \frac{3}{26} & \frac{4}{26} \end{array}$$

$\square$

3.3.2 Evidence Potentials

As indicated in Sect. 2.3 on page 24, evidence functions are actually potentials. To compute the joint posterior distribution resulting from incorporating a set of observations in the form of evidence functions, we simply extend the set of probability function constituents (possibly in the form of potentials) with corresponding evidence potentials, multiply, and normalize the product.

Before any evidence has been taken into account, the probability distribution $P(X')$ for a subset $X' \subseteq X$ of variables is referred to as the prior probability distribution for X'. The conditional probability distribution $P(X'|\varepsilon)$, where ε denotes evidence, is referred to as the posterior probability distribution for X' given ε. Given an evidence potential $\mathcal{E}_E$ on a subset $E \subseteq X \setminus X'$ (expressing ε), the posterior conditional distribution is obtained as

$$P(X'|\varepsilon) = \eta(P(X', \varepsilon)),$$

where

$$P(X', \varepsilon) = \sum_{X \setminus X'} P(X) * \mathcal{E}_E.$$

We define $P(X) * \mathcal{E}_E$ to have dimensionality $|X \setminus E|$. Thus, multiplication of $P(X)$ with $\mathcal{E}_E$ gives rise to a dimensionality decrease by $|E|$.

Example 3.9 (Example 3.8, cont.). Assume that we observe $B = \text{true}$, represented by the evidence potential $\mathcal{E}_B = (0, 1)$. Then the posterior marginal distribution, $P(A|\varepsilon)$, is given by

$$P(A|\varepsilon) = \eta(\phi * \psi * \mathcal{E}_B) =$$

$$\eta\left((1, 2) * \begin{pmatrix} 5 & 7 \\ 3 & 1 \end{pmatrix} * \begin{pmatrix} 0 \\ 1 \end{pmatrix}\right) = \eta(3, 2) = \left(\frac{3}{5}, \frac{2}{5}\right),$$

where the two-dimensional potential $\phi * \psi$ on A and B reduces to the one-dimensional potential $(3,2)$ on A when multiplied by the evidence potential $\mathcal{E}_B$.

□

3.3.3   Potential Calculus

To perform inference in probabilistic networks, we only need a few simple operations, namely, multiplication (combination), division, addition, and marginalization (projection). These are all defined very straightforwardly as follows.

Let ϕ and ψ be potentials defined on $\operatorname{dom}(X)$ and $\operatorname{dom}(Y)$, respectively, and let $z \in \operatorname{dom}(X \cup Y)$ be an arbitrary element (configuration).

We then define $\phi * \psi$ as

$$(\phi * \psi)(z) \triangleq \phi(z_X)\psi(z_Y), \tag{3.6}$$

where z_X and z_Y are projections of z onto $\operatorname{dom}(X)$ and $\operatorname{dom}(Y)$, respectively.[5] Addition is defined analogously. We need to take special care to avoid division by zero, but otherwise, division is defined in the obvious way:

$$(\phi/\psi)(z) \triangleq \begin{cases} 0 & \text{if } \phi(z_X) = 0 \\ \phi(z_X)/\psi(z_Y) & \text{if } \psi(z_Y) \neq 0 \\ \text{undefined} & \text{otherwise.} \end{cases} \tag{3.7}$$

As we shall see later, for all relevant operations involved in inference in probabilistic networks, $\phi(z_X) = 0$ implies $\psi(z_Y) = 0$ upon division of ϕ by ψ, and thus, defining $0/0 = 0$, the division operator is always defined.

Let $X' \subseteq X$ and let ϕ be a potential defined on $\operatorname{dom}(X)$. Then $\phi_{X'} = \sum_{X \setminus X'} \phi$ denotes the marginalization (or projection) of ϕ to $\operatorname{dom}(X')$ and is defined as

$$\phi_{X'}(x') \triangleq \sum_{z \in \operatorname{dom}(X \setminus X')} \phi(z.x'), \tag{3.8}$$

where $z.x'$ is the element in $\operatorname{dom}(X)$ for which $(z.x')_{X \setminus X'} = z$ and $(z.x')_{X'} = x'$. In other words, if, say, $X = (X_1, \ldots, X_n)$ and $X' = (X_k, \ldots, X_n)$, $1 < k < n$, then $z = (x_1, \ldots, x_{k-1})$ and $x' = (x_k, \ldots, x_n)$ are the projections of $z.x' = (x_1, \ldots, x_n)$ onto $\operatorname{dom}(X \setminus X')$ and $\operatorname{dom}(X')$, respectively.

Example 3.10 (Combination and Marginalization). Let $X = \{A, B\}$ and $Y = \{B, C, D\}$, where A, B, C, D are all binary variables with $\operatorname{dom}(A) = (a_1, a_2)$, etc. Let ϕ_X and ϕ_Y be potentials defined over $\operatorname{dom}(X)$ and $\operatorname{dom}(Y)$, respectively, where

---

[5]As defined in Sect. 2.2 on page 20.

$$\phi_X = \begin{array}{c|cc} & a_1 & a_2 \\ \hline b_1 & 0.1 & 0.9 \\ b_2 & 0.4 & 0.6 \end{array} \quad \text{and} \quad \phi_Y = \begin{array}{c|cccc} & \multicolumn{2}{c}{c_1} & \multicolumn{2}{c}{c_2} \\ & d_1 & d_2 & d_1 & d_2 \\ \hline b_1 & 0.11 & 0.14 & 0.06 & 0.09 \\ b_2 & 0.23 & 0.07 & 0.18 & 0.12 \end{array}$$

From (3.6), we get $\psi = \phi_X * \phi_Y$ to be

$$\psi = \begin{array}{c|cccc} & \multicolumn{2}{c}{c_1} & \multicolumn{2}{c}{c_2} \\ & d_1 & d_2 & d_1 & d_2 \\ \hline b_1 & (0.011, 0.099) & (0.014, 0.126) & (0.006, 0.054) & (0.009, 0.081) \\ b_2 & (0.092, 0.138) & (0.028, 0.042) & (0.072, 0.108) & (0.048, 0.072) \end{array}$$

where $(\phi_X * \phi_Y)(a_1, b_1, c_1, d_1) = \phi_X(a_1, b_1)\phi_Y(b_1, c_1, d_1) = 0.1 \cdot 0.11 = 0.011$, $(\phi_X * \phi_Y)(a_2, b_1, c_1, d_1) = \phi_X(a_2, b_1)\phi_Y(b_1, c_1, d_1) = 0.9 \cdot 0.11 = 0.099$, etc.

Now, if $Z = \{A, D\}$, then from (3.8), we get the marginal of ψ with respect to Z to be

$$\psi_Z = \begin{array}{c|cc} & d_1 & d_2 \\ \hline a_1 & 0.011 + 0.092 + 0.006 + 0.072 & 0.014 + 0.028 + 0.009 + 0.048 \\ a_2 & 0.099 + 0.138 + 0.054 + 0.108 & 0.126 + 0.042 + 0.081 + 0.072 \end{array}$$

$$= \begin{array}{c|cc} & d_1 & d_2 \\ \hline a_1 & 0.181 & 0.099 \\ a_2 & 0.399 & 0.321 \end{array}$$

where, for example, $\psi_Z(a_1, d_1) = \psi((b_1, c_1).(a_1, d_1)) + \psi((b_2, c_1).(a_1, d_1)) + \psi((b_1, c_2).(a_1, d_1)) + \psi((b_2, c_2).(a_1, d_1)) = 0.011 + 0.092 + 0.006 + 0.072 = 0.181$. Note that ψ (and hence also ψ_Z) is a probability distribution (i.e., $\sum_x \psi(x) = 1$), since ϕ_X is a conditional probability distribution for A given B and ϕ_Y is a joint probability distribution for $\{B, C, D\}$. $\qquad\square$

Distributive Law

Let ϕ and ψ be potentials defined on $\mathrm{dom}(X) = (x_1, \ldots, x_m)$ and $\mathrm{dom}(Y) = (y_1, \ldots, y_n)$. Assume that we wish to compute the marginal potential

$$(\phi * \psi)_{X' \cup Y'} = \sum_{X \setminus X'} \sum_{Y \setminus Y'} (\phi * \psi),$$

where $X' \subseteq X$ and $Y' \subseteq Y$ such that $X \setminus X' \cap Y = \emptyset$ and $Y \setminus Y' \cap X = \emptyset$ (i.e., the two subsets of variables to be marginalized out, namely, $X \setminus X'$ and $Y \setminus Y'$, do not intersect with the sets of variables on which ψ and ϕ, respectively, are defined).

Using the distributive law, we then get

$$\sum_{X \setminus X'} \sum_{Y \setminus Y'} (\phi * \psi) = \sum_{x \in \text{dom}(X \setminus X')} \sum_{y \in \text{dom}(Y \setminus Y')} \phi(x)\psi(y)$$

$$= \phi(x_1)\psi(y_1) + \cdots + \phi(x_1)\psi(y_n) + \cdots +$$

$$\phi(x_m)\psi(y_1) + \cdots + \phi(x_m)\psi(y_n)$$

$$= \phi(x_1)[\psi(y_1) + \cdots + \psi(y_n)] + \cdots +$$

$$\phi(x_m)[\psi(y_1) + \cdots + \psi(y_n)]$$

$$= \sum_{x \in \text{dom}(X \setminus X')} \phi(x) \sum_{y \in \text{dom}(Y \setminus Y')} \psi(y)$$

$$= \sum_{X \setminus X'} \phi \sum_{Y \setminus Y'} \psi, \tag{3.9}$$

where $\sum_X \phi \sum_Y \psi$ is short for $\sum_X (\phi * (\sum_Y \psi))$. Thus, if we wish to compute the marginal distribution $(\phi * \psi)_{X' \cup Y'}$, where $X \setminus X' \cap Y = \emptyset$ and $Y \setminus Y' \cap X = \emptyset$, then using the distributive law may help significantly in terms of reducing the computational complexity.

Example 3.11 (Distributive Law). Let ϕ, ψ, and ξ be potentials defined on dom (A, B, C), dom(B, D), and dom(C, D, E), respectively, and let $\mathcal{E}_E$ be an evidence potential defined on dom(E), where the variables $A, \ldots, E$ are all binary. Assume that we wish to compute $P(A|\varepsilon)$, where ε denotes the evidence provided through $\mathcal{E}_E$. A brute-force approach would be simply to combine all potentials, marginalize out variables $B, \ldots, E$, and normalize

$$P(A|\varepsilon) = \eta \left(\sum_B \sum_C \sum_D \sum_E (\phi * \psi * \xi * \mathcal{E}_E) \right).$$

Combining potentials ξ and $\mathcal{E}_E$ requires 8 multiplications. Next, combining ψ and $\xi * \mathcal{E}_E$ requires 16 multiplications, and, finally, combining ϕ and $\psi * \xi * \mathcal{E}_E$ requires 32 multiplications. Marginalizing out E, D, C, and B, respectively, requires 16, 8, 4, and 2 additions.

Alternatively, we could take advantage of the distributive law to compute the same thing:

$$P(A|\varepsilon) = \eta \left(\sum_B \sum_C \phi \sum_D \psi \sum_E (\xi * \mathcal{E}_E) \right).$$

First, combining ξ and $\mathcal{E}_E$ requires 8 multiplications. Then, marginalizing out E requires 4 additions. Multiplying the resulting potential by ψ requires 8 multiplications, and marginalizing out D requires 4 additions. Next, multiplying the resulting potential by ϕ requires 8 multiplications, and finally, marginalizing out C and B requires 4 and 2 additions, respectively.

Summing up the number of arithmetic operations used in the two computations, we find that the brute-force approach takes 56 multiplications and 30 additions, whereas the one exploiting the distributive law takes only 24 multiplications and 14 additions, less than half of what the brute-force approach requires. (On top of these numbers, we should also count the number of operations needed to normalize the final marginal, but that is the same in both cases.)

Note that the ordering (B, C, D, E) is just one out of $4! = 24$ different sequences in which we might marginalize out these four variables, and to each ordering is associated a certain number of arithmetic operations required to compute $P(A \,|\, \varepsilon)$.

$\square$

The single most important key to efficient inference in probabilistic networks is the ability to take advantage of the distributive law (i.e., to find optimal (or near optimal) sequences in which the variables are marginalized out). We shall return to this issue in Chap. 5.

3.3.4 Barren Variables

If a variable, X, of a probabilistic network is never observed, $P(X)$ is of no interest, and the same holds true for each of X's descendants (if any), then X is called a *barren variable* (Shachter 1986), as it provides no information relevant for the inference process. In fact, a barren variable provides "information" in the form of a vacuous potential (cf. (3.5)) and may hence be removed from the network.

Example 3.12 (Barren Variables). Consider $P(X, Y, Z) = P(X)P(Y \,|\, X)P(Z \,|\, Y)$ as a model over the variables X, Y, and Z. Following the discussion in Sect. 3.2.2 on page 46, this model can be represented graphically as indicated in Fig. 3.2a. Let $\mathcal{E}_Y$ and $\mathcal{E}_Z$ be evidence potentials for Y and Z, respectively, but where $\mathcal{E}_Z$ is always vacuous. Then the posterior probability distribution for X can be calculated as

$$P(X \,|\, \varepsilon) = \eta \left(P(X) \sum_Y P(Y \,|\, X) * \mathcal{E}_Y \sum_Z P(Z \,|\, Y) * \mathcal{E}_Z \right)$$

$$= \eta \left(P(X) \sum_Y P(Y \,|\, X) * \mathcal{E}_Y \sum_Z P(Z \,|\, Y) \right)$$

$$= \eta \left(P(X) \sum_Y P(Y \,|\, X) * \mathcal{E}_Y * 1_Y \right)$$

$$= \eta \left(P(X) \sum_Y P(Y \,|\, X) * \mathcal{E}_Y \right),$$

Fig. 3.2 (a) Model for
$P(X, Y, Z)$. (b) Equivalent
model when Z is barren

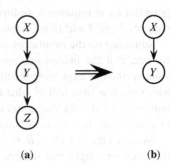

(a)   (b)

where $\sum_Z P(Z \mid Y) = 1_Y$ follows from (3.5) and ε denotes the evidence. If neither $P(Z)$ nor $P(Z \mid \varepsilon)$ are of any interest, then Z is barren, and the term $P(Z \mid Y)$ can be neglected in the inference process, and the model can be simplified to the one shown in Fig. 3.2b.   □

We shall return to the issue of barren variables in more detail in Sect. 5.1.1.

3.4   Fundamental Rule and Bayes' Rule

Generalizing Definition 3.1 on page 42 to arbitrary (random) variables X and Y, we get the *fundamental rule* of probability calculus:

$$P(X, Y) = P(X \mid Y)P(Y) = P(Y \mid X)P(X). \tag{3.10}$$

Bayes' rule follows immediately from (3.10):

$$P(Y \mid X) = \frac{P(X \mid Y)P(Y)}{P(X)}. \tag{3.11}$$

Using Definition 3.1 on page 42 and the rule of total probability, (3.11) can be rewritten as

$$P(Y \mid X) = \frac{P(X \mid Y)P(Y)}{P(X \mid Y = y_1)P(Y = y_1) + \cdots + P(X \mid Y = y_{\|Y\|})P(Y = y_{\|Y\|})}.$$

That is, the denominator in (3.11) can be derived from the numerator in (3.11). Since, furthermore, the denominator is obviously the same for all states of Y, we often write Bayes' rule as

$$P(Y \mid X) \propto P(X \mid Y)P(Y), \tag{3.12}$$

read as "$P(Y \mid X)$ is proportional to $P(X \mid Y)P(Y)$." Note that the proportionality factor $P(X)^{-1}$ is in fact a vector of proportionality constants, one for each state of X, determined in a normalization operation.

Division by zero in (3.11) is not a problem if we define $0/0 = 0$ as in (3.7), since for

$$P(x_i) = \sum_j P(x_i \mid y_j) P(y_j)$$

to be zero at least one of $P(x_i \mid y_j)$ and $P(y_j)$ must be zero for each j, and if this is the case, then both the numerator term, $P(x_i \mid y_j) P(y_j)$, and the denominator term, $P(x_i)$, of (3.11) will be zero.

Example 3.13 (Burglary or Earthquake, page 25). Given $P(E) = (0.01, 0.99)$, $P(B) = (0.1, 0.9)$, and the conditional probability table (CPT) for $P(A \mid B, E)$ from Example 3.7 on page 46, we can use Bayes' rule to compute $P(B \mid A)$, the conditional probability distribution for burglary (B) given alarm (A):

$$P(B \mid A) \propto \sum_E P(A \mid B, E) P(B) P(E) = P(A, B).$$

First, we compute the joint distribution for A, B, and E:

$$P(A, B, E) = P(A \mid B, E) P(B) P(E)$$

		$B = $ no		$B = $ yes	
A	$E = $ no	$E = $ yes	$E = $ no	$E = $ yes
no	0.88209	0.0009	0.0099	0.00001
yes	0.00891	0.0081	0.0891	0.00099

Next, we marginalize out E of $P(A, B, E)$ to obtain

$$P(A, B) = \sum_E P(A, B, E) =$$

A	$B = $ no	$B = $ yes
no	0.88299	0.00991
yes	0.00991	0.09009

Finally, we normalize $P(A, B)$ with respect to A, and get

$$P(B \mid A) = \eta_A(P(A, B)) =$$

B	$A = $ no	$A = $ yes
no	0.9889	0.1588
yes	0.0111	0.8412

$\square$

3.4.1 Interpretation of Bayes' Rule

Since Bayes' rule is so central to inference in Bayesian probability calculus, let us dwell a little on how Bayes' rule can be used and understood. Assume that we have

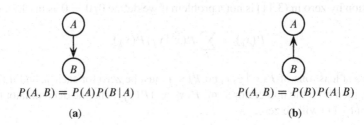

$$P(A, B) = P(A)P(B \mid A)$$

(a)

$$P(A, B) = P(B)P(A \mid B)$$

(b)

Fig. 3.3 Two equivalent models that can be obtained from each other through arc reversal

two (possibly, sets of) variables X and Y, a model $P(X, Y)$ given in the factorized form $P(X \mid Y)P(Y)$, and that we observe $X = x$. We would then typically want to compute $P(Y \mid x)$.

The prior distribution, $P(Y)$, expresses our initial belief about Y, and the posterior distribution, $P(Y \mid x)$, expresses our revised belief about Y in light of the observation $X = x$. Bayes' rule tells us how to obtain the posterior distribution by multiplying the prior $P(Y)$ by the ratio $P(x \mid Y)/P(x)$, known as the *normalized likelihood* of Y given x. Again, since $P(x)$ is a constant for each $y \in \text{dom}(Y)$, we get

$$P(Y \mid x) \propto P(Y)P(x \mid Y).$$

The quantity $P(x \mid Y) \triangleq L(Y \mid x)$ is called the *likelihood* of Y given x. Hence, we have

$$P(Y \mid x) \propto P(Y)L(Y \mid x). \tag{3.13}$$

In general,

$$\text{posterior} \propto \text{prior} \times \text{likelihood}.$$

In a machine learning context, Bayes' rule plays an important role. For example, consider a prior distribution, $P(M)$, for a random variable M, expressing a set of possible models. For any value d of another variable D, expressing data, the quantity $P(d \mid M)$—considered as a function of M—is the likelihood function for M given data d. The posterior distribution for M given the data is then

$$P(M \mid d) \propto P(M)P(d \mid M),$$

which provides a set of goodness-of-fit measures for models M (i.e., we obtain a conditional probability $P(m \mid d)$ for each $m \in \text{dom}(M)$).

3.4.1.1 Arc Reversal

Application of Bayes' rule can also be given a graphical interpretation. Consider, for example, two variables A and B and a model $P(A, B) = P(A)P(B \mid A)$. Again, following the discussion in Sect. 3.2.2 on page 46, this model can be represented graphically as indicated in Fig. 3.3a. Applying Bayes' rule on this model

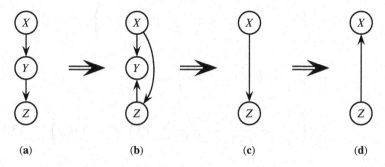

Fig. 3.4 (**a**) Model for $P(X, Y, Z)$. (**b**) Equivalent model obtained by reversing $Y \rightarrow Z$. (**c**) Equivalent model provided Y is barren. (**d**) Equivalent model obtained by reversing $X \rightarrow Z$

$$P(A \mid B) = \frac{P(A)P(B \mid A)}{\sum_A P(A)P(B \mid A)} = \frac{P(A, B)}{P(B)},$$

we obtain an equivalent model shown in Fig. 3.3b. Thus, one way of interpreting the application of Bayes' rule is through so-called *arc reversal*. As the typical inference task in probabilistic networks can be described as computing $P(X \mid \varepsilon)$, inference in probabilistic networks can be thought of as (repetitive) application of Bayes' rule or, in other words, as a sequence of arc reversals. Olmsted (1983) and Shachter (1990) have exploited this view of inference in an arc reversal algorithm for inference in probabilistic networks.

Example 3.14 (Arc Reversal). Consider the model in Fig. 3.4a, and assume that we wish to calculate the posterior marginal distribution for X given evidence, ε_Z, on Z. Using Shachter's arc reversal procedure (Shachter 1990) (not described in this book), we may proceed as follows:

$$P(X \mid \varepsilon) = \eta \left(\sum_Y \sum_Z P(X)P(Y \mid X)P(Z \mid Y)\varepsilon_Z \right)$$

$$= \eta \left(\sum_Y \sum_Z P(X)P(Y, Z \mid X)\varepsilon_Z \right) \tag{3.14}$$

$$= \eta \left(\sum_Y \sum_Z P(X)\frac{P(Y, Z \mid X)}{\sum_Y P(Y, Z \mid X)} \sum_Y P(Y, Z \mid X)\varepsilon_Z \right) \tag{3.15}$$

$$= \eta \left(\sum_Y \sum_Z P(X)P(Y \mid X, Z)P(Z \mid X)\varepsilon_Z \right) \tag{3.16}$$

$$= \eta \left(\sum_Z P(X) P(Z \mid X) \mathcal{E}_Z \sum_Y P(Y \mid X, Z) \right) \qquad (3.17)$$

$$= \eta \left(\sum_Z P(X) P(Z \mid X) \mathcal{E}_Z \right) \qquad (3.18)$$

$$= \eta \left(\sum_Z \frac{P(X) P(Z \mid X)}{\sum_X P(X) P(Z \mid X)} \sum_X P(X) P(Z \mid X) \mathcal{E}_Z \right) \qquad (3.19)$$

$$= \eta \left(\sum_Z P(X \mid Z) P(Z) \mathcal{E}_Z \right), \qquad (3.20)$$

where we combine $P(Y \mid X)$ and $P(Z \mid Y)$ into $P(Y, Z \mid X)$ (3.14), use Bayes' rule to reverse $Y \rightarrow Z$ (3.15), which induces a new edge $X \rightarrow Z$ (3.16), use the distributive law (3.17), eliminate barren variable Y (3.18), and finally use Bayes' rule to reverse $X \rightarrow Z$ (3.19). Now, if $\mathcal{E}_Z$ represent hard evidence (i.e., $Z = z$), (3.20) reduces to

$$P(X \mid \varepsilon) = P(X \mid Z = z),$$

that is, a simple lookup.   □

We shall return to the arc reversal approach in more detail in Sect. 5.1.1 on page 116.

3.5 Bayes' Factor

To calculate the relative support provided by an observation, $Y = y$, for two competing hypotheses, $H_0 : X = x_0$ and $H_1 : X = x_1$, the notion of *Bayes' factor* is useful:

$$B = \frac{\text{posterior odds}}{\text{prior odds}} = \frac{P(x_0 \mid y) / P(x_1 \mid y)}{P(x_0) / P(x_1)} = \frac{P(y \mid x_0)}{P(y \mid x_1)} = \frac{L(x_0 \mid y)}{L(x_1 \mid y)}; \qquad (3.21)$$

that is, the ratio of the likelihoods of the two hypotheses given the observation. Bayes' factor is also known as the *Bayesian likelihood ratio*.

From (3.21), we see that

$B > 1$ if the observation provides more support for H_0 than for H_1,

$B < 1$ if the observation provides less support for H_0 than for H_1, and

$B = 1$ if the observation does not provide useful information for differentiating between H_0 and H_1.

Example 3.15 (Balls in An Urn, page 42). Let hypotheses H_0 and H_1 be given as

$$H_0 : \text{The second ball drawn will be green: } X_2 = \text{green}$$
$$H_1 : \text{The second ball drawn will be blue: } X_2 = \text{blue,}$$

and assume we observe that the first ball drawn is blue (i.e., $X_1 = \text{blue}$). Now, using the numbers calculated in Example 3.5 on page 43, we get the Bayes' factor

$$B = \frac{P(X_2 = \text{green}| X_1 = \text{blue})/P(X_2 = \text{blue}| X_1 = \text{blue})}{P(X_2 = \text{green})/P(X_2 = \text{blue})} = \frac{\frac{3}{9}/\frac{4}{9}}{\frac{3}{10}/\frac{5}{10}} = \frac{5}{4}.$$

That is, since the posterior odds $(3/4)$ is greater than the prior odds $(3/5)$, the observation provides more support for H_0 than for H_1. Still, however, the probability that H_1 is going to be true is greater than the probability that H_0 is going to be true, as $P(H_0| X_1 = \text{blue}) = 3/9$ and $P(H_1| X_1 = \text{blue}) = 4/9$. $\square$

3.6 Independence

A variable X is *independent* of another variable Y with respect to a probability distribution P if

$$P(x|y) = P(x), \forall x \in \text{dom}(X), \forall y \in \text{dom}(Y). \tag{3.22}$$

Using standard notation, we express this property symbolically as $X \perp\!\!\!\perp_P Y$ or simply as $X \perp\!\!\!\perp Y$ when P is obvious from the context. Symmetry of independence (i.e., $X \perp\!\!\!\perp Y \equiv Y \perp\!\!\!\perp X$) can be verified from Bayes' rule:

$$P(x|y) = P(x) = \frac{P(y|x)P(x)}{P(y)} \quad \Leftrightarrow \quad P(y|x) = P(y).$$

The statement $X \perp\!\!\!\perp Y$ is often referred to as *marginal independence* between X and Y.

A variable X is *conditionally independent* of Y given Z (with respect to a probability distribution P) if

$$P(x|y, z) = P(x|z), \forall x \in \text{dom}(X), \forall y \in \text{dom}(Y), \forall z \in \text{dom}(Z). \tag{3.23}$$

The conditional independence statement expressed in (3.23) is indicated as $X \perp\!\!\!\perp Y | Z$ in the standard notation. With a slight misuse of notation, we shall also express this as $P(X|Y, Z) = P(X|Z).$[6]

---

[6]The misuse is concerned with differences in dimensionalities of $P(X|Y, Z)$ and $P(X|Z)$.

Example 3.16 (Conditional Independence). Consider the Burglary or Earthquake example from page 25. With $P(R \mid E)$ given as

$$P(R \mid E) = \begin{array}{c|cc} R \backslash E & = \text{no} & E = \text{yes} \\ \hline \text{no} & 0.999 & 0.01 \\ \text{yes} & 0.001 & 0.99 \end{array}$$

and

$$P(A, E, R) = \sum_B P(A \mid B, E) P(B) P(E) P(R \mid E)$$

$$= \sum_B P(A, B, E) P(R \mid E)$$

$$= P(A, E) P(R \mid E),$$

(see Example 3.13 on page 55), using the fundamental rule, we get

$$P(A \mid E, R) = \frac{P(A, E, R)}{\sum_A P(A, E, R)} = \frac{P(A, E) P(R \mid E)}{\sum_A P(A, E) P(R \mid E)}$$

$$= \begin{array}{c|cc|cc} & \multicolumn{2}{c|}{R = \text{no}} & \multicolumn{2}{c}{R = \text{yes}} \\ A & E = \text{no} & E = \text{yes} & E = \text{no} & E = \text{yes} \\ \hline \text{no} & 0.901 & 0.091 & 0.901 & 0.091 \\ \text{yes} & 0.099 & 0.909 & 0.099 & 0.909 \end{array}$$

and

$$P(A \mid E) = \frac{\sum_R P(A, E) P(R \mid E)}{\sum_{A,R} P(A, E) P(R \mid E)} = \begin{array}{c|cc} A & E = \text{no} & E = \text{yes} \\ \hline \text{no} & 0.901 & 0.091 \\ \text{yes} & 0.099 & 0.909 \end{array}$$

Obviously, $P(A \mid E, R) = P(A \mid E)$. Thus, we conclude that $A \perp\!\!\!\perp_P R \mid E$.   □

3.6.1 Independence and DAGs

Let P be a probability distribution over a set of variables V and let $\mathcal{G} = (V, E)$ be a DAG. Then $\mathcal{G}$ is said to be an *independency map* (or *I-map*) of P if $X_A \perp\!\!\!\perp_P X_B \mid X_S$ whenever $A \perp_{\mathcal{G}} B \mid S$ for subsets A, B, S of V. In other words, if for each pair of unconnected variables $u, v \in V$ in $\mathcal{G}$ (i.e., $u \not\sim_{\mathcal{G}} v$) it holds true that there is a set $S \subseteq V$ such that $X_u \perp\!\!\!\perp_P X_v \mid X_S$, then $\mathcal{G}$ is a representation of the independence properties of P. For brevity, we shall then say that "$\mathcal{G}$ is an I-map of P."

Note that a DAG that is an I-map of P does not necessarily represent all independence properties of P. In fact, the complete graph is an I-map of any distribution.

Fig. 3.5 Graphical
representations of
$X \perp\!\!\!\perp_P Y \,|\, Z$, representing,
respectively, (3.24)–(3.26)

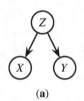 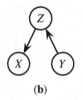 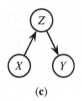

                      (a)   **(b)**   **(c)**

Definition 3.2 (I-map and D-map (Pearl 1988)). Let P be a probability distribution over a set of variables X_V, where $\mathcal{G} = (V, E)$ is a DAG over the vertices that indexes these variables. As defined by Pearl:

$$\mathcal{G} \text{ is a D-map of } P \text{ if } \quad A \perp_{\mathcal{G}} B \,|\, S \quad \Longleftarrow \quad X_A \perp\!\!\!\perp_P X_B \,|\, X_S.$$
$$\mathcal{G} \text{ is an I-map of } P \text{ if } \quad A \perp_{\mathcal{G}} B \,|\, S \quad \Longrightarrow \quad X_A \perp\!\!\!\perp_P X_B \,|\, X_S.$$

If $\mathcal{G}$ is both a D-map (or dependency map) and an I-map of P, then $\mathcal{G}$ is said to be a *perfect map* of P.

If $\mathcal{G}$ is an I-map of P, then P is said to be Markov with respect to $\mathcal{G}$, and if $\mathcal{G}$ is a D-map of P, then P is said to be faithful to $\mathcal{G}$ (Chickering & Meek 2003). We shall return to the issue of DAG faithfulness in Sect. 8.1.

Example 3.17 (Independence and DAGs). Let X, Y, and Z be three variables for which $X \perp\!\!\!\perp_P Y \,|\, Z$. Following the ordering (X, Y, Z) and using the fundamental rule (3.10) twice yields

$$P(X, Y, Z) = P(X \,|\, Y, Z) P(Y \,|\, Z) P(Z).$$

Since $X \perp\!\!\!\perp_P Y \,|\, Z$, this can be simplified to

$$P(X, Y, Z) = P(X \,|\, Z) P(Y \,|\, Z) P(Z). \tag{3.24}$$

Similarly, following the orderings (X, Z, Y) and (Y, Z, X), we get, respectively,

$$P(X, Y, Z) = P(X \,|\, Z) P(Z \,|\, Y) P(Y) \tag{3.25}$$

and

$$P(X, Y, Z) = P(Y \,|\, Z) P(Z \,|\, X) P(X). \tag{3.26}$$

Equations (3.24)–(3.26) have graphical representations as shown in Fig. 3.5a–c, respectively (see Sect. 3.2.2 on page 46). $\qquad\square$

Only a subset of the independence properties that can exist in a probability distribution can be represented by a DAG. That is, the DAG language is not rich enough to simultaneously capture all sets of independence statements.

Example 3.18 (No D-maps). Consider the following set of independence statements for a probability distribution P:

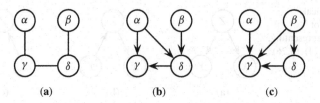

Fig. 3.6 (a) Preliminary skeleton for the independence statements of Example 3.18. (b) DAG representing CID_1 and CID_3. (c) DAG representing CID_1 and CID_2

$$CID_1 : \qquad X_\alpha \perp\!\!\!\perp_P X_\beta$$
$$CID_2 : X_\alpha \perp\!\!\!\perp_P X_\delta | \{X_\beta, X_\gamma\}$$
$$CID_3 : X_\beta \perp\!\!\!\perp_P X_\gamma | \{X_\alpha, X_\delta\}.$$

From these statements, we can conclude that a DAG, $\mathcal{G}$, over $\{\alpha, \beta, \gamma, \delta\}$ must include edges between each pair of vertices except (α, β), (α, δ), and (β, γ), as at least one independence statement has been specified for each of the variable pairs $\{X_\alpha, X_\beta\}$, $\{X_\alpha, X_\delta\}$, and $\{X_\beta, X_\gamma\}$, respectively. A preliminary skeleton of the possible DAGs therefore appears as shown in Fig. 3.6a.

Recalling the d-separation criterion or the directed global Markov criterion (see Sect. 2.6 on page 32), we see that for CID_1 to hold true, there must be a converging connection at γ or δ. However, a converging connection at, for example, γ implies $\alpha \perp_{\mathcal{G}} \delta$, making $\mathcal{G}$ not being an I-map of P. To remedy that, we will have to include an edge between α and δ. Now, to ensure $\alpha \perp_{\mathcal{G}} \beta$, the edges $\alpha - \delta$ and $\beta - \delta$ must meet head-to-head at δ (i.e., must converge at δ). The resulting DAG in Fig. 3.6b is an I-map of P but not a D-map as CID_2 is not represented. Similarly, the DAG in Fig. 3.6c represents CID_1 and CID_2 but not CID_3.

Each DAG in Fig. 3.6b, c is an I-map of P, but neither of them are D-maps of P, as they both fail to represent all three independence statements. $\square$

3.7 Chain Rule

For a probability distribution, $P(X)$, over a set of variables $X = \{X_1, \ldots, X_n\}$, we can use the fundamental rule repetitively to decompose it into a product of conditional probability distributions:

$$P(X) = P(X_1 | X_2, \ldots, X_n) P(X_2, \ldots, X_n)$$
$$= P(X_1 | X_2, \ldots, X_n) P(X_2 | X_3, \ldots, X_n) \cdots P(X_{n-1} | P_n) P(X_n)$$

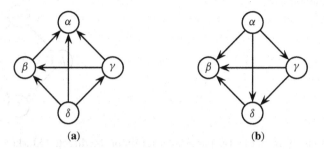

Fig. 3.7 (**a**) DAG corresponding to (3.28). (**b**) DAG corresponding to (3.29)

$$= \prod_{i=1}^{n} P(X_i \mid X_{i+1}, \dots, X_n). \tag{3.27}$$

Notice that the actual conditional distributions that comprise the factors of the decomposition are determined by the order in which we select the head variables of the conditional distributions. Thus, there are n factorial different factorizations of $P(X)$, and to each factorization corresponds a unique DAG, but all of these DAGs are equivalent in terms of dependence and independence properties, as they are all complete graphs, and hence represent no independence statements.[7]

Example 3.19 (Chain Decomposition and DAGs). Let $V = \{\alpha, \beta, \gamma, \delta\}$. Then $P(X_V)$ factorizes as

$$P(X_V) = P(X_\alpha, X_\beta, X_\gamma, X_\delta) = P(X_\alpha \mid X_\beta, X_\gamma, X_\delta) P(X_\beta, X_\gamma, X_\delta)$$

$$= P(X_\alpha \mid X_\beta, X_\gamma, X_\delta) P(X_\beta \mid X_\gamma, X_\delta) P(X_\gamma, X_\delta)$$

$$= P(X_\alpha \mid X_\beta, X_\gamma, X_\delta) P(X_\beta \mid X_\gamma, X_\delta) P(X_\gamma \mid X_\delta) P(X_\delta) \tag{3.28}$$

$$= P(X_\beta \mid X_\alpha, X_\gamma, X_\delta) P(X_\delta \mid X_\alpha, X_\gamma) P(X_\gamma \mid X_\alpha) P(X_\alpha) \tag{3.29}$$

$$= \cdots$$

The DAGs corresponding to (3.28) and (3.29) appear in Fig. 3.7a, b, respectively. □

Assume that DAG $\mathcal{G}$ is a perfect map of P and that the order in which we select the head variables of the conditional distributions respect a topological ordering $(X_{v_1}, \dots, X_{v_n})$[8] with respect to $\mathcal{G}$: $\text{pa}(v_i) \subseteq \{v_1, \dots, v_{i-1}\}$ for all $i = 1, \dots, n$ (i.e., the parents of each variable are selected before the variable itself). That is, the tail variables always include the parents.

---

[7]See (8.1) for the number of possible DAGs on n vertices.

[8]For notational convenience, we assume (without loss of generality) that $v_1, \dots, v_n$ is a topological ordering.

Fig. 3.8 A perfect map of a distribution P; see Example 3.20

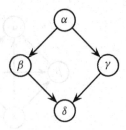

It follows easily from the d-separation criterion or the directed Markov criterion (Sect. 2.6) that for any vertex v of a DAG, $\mathcal{G}$, $v \perp_{\mathcal{G}} \mathrm{nd}(v) \,|\, \mathrm{pa}(v)$.[9] Since $\mathcal{G}$ is a perfect map of P, it follows that $X_v \perp\!\!\!\perp_P X_{\mathrm{nd}(v)} \,|\, X_{\mathrm{pa}(v)}$. Therefore, each term $P(X_{v_i} \,|\, X_{v_1}, \ldots, X_{v_i})$ can be reduced to $P(X_{v_i} \,|\, X_{\mathrm{pa}(v_i)})$. The product in (3.27) then simplifies to the *chain rule*:

$$P(X_V) = \prod_{i=1}^{n} P(X_{v_i} \,|\, X_{\mathrm{pa}(v_i)}). \tag{3.30}$$

Example 3.20 (Example 3.19, cont.). Assume that the complete set of independence properties that P satisfies comprises $\{X_\beta \perp\!\!\!\perp_P X_\gamma \,|\, X_\alpha, X_\alpha \perp\!\!\!\perp_P X_\delta \,|\, \{X_\beta, X_\gamma\}\}$. Then the DAG in Fig. 3.8 is a perfect map of P. From the chain rule, we can therefore write the joint distribution as

$$P(X_V) = P(X_\alpha) P(X_\beta \,|\, X_\alpha) P(X_\gamma \,|\, X_\alpha) P(X_\delta \,|\, X_\beta, X_\gamma),$$

where the tail variables are exactly the set of parents for each head variable. $\Box$

3.8 Summary

Bayesian probability calculus is based on a few very simple and intuitive axioms that express ground statements of probability about the occurrence of a single event, the occurrence of mutually exclusive events, and the co-occurrence of events. Being defined, basically, as exhaustive lists of mutually exclusive events, (discrete) variables hence provide an excellent concept on which to base a calculus of distributions of probabilities. For example, we saw how the axiom on mutually exclusive events implies the rule of total probability, which is the basis for computing a (lower-dimensional) marginal probability distribution through projection from a (higher-dimensional) distribution. Together with the straightforward operations of multiplication (combination), division, and addition, the projection

---

[9] This result is probably easiest to acknowledge from the directed Markov criterion: The graph $(\mathcal{G}_{\mathrm{An}(\{v\} \cup \mathrm{nd}(v) \cup \mathrm{pa}(v))})^m$ obviously excludes all descendants of v forcing all paths to v to go through $\mathrm{pa}(v)$.

(or marginalization) operation provide a complete set of operations for probabilistic inference.

As a matter of convenience, the notion of probability potentials was introduced as a generalization of probability distributions in the sense that the elements of a probability potential do not necessarily sum to 1. Probability distributions are restored through the operation of normalization, where the normalization constant expresses the reciprocal value of the probability of the evidence.

The fundamental trick in making efficient probabilistic inference with a joint probability distribution over a (possibly large) collection of discrete random variables lies in the ability to exploit the (conditional) independence properties of the distribution. The extent to which these properties allow a factorization of the distribution into lower-dimensional distributions (i.e., distributions defined on subsets of the variables) determines how efficiently inference can be performed. Basically, this gain in efficiency is realized through the exploitation of the distributive law, which implies interleaving of the operations of combination and marginalization. Allowing a marginalization operation (dimensionality decrease) to be performed before a combination operation (dimensionality increase) reduces the total amount of arithmetic operations needed.

The ability to perform abductive reasoning in probabilistic inference (e.g., to compute $P(X \mid Y = y)$ given $P(Y \mid X)$ and $P(X)$) follows from Bayes' rule, which in turn follows from the fundamental rule of probability calculus that is a generalization of the axiom on the co-occurrence of events. An often-used interpretation of Bayes' rule states that the posterior probability of an event given some observation, say $P(X = x \mid Y = y)$ (i.e., our belief about the probability of the occurrence of the event $X = x$ *after* the event $Y = y$ has been observed), is proportional to the prior probability of the event, $P(X = x)$ (i.e., our belief about the probability of the occurrence of the event $X = x$ *before* observing $Y = y$), times the likelihood of the event given the observation, defined as $L(X = x \mid Y = y) = P(Y = y \mid X = x)$.

We saw how to establish the important connection between the notion of d-separation (and the equivalent directed, global Markov property) defined on a DAG, $\mathcal{G} = (V, E)$, as discussed in Chap. 2, and the independence properties of a joint probability distribution, P, defined over variables represented by V, where the directed edges, E, lead from nodes representing tail variables to nodes representing head variables of conditional probability distributions constituting a factorization of P. In fact, with $\mathcal{G}$ so defined, there is a one-to-one correspondence between the statements of d-separation in $\mathcal{G}$ and the (conditional) independence statements of P (i.e., $\mathcal{G}$ is a perfect map of P). This correspondence between d-separation and (conditional) independence is expressed in the chain rule on page 64.[10]

---

[10]Special cases in which variables are independent only for particular values of some other variable(s) might exist. Such *context-specific independence* properties obviously cannot be captured by a DAG. Thus, the one-to-one correspondence should be understood with respect to independence statements on the "level of variables."

On the basis of the fundamental concepts introduced in Chaps. 2 and 3, we shall see in Chaps. 4 and 5, respectively, how different kinds of probabilistic networks can be defined and how inference in these can be performed.

Exercises

Exercise 3.1. Assume that a non-red ball is removed from the urn in Example 3.4 on page 42. What is then the probability of picking a blue ball from the urn?

Exercise 3.2. Assume that smoking (S = true) causes lung cancer (L = true) in one out every ten cases and that nonsmokers (S = false) get lung cancer in one out of 500 cases.

(a) Specify the probability table for $P(L \mid S)$.
(b) Assuming that smoking is the only cause of lung cancer, what is then the frequency of lung cancer in a population, where one third of the population are smokers?

Exercise 3.3. Suppose we have the simple model $X \rightarrow Y$ and are given $P(X)$, $P(Y \mid X)$, and evidence $Y = y$.

(a) Indicate a minimal-cost procedure for computing $P(X \mid Y = y)$.
(b) Use the procedure to compute $P(S \mid L = \text{true})$ in Exercise 3.2.

Exercise 3.4. Normally, John arrives on time at his office. If, however, the roads are icy, there is a chance that he will be late. Let I = icy denote the event that the roads are icy, and let L = late denote the event that John is late. Assume that our prior knowledge about road conditions (i.e., without knowing when John arrives) is given by $P(I) = P(I = \text{icy}, I = \neg\text{icy}) = (0.01, 0.99)$. Also, assume our experience tells us that $P(L = \text{late} \mid \text{icy}) = 0.9$ and $P(L = \text{late} \mid \neg\text{icy}) = 0.2$.

(a) What is the likelihood for icy roads given that John arrives late?
(b) What is the probability of icy roads given that John arrives late?

Exercise 3.5. Assume that the complete list of conditional independence statements satisfied by a probability distribution $P(A, B, C, D, E)$ is given by:

$A \perp\!\!\!\perp D \mid \{B, C\}$	$B \perp\!\!\!\perp C \mid A$	$D \perp\!\!\!\perp E \mid C$
$A \perp\!\!\!\perp D \mid \{B, C, E\}$	$B \perp\!\!\!\perp C \mid \{A, E\}$	$D \perp\!\!\!\perp E \mid \{B, C\}$
$A \perp\!\!\!\perp E \mid C$	$B \perp\!\!\!\perp E \mid A$	$D \perp\!\!\!\perp E \mid \{A, C\}$
$A \perp\!\!\!\perp E \mid \{B, C\}$	$B \perp\!\!\!\perp E \mid C$	$D \perp\!\!\!\perp E \mid \{A, B, C\}$
$A \perp\!\!\!\perp E \mid \{C, D\}$	$B \perp\!\!\!\perp E \mid \{A, C\}$	
$A \perp\!\!\!\perp E \mid \{B, C, D\}$	$B \perp\!\!\!\perp E \mid \{C, D\}$	
	$B \perp\!\!\!\perp E \mid \{A, C, D\}$	

(a) Draw a DAG fulfilling the assumptions.
(b) How many DAGs fulfill the assumptions?
(c) Make factorizations of P corresponding to the DAGs.

Exercise 3.6. Let $W \subseteq U$, and let $\phi \equiv \phi_U$ be a potential defined on $\text{dom}(U)$, where $U = \{A, B, C, D\}$, $W = \{A, C\}$, and all four variables are binary, where $\text{dom}(A) = (a_1, a_2), \text{dom}(B) = (b_1, b_2)$, etc. Let ϕ_U be given by the following table:

	c_1 d_1	c_1 d_2	c_2 d_1	c_2 d_2
$a_1\ b_1$	0.0957	0.0672	0.0341	0.0513
$a_1\ b_2$	0.1021	0.0162	0.0634	0.1287
$a_2\ b_1$	0.0174	0.1297	0.0040	0.1089
$a_2\ b_2$	0.0624	0.0776	0.0307	0.0107

(a) Compute ϕ_W.
(b) Indicate the table for the evidence function $\mathcal{E}_D$ (defined on $\text{dom}(D)$), representing the evidence "D is in state d_1."
(c) Compute $\phi_U * \mathcal{E}_D$.
(d) Compute the normalization constant, μ, in $P(U \mid e) = \mu * \phi_U * \mathcal{E}_D$.

Chapter 4
Probabilistic Networks

In this chapter we introduce probabilistic networks for belief update and decision making under uncertainty.

Many real-life situations can be modeled as a domain of entities represented as random variables in a probabilistic network. A probabilistic network is a clever graphical representation of dependence and independence relations between random variables. A domain of random variables can, for instance, form the basis of a decision support system to help decision makers identify the most beneficial decision in a given situation.

A probabilistic network represents and processes probabilistic knowledge. The representational components of a probabilistic network are a qualitative and a quantitative component. The qualitative component encodes a set of (conditional) dependence and independence statements among a set of random variables, informational precedence, and preference relations. The statements of (conditional) dependence and independence, information precedence, and preference relations are visually encoded using a graphical language. The quantitative component, on the other hand, specifies the strengths of dependence relations using probability theory and preference relations using utility theory.

The graphical representation of a probabilistic network describes knowledge of a problem domain in a precise manner. The graphical representation is intuitive and easy to comprehend, making it an ideal tool for communication of domain knowledge between experts, users, and systems. For these reasons, the formalism of probabilistic networks is becoming an increasingly popular knowledge representation for belief update and decision making under uncertainty.

Since a probabilistic network consists of two components, it is customary to consider its construction as a two-phase process: the construction of the qualitative component and subsequently the construction of the quantitative component. The qualitative component defines the structure of the quantitative component. As the first step, the qualitative structure of the model is constructed using a graphical language. This step consists of identifying variables and relations between variables. As the second step, the parameters of the quantitative part as defined by the qualitative part are assessed.

U.B. Kjærulff and A.L. Madsen, *Bayesian Networks and Influence Diagrams: A Guide to Construction and Analysis*, ISS 22, DOI 10.1007/978-1-4614-5104-4_4, © Springer Science+Business Media New York 2013

In this book, we consider the subclass of probabilistic networks known as Bayesian networks and influence diagrams. Bayesian networks and influence diagrams are ideal knowledge representations for use in many situations involving belief update and decision making under uncertainty. These models are often characterized as normative expert systems as they provide model-based domain descriptions, where the model is reflecting properties of the problem domain (rather than the domain expert) and probability calculus is used as the calculus for uncertainty.

A Bayesian network model representation of a problem domain can be used as the basis for performing inference and analysis about the domain. Decision options and utilities associated with these options can be incorporated explicitly into the model, in which case the model becomes an influence diagram, capable of computing expected utilities of all decision options given the information known at the time of decision. Bayesian networks and influence diagrams are applicable for a large range of domain areas with inherent uncertainty.

Section 4.1 considers Bayesian networks as probabilistic models for belief update. We consider Bayesian network models containing discrete variables only and models containing a mixture of continuous and discrete variables. Section 4.2 considers influence diagrams as probabilistic networks for decision making under uncertainty. The influence diagram is a Bayesian network augmented with decision variables, informational precedence relations, and preference relations. We consider influence diagram models containing discrete variables only and models containing a mixture of continuous and discrete variables. In Sect. 4.3 object-oriented probabilistic networks are considered. An object-oriented probabilistic network is a flexible framework for building hierarchical knowledge representations using the notions of classes and instances. In Sect. 4.4 dynamic probabilistic networks are considered. A dynamic probabilistic network is a method for representing dynamic systems that are changing over time.

4.1 Belief Update

A probabilistic interaction model between a set of random variables may be represented as a joint probability distribution. Considering the case where random variables are discrete, it is obvious that the size of the joint probability distribution will grow exponentially with the number of variables as the joint distribution must contain one probability for each configuration of the random variables. Therefore, we need a more compact representation for reasoning about the state of large and complex systems involving a large number of variables.

To facilitate an efficient representation of a large and complex domain with many random variables, the framework of Bayesian networks uses a graphical representation to encode dependence and independence relations among the random variables. The dependence and independence relations induce a compact representation of the joint probability distribution.

4.1.1 Discrete Bayesian Networks

A (discrete) Bayesian network, $\mathcal{N} = (\mathcal{X}, \mathcal{G}, \mathcal{P})$, over variables, $\mathcal{X}$, consists of a DAG $\mathcal{G} = (V, E)$ and a set of conditional probability distributions $\mathcal{P}$. Each node v in G corresponds one-to-one with a discrete random variable $X_v \in \mathcal{X}$ with a finite set of mutually exclusive states. The directed links $E \subseteq V \times V$ of $\mathcal{G}$ specify assumptions of conditional dependence and independence between random variables according to the d-separation criterion (see Proposition 2.4 on page 33).

There is a conditional probability distribution, $P(X_v \mid X_{\mathrm{pa}(v)}) \in \mathcal{P}$, for each variable $X_v \in \mathcal{X}$. The set of variables represented by the parents, pa(v), of $v \in V$ in $\mathcal{G} = (V, E)$ is sometimes called the *conditioning variables* of X_v — the *conditioned* variable.

Definition 4.1 (Jensen 2001). A (discrete) Bayesian network $\mathcal{N} = (\mathcal{X}, \mathcal{G}, \mathcal{P})$ consists of:

- A DAG $\mathcal{G} = (V, E)$ with nodes $V = \{v_1, \ldots, v_n\}$ and directed links E
- A set of discrete random variables, $\mathcal{X}$, represented by the nodes of $\mathcal{G}$
- A set of conditional probability distributions, $\mathcal{P}$, containing one distribution, $P(X_v \mid X_{\mathrm{pa}(v)})$, for each random variable $X_v \in \mathcal{X}$

A Bayesian network encodes a joint probability distribution over a set of random variables, $\mathcal{X}$, of a problem domain. The set of conditional probability distributions, $\mathcal{P}$, specifies a multiplicative factorization of the joint probability distribution over $\mathcal{X}$ as represented by the chain rule of Bayesian networks (see Sect. 3.7 on page 62):

$$P(\mathcal{X}) = \prod_{v \in V} P(X_v \mid X_{\mathrm{pa}(v)}). \tag{4.1}$$

Even though the joint probability distribution specified by a Bayesian network is defined in terms of conditional independence, a Bayesian network is most often constructed using the notion of cause–effect relations, see Sect. 2.4. In practice, cause–effect relations between entities of a problem domain can be represented in a Bayesian network using a graph of nodes representing random variables and links representing cause–effect relations between the entities. Usually, the construction of a Bayesian network (or any probabilistic network for that matter) proceeds according to an iterative procedure where the set of nodes and their states and the set of links are updated iteratively as the model becomes more and more refined. In Chaps. 6 and 7, we consider in detail the art of building efficient probabilistic network representations of a problem domain.

To solve a Bayesian network $\mathcal{N} = (\mathcal{X}, \mathcal{G}, \mathcal{P})$ is to compute all posterior marginals given a set of evidence ε, that is, $P(X \mid \varepsilon)$ for all $X \in \mathcal{X}$. If the evidence set is empty, that is, $\varepsilon = \emptyset$, then the task is to compute all prior marginals, that is, $P(X)$ for all $X \in \mathcal{X}$.

Example 4.1 (Apple Jack (Madsen, Nielsen & Jensen 1998)). Let us consider the small orchard belonging to Jack Fletcher (also called Apple Jack). One day, Apple

Fig. 4.1 The Apple Jack
network

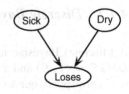

Jack discovers that his finest apple tree is *losing its leaves*. Apple Jack wants to know why this is happening. He knows that if the tree is *dry* (for instance, caused by a drought), there is no mystery as it is common for trees to lose their leaves during a drought. On the other hand, the loss of leaves can be *an indication of a disease*.

The qualitative knowledge about the cause–effect relations of this situation can be modeled by the DAG $\mathcal{G}$ shown in Fig. 4.1. The graph consists of three nodes: Sick, Dry, and Loses that represent variables of the same names. Each variable may be in one of two states: no and yes, that is, dom(Dry) = dom(Loses) = dom(Sick) = (no, yes). The variable Sick tells us that the apple tree is sick by being in state yes. Otherwise, it will be in state no. The variables Dry and Loses tell us whether or not the tree is dry and whether or not the tree is losing its leaves, respectively.

The graph, $\mathcal{G}$, shown in Fig. 4.1 models the cause–effect relations between variables Sick and Loses as well as between Dry and Loses. This is represented by the two (causal) links (Sick, Loses) and (Dry, Loses). In this way, Sick and Dry are common causes of the effect Loses.

Let us return to the discussion of causality considered previously in Sect. 2.4. When there is a causal dependence relation going from a variable A to another variable B, we expect that when A is in a certain state, this has an impact on the state of B. One should be careful when modeling causal dependence relations in a Bayesian network. Sometimes it is not quite obvious in which direction a link should point. In the Apple Jack example, we say that there is a causal impact from Sick on Loses, because when a tree is sick, this might cause the tree to lose its leaves. Why can we not say that when the tree loses its leaves, it might be sick and turn the link in the other direction? The reason is that it is the sickness that causes the tree to lose its leaves and not the lost leaves that causes the sickness.

Figure 4.1 shows the graphical representation of the Bayesian network model. This is referred to as the qualitative representation. To have a complete Bayesian network, we need to specify the quantitative representation. Recall that each variable has two states, no and yes.

The quantitative representation of a Bayesian network is the set of conditional probability distributions, $\mathcal{P}$, defined by the structure of $\mathcal{G}$. Table 4.1 shows the conditional probability distribution of Loses given Sick and Dry, that is, $P(\text{Loses} \mid \text{Dry}, \text{Sick})$. For variables Sick and Dry, we assume that $P(S) = (0.9, 0.1)$ and $P(D) = (0.9, 0.1)$ (we use D as short for Dry, S as short for Sick, and L as short for Loses).

Table 4.1 The conditional			L	
probability distribution $P(L \mid D, S)$	D	S	no	yes
	no	no	0.98	0.02
	no	yes	0.1	0.9
	yes	no	0.15	0.85
	yes	yes	0.05	0.95

Note that all distributions specify the probability of a variable being in a specific state depending on the configuration of its parent variables, but since Sick and Dry do not have any parent variables, their distributions are marginal distributions.

The model may be used to compute all prior marginals and the posterior distribution of each nonevidence variable given evidence in the form of observations on a subset of the variables in the model. The priors for D and S equal the specified marginal distributions, that is, $P(D) = P(S) = (0.9, 0.1)$, while the prior distribution for L is computed through combination of the distributions specified for the three variables, followed by marginalization, where variables D and S are marginalized out. This yields $P(L) = (0.82, 0.18)$ (see Example 3.10 on page 50 for details on combination and marginalization). Following a similar procedure, the posteriors of D and S given $L = $ yes can be computed to be $P(D \mid L = $ yes$) = (0.53, 0.47)$ and $P(S \mid L = $ yes$) = (0.51, 0.49)$. Thus, according to the model, the tree being sick is the most likely cause of the loss of leaves. □

The specification of a conditional probability distribution $P(X_v \mid X_{\mathrm{pa}(v)})$ can be a labor-intensive knowledge acquisition task as the number of parameters grows exponentially with the size of $\mathrm{dom}(X_{\mathrm{fa}(v)})$, where $\mathrm{fa}(v) = \mathrm{pa}(v) \cup \{v\}$. Different techniques can be used to simplify the knowledge acquisition task, assumptions can be made, or the parameters can be estimated from data.

The complexity of a Bayesian network is defined in terms of the family $\mathrm{fa}(v)$ with the largest state space size $\|X_{\mathrm{fa}(v)}\| \triangleq |\mathrm{dom}(X_{\mathrm{fa}(v)})|$. As the state space size of a family of variables grows exponentially with the size of the family, we seek to reduce the size of the parent sets to a minimum. Another useful measure of the complexity of a Bayesian network is the number of cycles and the length of cycles in its graph.

Definition 4.2. A Bayesian network $\mathcal{N} = (\mathcal{X}, \mathcal{G}, \mathcal{P})$ is minimal if and only if for every variable $X_v \in \mathcal{X}$ and for every parent $Y \in X_{\mathrm{pa}(v)}$, X_v is not independent of Y given $X_{\mathrm{pa}(v)} \setminus \{Y\}$.

Definition 4.2 says that the parent set $X_{\mathrm{pa}(v)}$ of X_v should be limited to the set of variables with a direct impact on X_v.

Example 4.2 (Chest Clinic (Lauritzen & Spiegelhalter 1988)). A physician at a chest clinic wants to diagnose her patients with respect to three diseases based on observations of symptoms and possible causes of the diseases. The fictitious qualitative medical knowledge is the following.

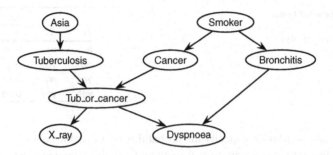

Fig. 4.2 A graph specifying the independence and dependence relations of the Asia example

The physician is trying to diagnose a patient who may be suffering from one or more of *tuberculosis, lung cancer,* or *bronchitis. Shortness of breath* (dyspnoea) may be due to tuberculosis, lung cancer, bronchitis, none of them, or more than one of them. *A recent visit to Asia* increases the chances of tuberculosis, while *smoking* is known to be a risk factor for both lung cancer and bronchitis. *The results of a single chest X-ray* do not discriminate between lung cancer and tuberculosis, as neither does the presence nor absence of dyspnoea.

From the description of the situation, it is clear that there are three possible diseases to consider (lung cancer, tuberculosis, and bronchitis). The three diseases produce three variables Tuberculosis (T), Cancer (L), and Bronchitis (B) with states no and yes. These variables are the targets of the reasoning and may, for this reason, be referred to as *hypothesis variables.*The diseases may be manifested in two symptoms (results of the X-ray and shortness of breath). The two symptoms produce two variables X_ray (X), and Dyspnoea (D) with states no and yes. In addition, there are two causes or risk factors (smoking and a visit to Asia) to consider. The two risk factors produce variables Asia (A) and Smoker (S) with states no and yes.

An acyclic, directed graph, $\mathcal{G}$, encoding the above medical qualitative knowledge is shown in Fig. 4.2, where the variable Tub_or_cancer (E) is a mediating variable (modeling trick, see Sect. 6.2.2 on page 152) specifying whether or not the patient has tuberculosis or lung cancer (or both).

Using the structure of $\mathcal{G}$, we may perform an analysis of dependence and independence properties between variables in order to ensure that the qualitative structure encodes the domain knowledge correctly. This analysis would be based on an application of the d-separation criterion.

Figure 4.2 only presents the qualitative structure $\mathcal{G}$ (and the variables) of $\mathcal{N} = (\mathcal{X}, \mathcal{G}, \mathcal{P})$. In order to have a fully specified Bayesian network, it is necessary to specify the quantitative part, $\mathcal{P}$, too.

The quantitative domain knowledge is specified in the following set of (conditional) probability distributions $P(A) = (0.99, 0.01)$, $P(S) = (0.5, 0.5)$, and the remaining conditional probability distributions, except $P(E\,|\,L, T)$, are shown in Tables 4.2 and 4.3.

Table 4.2 The conditional probability distributions $P(L|S)$, $P(B|S)$, $P(T|A)$, and $P(X|E)$

$P(L|S)$	$S =$ no	$S =$ yes		$P(B|S)$	$S =$ no	$S =$ yes
$L =$ no	0.99	0.9		$B =$ no	0.7	0.4
$L =$ yes	0.01	0.1		$B =$ yes	0.3	0.6

$P(T|A)$	$A =$ no	$A =$ yes		$P(X|E)$	$E =$ no	$E =$ yes
$T =$ no	0.99	0.95		$X =$ no	0.95	0.02
$T =$ yes	0.01	0.05		$X =$ yes	0.05	0.98

Table 4.3 The conditional probability distribution $P(D|B, E)$

	$B =$ no		$B =$ yes	
	$E =$ no	$E =$ yes	$E =$ no	$E =$ yes
$D =$ no	0.9	0.3	0.2	0.1
$D =$ yes	0.3	0.7	0.8	0.9

Table 4.4 Posterior distributions of the disease variables given various evidence scenarios

ε	$P(B =$ yes$|\varepsilon)$	$P(L =$ yes$|\varepsilon)$	$P(T =$ yes$|\varepsilon)$
$\emptyset$	0.45	0.055	0.01
$\{S =$ yes$\}$	0.6	0.1	0.01
$\{S =$ yes$, D =$ yes$\}$	0.88	0.15	0.015
$\{S =$ yes$, D =$ yes$, X =$ yes$\}$	0.71	0.72	0.08

The conditional probability table of the random variable E can be generated from a mathematical expression. From our domain knowledge of the diagnosis problem, we know that E represents the disjunction of L and T. That is, E represents whether or not the patient has tuberculosis or lung cancer. From this, we can express E as $E = T \vee L$. This produces the conditional probability $P(E =$ yes$|L = l, T = t) = 1$ whenever l or t is yes and 0 otherwise.

We will, in a later section, consider in more detail how to build mathematical expressions for the generation of conditional probability distributions (see Sect. 6.5.3 on page 180).

Using the Bayesian network model just developed, we may compute the posterior probability of the three diseases given various subsets of evidence on the causes and symptoms as shown in Table 4.4. □

4.1.2 Conditional Linear Gaussian Bayesian Networks

Up until now, we have considered Bayesian networks over discrete random variables only. However, there are many reasons for extending our considerations to include continuous variables. In this section we will consider Bayesian networks consisting of both continuous and discrete variables. For reasons to become clear later, we restrict our attention to the case of conditional linear Gaussian (also known as normal) distributions and the case of conditional linear Gaussian Bayesian networks. We refer to a conditional linear Gaussian Bayesian network as a CLG Bayesian network.

A CLG Bayesian network $\mathcal{N} = (\mathcal{X}, \mathcal{G}, \mathcal{P}, \mathcal{F})$ consists of an acyclic, directed graph $\mathcal{G} = (V, E)$, a set of conditional probability distributions $\mathcal{P}$, and a set of density functions $\mathcal{F}$. There will be one conditional probability distribution for each discrete random variable X of $\mathcal{X}$ and one density function for each continuous random variable Y of $\mathcal{X}$.

A CLG Bayesian network specifies a distribution over a mixture of discrete and continuous variables (Lauritzen 1992b, Lauritzen & Jensen 2001). The variables, $\mathcal{X}$, are partitioned into the set of continuous variables, $\mathcal{X}_\Gamma$, and the set of discrete variables, $\mathcal{X}_\Delta$. Each node of $\mathcal{G}$ represents either a discrete random variable with a finite set of mutually exclusive and exhaustive states or a continuous random variable with a conditional linear Gaussian distribution conditional on the configuration of its discrete parent variables. This implies an important constraint on the structure of $\mathcal{G}$, namely, that a discrete random variable X_v may only have discrete parents, that is, $X_{\mathrm{pa}(v)} \subseteq \mathcal{X}_\Delta$ for any $X_v \in \mathcal{X}_\Delta$.

Any Gaussian distribution function can be specified by its mean and variance parameter. As mentioned above, we consider the case where a continuous random variable can have a single Gaussian distribution function for each configuration of its discrete parent variables. If a continuous variable has one or more continuous variables as parents, the mean may depend linearly on the state of the continuous parent variables. Continuous parent variables of discrete variables are disallowed.

A random variable, X, has a continuous distribution if there exists a nonnegative function p, defined on the real line, such that for any interval J,

$$P(X \in J) = \int_J p(x)dx,$$

where the function p is the probability density function of X (DeGroot 1986). The probability density function of a *Gaussian* (or *normal*- *distributed variable*, X, with a mean value, μ, and a positive variance, σ^2, is (i.e., $X \sim \mathbb{N}(\mu, \sigma^2)$ or $\mathcal{L}(X) = \mathbb{N}(\mu, \sigma^2)$)

$$p(x; \mu, \sigma^2) = \mathbb{N}(\mu, \sigma^2) = \frac{1}{\sqrt{(2\pi\sigma^2)}} \exp\left[-\frac{(x-\mu)^2}{2\sigma^2}\right],$$

where $x \in \mathbb{R}$.[1]

A continuous random variable, X, has a *conditional linear Gaussian distribution* (or CLG distribution), conditional on the configuration of the parent variables ($Z \subseteq \mathcal{X}_\Gamma, I \subseteq \mathcal{X}_\Delta$) if

$$\mathcal{L}(X \mid Z = z, I = i) = \mathbb{N}(A(i) + B(i)^T z, C(i)), \tag{4.2}$$

---

[1] $\mathcal{L}(X)$ should be read as "the law of X."

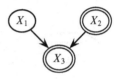

Fig. 4.3 CLG Bayesian
network with X_1 discrete
and X_2 and X_3 continuous

where A is a table of mean values (one value for each configuration i of the discrete
parent variables I), B is a table of regression coefficient vectors (one vector for
each configuration i of I with one regression coefficient for each continuous parent
variable), and C is a table of variances (one for each configuration i of I). Notice
that the mean value $A(i) + B(i)^T z$ of X depends linearly on the values of the
continuous parent variables Z, while the variance is independent of Z. We allow
for the situation where the variance is zero such that deterministic relations between
continuous variables can be represented.

The quantitative part of a CLG Bayesian network consists of a conditional
probability distribution for each $X \in \mathcal{X}_\Delta$ and a conditional Gaussian distribution
for each $X \in \mathcal{X}_\Gamma$. For each $X \in \mathcal{X}_\Gamma$ with discrete parents, I, and continuous
parents, Z, we need to specify a one-dimensional Gaussian probability distribution
for each configuration i of I as shown in (4.2).

Definition 4.3. A CLG Bayesian network $\mathcal{N} = (\mathcal{X}, \mathcal{G}, \mathcal{P}, \mathcal{F})$ consists of:

- A DAG $\mathcal{G} = (V, E)$ with nodes V and directed links E
- A set of random variables, $\mathcal{X}$, represented by the nodes of $\mathcal{G}$
- A set of conditional probability distributions, $\mathcal{P}$, containing one distribution,
  $P(X_v \,|\, X_{\mathrm{pa}(v)})$, for each discrete random variable X_v
- A set of conditional linear Gaussian probability density functions, $\mathcal{F}$, containing
  one density function, $p(Y_v \,|\, X_{\mathrm{pa}(v)})$, for each continuous random variable Y_v

The joint distribution over all the variables in a CLG Bayesian network has the
form $P(\mathcal{X}_\Delta = i) * \mathbb{N}_{|\mathcal{X}_\Gamma|}(\mu(i), \sigma^2(i))$, where $\mathbb{N}_k(\mu, \sigma^2)$ denotes a k-dimensional
Gaussian distribution. The chain rule of CLG Bayesian networks is

$$P(\mathcal{X}_\Delta = i) * \mathbb{N}_{|\mathcal{X}_\Gamma|}(\mu(i), \sigma^2(i)) = \prod_{v \in V_\Delta} P(i_v \,|\, i_{\mathrm{pa}(v)}) * \prod_{w \in V_\Gamma} p(y_w \,|\, X_{\mathrm{pa}(w)}),$$

for each configuration i of $\mathcal{X}_\Delta$.

Recall from Table 2.2 that in the graphical representation of a CLG Bayesian
network, continuous variables are represented by double ovals.

Example 4.3 (CLG Bayesian Network). Figure 4.3 shows an example of the qualitative specification of a CLG Bayesian network, $\mathcal{N}$, with three variables, that is,
$\mathcal{X} = \{X_1, X_2, X_3\}$, where $\mathcal{X}_\Delta = \{X_1\}$ and $\mathcal{X}_\Gamma = \{X_2, X_3\}$. Hence, $\mathcal{N}$ consists of
a continuous random variable X_3 having one discrete random variable X_1 (binary
with states false and true) and one continuous random variable X_2 as parents.

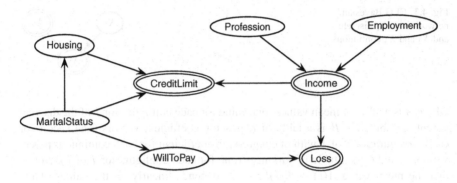

Fig. 4.4 CLG Bayesian network for credit account management

To complete the model, we need to specify the relevant conditional probability distribution and density functions. The quantitative specification could, for instance, consist of the following conditional linear Gaussian distribution functions for X_3:

$$\mathcal{L}(X_3 \mid \text{false}, x_2) = \mathbb{N}(-5 + (-2 * x_2), 1.1)$$

$$\mathcal{L}(X_3 \mid \text{true}, x_2) = \mathbb{N}(5 + (2 * x_2), 1.2).$$

The quantitative specification is completed by letting X_2 have a standard normal distribution (i.e., $X_2 \sim \mathbb{N}(0, 1)$) and $P(X_1) = (0.75, 0.25)$.

The qualitative and quantitative specifications complete the specification of $\mathcal{N}$. The joint distribution induced by $\mathcal{N}$ is

$$P(X_1 = \text{false}) * p(X_2, X_3) = 0.75 * \mathbb{N}\left(\begin{pmatrix} 0 \\ -5 \end{pmatrix}, \begin{pmatrix} 1 & 10 \\ 10 & 5.1 \end{pmatrix}\right),$$

$$P(X_1 = \text{true}) * p(X_2, X_3) = 0.25 * \mathbb{N}\left(\begin{pmatrix} 0 \\ 5 \end{pmatrix}, \begin{pmatrix} 1 & 10 \\ 10 & 5.2 \end{pmatrix}\right).$$

□

Determining the joint distribution induced by $\mathcal{N}$ requires a series of nontrivial computations. We refer the reader to the next chapter for a brief treatment of inference in CLG Bayesian networks. A detailed treatment of these computations is beyond the scope of this book.

Example 4.4 (Adapted from Lauritzen (1992a)). Consider a banker monitoring her clients in order to limit future loss from each client account. The task of the banker is to identify clients who may have problems repaying their loans by predicting potential future loss originating from each individual customer based on demographic information and credit limit.

Figure 4.4 shows a simple CLG Bayesian network model for this scenario. Loss is a linear function of variables Income (I) given variable WillToPay (W).

CreditLimit (C) is a linear function of Income given Housing (H) and Marital Status (M). In addition MaritalStatus is also a causal factor of Housing and WillToPay, while Profession and Employment are causal factors of Income.

With the model, the banker may enter observations on each client and compute an expected loss for that client. The model may be extended to include various risk indicators and controls in order to facilitate a scenario-based analysis on each client. ☐

The reason for restricting our attention to the case of conditional linear Gaussian distributions is that only for this case is exact probabilistic inference feasible by local computations. For most other cases, it is necessary to resort to approximate algorithms.

4.2   Decision Making Under Uncertainty

The framework of influence diagrams (Howard & Matheson 1981) is an effective modeling framework for representation and analysis of (Bayesian) decision making under uncertainty. Influence diagrams provide a natural representation for capturing the semantics of decision making with a minimum of clutter and confusion for the decision maker (Shachter & Peot 1992). Solving a decision problem amounts to (1) determining an (optimal) strategy that maximizes the expected utility for the decision maker and (2) computing the expected utility of adhering to this strategy.

An influence diagram is a type of causal model that differs from a Bayesian network. A Bayesian network is a probabilistic network for belief update, whereas an influence diagram is a probabilistic network for reasoning about decision making under uncertainty. An influence diagram is a graphical representation of a decision problem involving a sequence of interleaved decisions and observations. Similar to Bayesian networks, an influence diagram is a compact and intuitive probabilistic knowledge representation (a probabilistic network). It consists of a graphical representation describing dependence relations between entities of a problem domain, points in time where decisions are to be made, and a precedence ordering specifying the order on decisions and observations. It also consists of a quantification of the strengths of the dependence relations and the preferences of the decision maker. As such, an influence diagram can be considered as a Bayesian network augmented with decision variables, utility functions specifying the preferences of the decision maker, and a precedence ordering.

As decision makers we are interested in making the best possible decisions given our model of the problem domain. Therefore, we associate utilities with state configurations of the network. These utilities are represented by *utility functions* (also known as *value functions*). Each utility function associates a utility value with each configuration of its domain variables. The objective of decision analysis is to identify the decision options that produce the highest expected utility.

By making decisions, we influence the probabilities of the configurations of the network. To identify the decision option with the highest expected utility, we compute the expected utility of each decision alternative. If A is a decision variable with options $a_1, \ldots, a_m$, H is a hypothesis with states $h_1, \ldots, h_n$, and ε is a set of observations in the form of evidence, then we can compute the utility of each outcome of the hypothesis and the expected utility of each action. The utility of an outcome (a_i, h_j) is $U(a_i, h_j)$ where $U(\cdot)$ is our utility function. The expected utility of performing action a_i is

$$\mathrm{EU}(a_i) = \sum_j U(a_i, h_j) P(h_j \mid \varepsilon),$$

where $P(\cdot)$ represents our belief in H given ε. The utility function $U(\cdot)$ encodes the preferences of the decision maker on a numerical scale.

We shall choose the alternative with the highest expected utility; this is known as the (maximum) expected utility principle. Choosing the action, which maximizes the expected utility, amounts to selecting an option a^* such that

$$a^* = \arg\max_{a \in A} \mathrm{EU}(a).$$

There is an important difference between observations and actions. An observation of an event is passive in the sense that we assume that an observation does not affect the state of the world, whereas the decision on an action is active in the sense that an action enforces a certain event. The event enforced by a decision may or may not be included in the model depending on whether or not the event is relevant for the reasoning. If the event enforced by an action A is represented in our model, then A is referred to as an intervening action, otherwise it is referred to as a nonintervening action.

4.2.1   Discrete Influence Diagrams

An (discrete) influence diagram $\mathcal{N} = (\mathcal{X}, \mathcal{G}, \mathcal{P}, \mathcal{U})$ is a four-tuple consisting of a set, $\mathcal{X}$, of discrete random variables and discrete decision variables, an acyclic, directed graph $\mathcal{G}$, a set of conditional probability distributions $\mathcal{P}$, and a set of utility functions $\mathcal{U}$. The acyclic, directed graph, $\mathcal{G} = (V, E)$, contains nodes representing random variables, decision variables, and utility functions (also known as value or utility nodes).

Each decision variable, D, represents a specific point in time under the model of the problem domain where the decision maker has to make a decision. The decision options or alternatives are the states $(d_1, \ldots, d_n)$ of D where $n = \|D\|$. The usefulness of each decision option is measured by the local utility functions associated with D or one of its descendants in $\mathcal{G}$. Each local utility function $u(X_{\mathrm{pa}(v)}) \in \mathcal{U}$, where $v \in V_U$ is a utility node, represents an additive contribution to the total utility function $u(\mathcal{X})$ in $\mathcal{N}$. Thus, the total utility function is the sum of all the utility functions in the influence diagram, that is, $u(\mathcal{X}) = \sum_{v \in V_U} u(X_{pa(v)})$.

Definition 4.4. A (discrete) influence diagram $\mathcal{N} = (\mathcal{X}, \mathcal{G}, \mathcal{P}, \mathcal{U})$ consists of:

- A DAG $\mathcal{G} = (V, E)$ with nodes, V, and directed links, E, encoding dependence relations and information precedence including a total order on decisions
- A set of discrete random variables, $\mathcal{X}_C$, and discrete decision variables, $\mathcal{X}_D$, such that $\mathcal{X} = \mathcal{X}_C \cup \mathcal{X}_D$ represented by nodes of $\mathcal{G}$
- A set of conditional probability distributions, $\mathcal{P}$, containing one distribution, $P(X_v | X_{pa(v)})$, for each discrete random variable X_v
- A set of utility functions, $\mathcal{U}$, containing one utility function, $u(X_{pa(v)})$, for each node v in the subset $V_U \subset V$ of utility nodes

An influence diagram supports the representation and solution of sequential decision problems with multiple local utility functions under the *no-forgetting assumption* (Howard & Matheson 1981), that is, perfect recall is assumed of all observations and decisions made in the past.

An influence diagram, $\mathcal{N} = (\mathcal{X}, \mathcal{G}, \mathcal{P}, \mathcal{U})$, should be constructed such that one can determine exactly which variables are known prior to making each decision. If the state of a variable $X_v \in \mathcal{X}_C$ will be known at the time of making a decision $D_w \in \mathcal{X}_D$, this will (probably) have an impact on the choice of alternative at D. An observation on X_v made prior to decision D_w is represented in $\mathcal{N}$ by making v a parent of w in $\mathcal{G}$. If v is a parent of w in $\mathcal{G} = (V, E)$ (i.e., $(v, w) \in E$, implying $X_v \in X_{pa(w)}$), then it is assumed that X_v is observed prior to making the decision represented by D_w. The link (v, w) is then referred to as an *informational link*.

In an (perfect recall) influence diagram, there must also be a total order on the decision variables $\mathcal{X}_D = \{D_1, \ldots, D_n\} \subseteq \mathcal{X}$. This is referred to as the *regularity* constraint. That is, there can be only one sequence in which the decisions are made. We add informational links to specify a total order $(D_1, \ldots, D_n)$ on $\mathcal{X}_D = \{D_1, \ldots, D_n\}$. There need only be a directed path from one decision variable to the next one in the decision sequence in order to enforce a total order on the decisions.

In short, a link, (w, v), into a node representing a random variable, X_v, denotes a possible probabilistic dependence relation of X_v on Y_w, while a link from a node representing a variable, X, into a node representing a decision variable, D, denotes that the state of X is known when decision D is to be made. A link, (w, v), into a node representing a local utility function, u, denotes functional dependence of u on $X_v \in \mathcal{X}$.

The chain rule of influence diagrams is

$$EU(\mathcal{X}) = \prod_{X_v \in \mathcal{X}_C} P(X_v | X_{pa(v)}) \sum_{w \in V_U} u(X_{pa(w)}).$$

An influence diagram is a compact representation of a joint expected utility function.

Fig. 4.5 The oil wildcatter
network

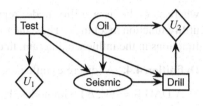

Table 4.5 The conditional
probability distribu-
tion P(Seismic | Oil, Test = yes)

	Seismic		
Oil	diffuse	open	closed
dry	0.6	0.3	0.1
wet	0.3	0.4	0.3
soaking	0.1	0.4	0.5

In the graphical representation of an influence diagram, utility functions are represented by rhombuses (diamond-shaped nodes), whereas decision variables are represented as rectangles, see Table 2.2.

Example 4.5 (Oil Wildcatter (Raiffa 1968)).   Consider the fictitious example of an oil wildcatter about to decide whether or not to drill for oil at a specific site.

The situation of the oil wildcatter is the following. The oil wildcatter must decide either to *drill* or *not to drill*. He is uncertain whether the *hole* will be dry, wet, or soaking with oil. The wildcatter could *take seismic soundings* that will help determine the *geological structure* of the site. The soundings will give a closed reflection pattern (indication of much oil), an open pattern (indication of some oil), or a diffuse pattern (almost no hope of oil).

The qualitative domain knowledge extracted from the above description can be formulated as the DAG shown in Fig. 4.5. The state spaces of the variables are as follows dom(Drill) = (no, yes), dom(Oil) = (dry, wet, soaking), dom(Seismic) = (closed, open, diffuse), and dom(Test) = (no, yes).

Figure 4.5 shows how the qualitative knowledge of the example can be compactly specified in the structure of an influence diagram $\mathcal{N} = (\mathcal{X}, \mathcal{G}, \mathcal{P}, \mathcal{U})$.

The quantitative probabilistic knowledge as defined by the structure of $\mathcal{G}$ consists of P(Oil) and P(Seismic | Oil, Test), while the quantitative utility knowledge consists of U_1(Test) and U_2(Drill, Oil).

The cost of testing is $10k$, whereas the cost of drilling is $70k$. The utility of drilling is $0k$, $120k$, and $270k$ for a dry, wet, and soaking hole, respectively. Hence, U_1(Test) = $(0, -10)$ and U_2(Drill = yes, Oil) = $(-70, 50, 200)$. The test result Seismic depends on the amount of oil represented by variable Oil as specified in Table 4.5. The prior belief of the oil wildcatter on the amount of oil at the site is P(Oil) = $(0.5, 0.3, 0.2)$.

This produces a completely specified influence diagram representation of the oil wildcatter decision problem. The decision strategy of the oil wildcatter will be considered in Example 4.7 on the following page. □

As a consequence of the total order on decisions and the set of informational links, the set of discrete random variables and decision variables is subjected to a partial ordering. The random variables are partitioned into disjoint *information sets* $\mathcal{I}_0, \ldots, \mathcal{I}_n$ (i.e., $\mathcal{I}_i \cap \mathcal{I}_j = \emptyset$ for $i \neq j$) relative to the decision variables specifying the precedence order. The information set $\mathcal{I}_i$ is the set of variables observed after decision D_i and before decision D_{i+1}. The partition induces a partial ordering, $\prec$, on the variables $\mathcal{X}$. The set of variables observed between decisions D_i and D_{i+1} precedes D_{i+1} and succeeds D_i in the ordering

$$\mathcal{I}_0 \prec D_1 \prec \mathcal{I}_1 \prec \cdots \prec D_n \prec \mathcal{I}_n,$$

where $\mathcal{I}_0$ is the set of discrete random variables observed before the first decision, $\mathcal{I}_i$ is the set of discrete random variables observed after making decision D_i and before making decision D_{i+1}, for all $i = 1, \ldots, n-1$, and $\mathcal{I}_n$ is the set of discrete random variables never observed or observed after the last decision D_n has been made. If the influence diagram is not constructed or used according to this constraint, the computed expected utilities will (of course) not be correct.

Example 4.6 (Partial Order of Information Set). The total order on decisions and the informational links of Example 4.5 on the preceding page induce the following partial order:

$$\{\} \prec \text{Test} \prec \{\text{Seismic}\} \prec \text{Drill} \prec \{\text{Oil}\}.$$

This partial order turns out to be a total order. In general, this is not the case. The total order specifies the flow of information in the decision problem. No observations are made prior to the decision on whether or not to Test. After testing and before deciding on whether or not to Drill, the oil wildcatter will make an observation on Seismic, that is, the test result is available before the Drill decision. After drilling Oil is observed. □

To solve an influence diagram $\mathcal{N} = (\mathcal{X}, \mathcal{G}, \mathcal{P}, \mathcal{U})$ with decision variables, $\mathcal{X}_D$, is to identify an optimal strategy, $\hat{\Delta}$, over $\mathcal{X}_D$ maximizing the expected utility for the decision maker and to compute the *(maximum) expected utility* $\text{EU}(\hat{\Delta})$ of $\hat{\Delta}$. A *strategy*, Δ, is an ordered set of decision policies $\Delta = (\delta_1, \ldots, \delta_n)$ including one decision policy for each decision $D \in \mathcal{X}_D$. An *optimal strategy* $\hat{\Delta} = (\hat{\delta}_1, \ldots, \hat{\delta}_n)$, maximizes the expected utility over all possible strategies, that is, it satisfies

$$\text{EU}(\hat{\Delta}) \geq \text{EU}(\Delta),$$

for all strategies Δ.

The *decision history* of D_i, denoted $\mathcal{H}(D_i)$, is the set of previous decisions and their parent variables

$$\mathcal{H}(D_i) = \bigcup_{j=1}^{i-1}(\{D_j\} \cup X_{\mathrm{pa}(v_j)}) = \{D_1, \ldots, D_{i-1}\} \cup \bigcup_{j=0}^{i-2} \mathcal{I}_j,$$

where v_j denotes the node that represents D_j.

The *decision past* of D_j, denoted $\mathcal{I}(D_i)$, is the set of its parent variables and the decision history $\mathcal{H}(D_i)$

$$\mathcal{I}(D_i) = X_{\mathrm{pa}(v_i)} \cup \mathcal{H}(D_i)$$

$$= X_{\mathrm{pa}(v_i)} \cup \bigcup_{j=1}^{i-1}(\{D_j\} \cup X_{\mathrm{pa}(v_j)})$$

$$= \{D_1, \ldots, D_{i-1}\} \cup \bigcup_{j=1}^{i-1} \mathcal{I}_j.$$

Hence, $\mathcal{I}(D_i) \setminus \mathcal{H}(D_i) = \mathcal{I}_{i-1}$ are the variables observed between D_{i-1} and D_i. The *decision future* of D_i, denoted $\mathcal{F}(D_i)$ is the set of its descendant variables

$$\mathcal{F}(D_i) = \mathcal{I}_i \cup \left(\bigcup_{j=i+1}^{n}(\{D_j\} \cup X_{\mathrm{pa}(v_j)}) \right)$$

$$= \{D_{i+1}, \ldots, D_n\} \cup \bigcup_{j=i}^{n} \mathcal{I}_j.$$

A *policy* δ_i is a mapping from the information set $\mathcal{I}(D_i)$ of D_i to the state space $\mathrm{dom}(D_i)$ of D_i such that $\delta_i : \mathcal{I}(D_i) \to \mathrm{dom}(D_i)$. A policy for decision D specifies the (optimal) action for the decision maker for all possible observations made prior to making decision D.

It is only necessary to consider δ_i as a function from relevant observations on $\mathcal{I}(D_i)$ to $\mathrm{dom}(D_i)$, that is, observations with an unblocked path to a utility descendant of D_i. Relevance of an observation with respect to a decision is defined in Sect. 4.2.3 on page 93.

Example 4.7 (Oil Wildcatter Strategy). After solving the influence diagram, we obtain an optimal strategy $\hat{\Delta} = \{\hat{\delta}_{\mathsf{Test}}, \hat{\delta}_{\mathsf{Drill}}\}$. Hence, the optimal strategy $\hat{\Delta}$ (we show how to identify the optimal strategy for this example in Example 5.11 on page 129) consists of a policy $\hat{\delta}_{\mathsf{Test}}$ for Test and a policy $\hat{\delta}_{\mathsf{Drill}}$ for Drill given Test and Seismic

$$\hat{\delta}_{\mathsf{Test}} = \mathsf{yes}$$

$$\hat{\delta}_{\text{Drill}}(\text{Seismic}, \text{Test}) = \begin{cases} \text{yes} & \text{Seismic} = \text{closed}, \text{Test} = \text{no} \\ \text{yes} & \text{Seismic} = \text{open}, \text{Test} = \text{no} \\ \text{yes} & \text{Seismic} = \text{diffuse}, \text{Test} = \text{no} \\ \text{yes} & \text{Seismic} = \text{closed}, \text{Test} = \text{yes} \\ \text{yes} & \text{Seismic} = \text{open}, \text{Test} = \text{yes} \\ \text{no} & \text{Seismic} = \text{diffuse}, \text{Test} = \text{yes} \end{cases}$$

The policy for Test says that we should always test, while the policy for Drill says that we should drill except when the test produces a diffuse pattern indicating almost no hope of oil.   □

An intervening decision D of an influence diagram is a decision that may impact the state or value of another variable X represented in the model. In order for D to potentially impact the value of X, X must be a descendant of D in G. This can be realized by considering the d-separation criterion (consider the information blocking properties of the converging connection) and the set of evidence available when making the decision D. Consider, for instance, the influence diagram shown in Fig. 4.5. The decision Test is an intervening decision as it impacts the value of Seismic. It cannot, however, impact the value of Oil as Oil is a non-descendant of Test, and we have no *down-stream* evidence when making the decision on Test. Since decision D may only have a potential impact on its descendants, the usefulness of D can only be measured by the utility descendants of D.

A total ordering on the decision variables is usually assumed. This assumption can, however, be relaxed. Nielsen & Jensen (1999) describe when decision problems with only a partial ordering on the decision variables are *well defined*. In addition, the limited memory influence diagram (Lauritzen & Nilsson 2001), see Sect. 4.2.3, and the unconstrained influence diagram (Vomlelová & Jensen 2002) support the use of unordered decision variables.

Example 4.8 (Apple Jack). We consider once again the problems of Apple Jack from Example 4.1 on page 71. A Bayesian network for reasoning about the causes of the apple tree losing its leaves was shown in Fig. 4.1 on page 72.

We continue the example by assuming that Apple Jack wants to decide whether or not to invest resources in giving the tree some treatment against a possible disease. Since this involves a decision through time, we have to extend the Bayesian network to capture the impact of the treatment on the development of the disease. We first add three variables similar to those already in the network. The new variables Sick*, Dry*, and Loses* correspond to the original variables, except that they represent the situation at the time of harvest, that is, after the treatment decision. These variables have been added in Fig. 4.6.

The additional variables have the same states as the original variables: Sick*, Dry*, and Loses* all have states no and yes. In the extended model, we expect a causal influence from the original Sick variable on the Sick* variable and from the original Dry variable on the Dry* variable. The reason is the following. If, for

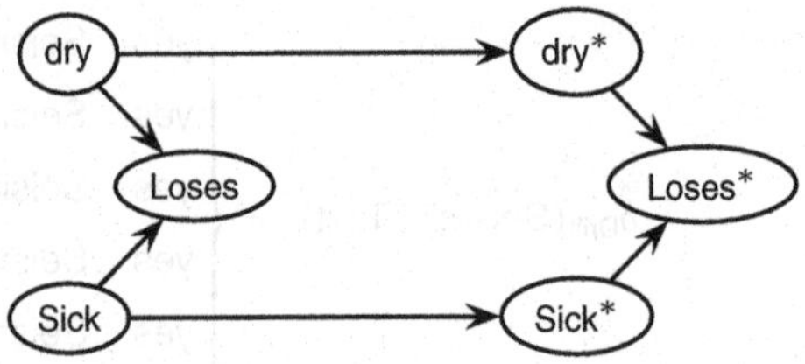

Fig. 4.6 We model the
system at two different points
in time (before and after a
decision) by replicating the
structure

Fig. 4.7 Addition of a
decision variable for
treatment to the Bayesian
network in Fig. 4.6

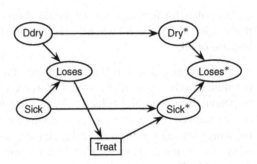

example, we expect the tree to be sick now, then this is also likely to be the case
in the future and especially at the time of harvest. Of course, the strength of the
influence depends on how far out in the future we look. Perhaps one could also have
a causal influence from Loses on Loses*, but we have chosen not to model such a
possible dependence relation in this model.

Apple Jack may try to heal the tree with a treatment to get rid of the possible
disease. If he expects that the loss of leaves is caused by drought, he might save
his money and just wait for rain. The action of giving the tree a treatment is now
added as a decision variable to the Bayesian network, which will then no longer be
a Bayesian network. Instead, it becomes the influence diagram shown in Fig. 4.7.

The treat decision variable has the states no and yes. There is a causal
link (Treat, Sick*) from the decision Treat to Sick* as we expect the treatment to
have a causal impact on the future health of the tree. There is an informational link
from Loses to Treat as we expect Apple Jack to observe whether or not the apple
tree is losing its leaves prior to making the decision on treatment.

We need to specify the utility functions enabling us to compute the expected
utility of the decision options. This is done by adding utility functions to the
influence diagram. Each utility function will represent a term of an additively
decomposing utility function, and each term will contribute to the total utility. The
utility functions are shown in Fig. 4.8.

The utility function C specifies the cost of the treatment, while utility function H
specifies the reward of the harvest. The latter depends on the state of Sick*,
indicating that the production of apples depends on the health of the tree.

Fig. 4.8 A complete qualitative representation of the influence diagram used for decision making in Apple Jack's orchard

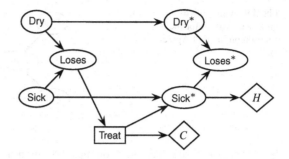

Table 4.6 The conditional probability distribution $P(\text{Sick}^* \mid \text{Treat}, \text{Sick})$

		Sick*	
Treat	Sick	no	yes
no	no	0.98	0.02
no	yes	0.01	0.99
yes	no	0.99	0.01
yes	yes	0.8	0.2

Table 4.7 The conditional probability distribution $P(\text{Dry}^* \mid \text{Dry})$

	Dry*	
Dry	no	yes
no	0.95	0.05
yes	0.4	0.6

Table 4.8 The conditional probability distribution $P(\text{Loses}^* \mid \text{Dry}^*, \text{Sick}^*)$

		Loses*	
Dry*	Sick*	no	yes
no	no	0.98	0.02
no	yes	0.1	0.9
yes	no	0.15	0.85
yes	yes	0.05	0.95

Figure 4.8 shows the complete qualitative representation of the influence diagram $N = (X, G, P, U)$. To complete the quantitative representation as well, we need to specify the conditional probability distributions, P, and utility functions, U, of N. Recall that a decision variable does not have any distribution. The appropriate probability distributions are specified in Tables 4.6–4.8.

If we have a healthy tree (Sick* is in state no), then Apple Jack will get an income of € 200, while if the tree is sick (Sick* is in state yes), his income is only € 30, that is, $H(\text{Sick}^*) = (200, 30)$. To treat the tree, he has to spend € 80, that is, $C(\text{Treat}) = (0, -80)$.

Since Dry* and Loses* are not relevant for the decision on whether or not to treat and since we do not care about their distribution, we remove them from our model producing the final model shown in Fig. 4.9. Variables Dry* and Loses* are in fact

Fig. 4.9 A simplified
influence diagram for the
decision problem of Apple
Jack

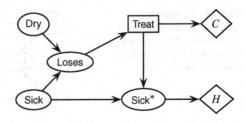

barren variables, see Sect. 3.3.4 on page 53. In an influence diagram, a variable is
a barren variable when none of its descendants are utility nodes, and none of its
descendants are ever observed.

The purpose of our influence diagram is to be able to determine the optimal
strategy for Apple Jack. After solving $\mathcal{N}$, we obtain the following policy (δ_{Treat} :
Loses $\rightarrow$ dom(Treat)) for Treat:

$$\delta_{\text{Treat}}(\text{Loses}) = \begin{cases} \text{no} & \text{Loses} = \text{no} \\ \text{yes} & \text{Loses} = \text{yes} \end{cases}$$

Hence, we should only treat the tree when it loses its leaves. In Sect. 5.2, we describe
how to solve an influence diagram.   □

Notice that since a policy is a mapping from all possible observations to decision
options, it is sufficient to solve an influence diagram once. Hence, the computed
strategy can be used by the decision maker each time she or he is faced with the
decision problem.

Implications of Perfect Recall

As mentioned above, when using influence diagrams to represent decision problems,
we assume perfect recall. This assumption states that at the time of any decision, the
decision maker remembers all past decisions and all previously known information
(as enforced by the informational links). This implies that a decision variable and all
of its parent variables are informational parents of all subsequent decision variables.
Due to this assumption, it is not necessary to include no-forgetting links in the DAG
of the influence diagram as they—if missing—will implicitly be assumed present.

Example 4.9 (Jensen, Jensen & Dittmer (1994)). Let $\mathcal{N}$ be the influence diagram in
Fig. 4.10 on the facing page. This influence diagram represents a decision problem
involving four decisions D_1, D_2, D_3, and D_4 in that order.

From the structure of $\mathcal{N}$, the following partial ordering on the random and
decision variables can be read:

$$\{B\} \prec D_1 \prec \{E, F\} \prec D_2 \prec \{\} \prec D_3 \prec \{G\} \prec D_4 \prec \{A, C, D, H, I, J, K, L\}.$$

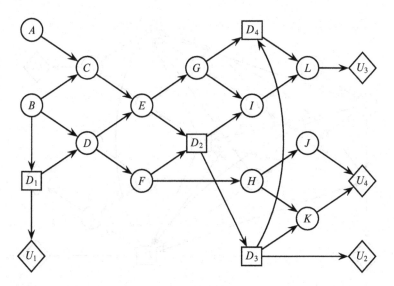

Fig. 4.10 An influence diagram representing the sequence of decisions D_1, D_2, D_3, D_4

This partial ordering specifies the flow of information in the decision problem represented by $\mathcal{N}$. Thus, the initial (relevant) information available to the decision maker is an observation of B. After making a decision on D_1, the decision maker observes E and F. After the observations of E and F, a decision on D_2 is made, and so on.

Notice that no-forgetting links have been left out, for example, there are no links from B to D_2, D_3, or D_4. These links are included in Fig. 4.11. The difference in complexity of reading the graph is apparent.

As this example shows, a rather informative analysis can be performed by reading only the structure of the graph of $\mathcal{N}$.   □

4.2.2 Conditional LQG Influence Diagrams

Conditional linear–quadratic Gaussian influence diagrams combine conditional linear Gaussian Bayesian networks, discrete influence diagrams, and quadratic utility functions into a single framework supporting decision making under uncertainty with both continuous and discrete variables (Madsen & Jensen 2005).

Definition 4.5. A CLQG influence diagram $\mathcal{N} = (\mathcal{X}, \mathcal{G}, \mathcal{P}, \mathcal{F}, \mathcal{U})$ consists of:

- A DAG $\mathcal{G} = (V, E)$ with nodes, V, and directed links, E, encoding dependence relations and information precedence including a total order on decisions

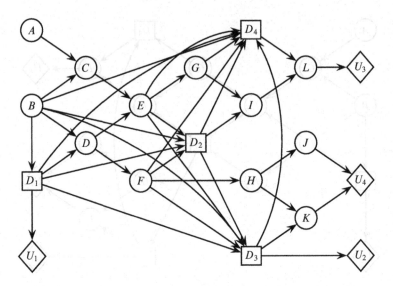

Fig. 4.11 The influence diagram of Fig. 4.10 with no-forgetting links

- A set of random variables, $\mathfrak{X}_C$, and decision variables, $\mathfrak{X}_D$, such that $\mathfrak{X} = \mathfrak{X}_C \cup \mathfrak{X}_D$ represented by nodes of $\mathcal{G}$
- A set of conditional probability distributions, $\mathcal{P}$, containing one distribution, $P(X_v \mid X_{\mathrm{pa}(v)})$, for each discrete random variable X_v
- A set of conditional linear Gaussian probability density functions, $\mathcal{F}$, containing one density function, $p(Y_w \mid X_{\mathrm{pa}(w)})$, for each continuous random variable Y_w
- A set of linear–quadratic utility functions, $\mathcal{U}$, containing one utility function, $u(X_{\mathrm{pa}(v)})$, for each node v in the subset $V_U \subset V$ of utility nodes

We refer to a conditional linear–quadratic Gaussian influence diagram as a CLQG influence diagram. The chain rule of CLQG influence diagrams is

$$\mathrm{EU}(\mathfrak{X}_\Delta = i, \mathfrak{X}_\Gamma) = P(\mathfrak{X}_\Delta = i) * \mathbb{N}_{|\mathfrak{X}_\Gamma|}\big(\mu(i), \sigma^2(i)\big) * \sum_{z \in V_U} u(X_{\mathrm{pa}(z)})$$

$$= \prod_{v \in V_\Delta} P(i_v \mid i_{\mathrm{pa}(v)}) * \prod_{w \in V_\Gamma} p(y_w \mid X_{\mathrm{pa}(w)}) *$$

$$\sum_{z \in V_U} u(X_{\mathrm{pa}(z)})$$

for each configuration i of $\mathfrak{X}_\Delta$.

Fig. 4.12 A CLQG influence diagram for a simple guessing game

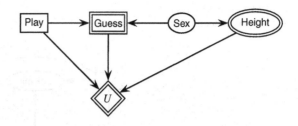

Recall that in the graphical representation of a CLQG influence diagram, continuous utility functions are represented by double rhombuses and continuous decision variables as double rectangles, see Table 2.2 on page 23 for an overview of vertex symbols.

A CLQG influence diagram is a compact representation of a joint expected utility function over continuous and discrete variables, where continuous variables are assumed to follow a linear Gaussian distribution conditional on a subset of discrete variables, while utility functions are assumed to be linear–quadratic in the continuous variables (and constant in the discrete). This may seem a severe assumption which could be limiting to the usefulness of the CLQG influence diagram. The assumption seems to indicate that all local utility functions specified in a CLQG influence diagram should be linear–quadratic in the continuous variables. This is not the case, however, as the following examples show. We will consider the assumption in more detail in Sect. 5.2 on solving decision models.

Example 4.10 (Guessing Game (Madsen & Jensen 2005)). Figure 4.12 illustrates a CLQG influence diagram, $\mathcal{N}$, representation of a simple guessing game with two decisions.

The first decision, represented by the discrete decision variable Play with states reward and Play, is to either accept an immediate reward or to play a game where you will receive a payoff determined by how good you are at guessing the height of a person, represented by the continuous random variable Height, based on knowledge about the sex of the person, represented by the discrete random variable Sex with states female and male. The second decision, represented by the real-valued decision variable Guess, is your guess on the height of the person given knowledge about the sex of the person.

The payoff is a constant (higher than the reward) minus the distance of your guess from the true height of the person measured as height minus guess squared.

To quantify $\mathcal{N}$, we need to specify a prior probability distribution for Sex, a conditional Gaussian distribution for Height and a utility function over Play, Guess, and Height. Assume the prior distribution on Sex is $P(\mathsf{Sex}) = (0.5, 0.5)$, whereas the distribution for Height is

$$\mathcal{L}(\mathsf{Height}|\mathsf{female}) = \mathbb{N}(170, 400)$$

$$\mathcal{L}(\mathsf{Height}|\mathsf{male}) = \mathbb{N}(180, 100).$$

Fig. 4.13 A revised version
of the oil wildcatter problem

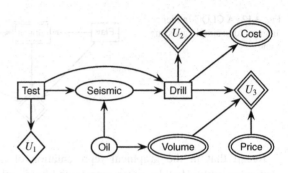

We assume the average height of a female to be 170 cm with a standard deviation
of 20 cm and average height of a male to be 180 cm with a standard deviation
of 10 cm. The utility function over Play, Guess, *and* Height is

$$u(\text{play}, d_2, h) = 150 - (h - d_2)^2$$
$$u(\text{reward}, d_2, h) = 100.$$

We assume the immediate reward is 100. After solving $\mathcal{N}$, we obtain an optimal
strategy $\Delta = \{\delta_{\text{Play}}, \delta_{\text{Guess}}\}$

$$\delta_{\text{Play}} = \text{play}$$
$$\delta_{\text{Guess}}(\text{play}, \text{female}) = 170$$
$$\delta_{\text{Guess}}(\text{play}, \text{male}) = 180.$$

The optimal strategy is to guess that the height of a female person is 170 cm and the
height of a male person is 180 cm.

In this example the policy for Guess reduces to a constant for each configuration
of its parent variables. In the general case, the policy for a continuous decision
variable is a multilinear function in its continuous parent variables given the discrete
parent variables.   □

As another example of a CLQG influence diagram, consider a revised extension
of the oil wildcatter problem of Raiffa (1968) (Example 4.5 on page 82). The revised
Oil Wildcatter problem, which is further revised here, is due to Cobb & Shenoy
(2004).

Example 4.11 (Oil Wildcatter (Madsen & Jensen 2005)). The network of the
revised version of the Oil Wildcatter problem is shown in Fig. 4.13. First, the
decision maker makes a decision on whether or not to perform a test Test of the
geological structure of the site under consideration. When performed, this test
will produce a test result, Seismic depending on the amount of oil Oil. Next, a
decision Drill on whether or not to drill is made. There is a cost Cost associated with
drilling, while the revenue is a function of oil volume Volume and oil price Price.

We assume the continuous random variables (i.e., cost of drilling, oil price, and oil volume) to follow (conditional) Gaussian distributions. The utility function can be stated in thousands of euros as $U_1(\text{Test} = \text{yes}) = -10$, $U_2(\text{Cost} = c, \text{Drill} = \text{yes}) = -c$, $U_3(\text{Volume} = v, \text{Price} = p, \text{Drill} = \text{yes}) = v * p$, and zero for the no drill and no test situations.

If the hole is dry, then no oil is extracted: $\mathcal{L}(\text{Volume} | \text{Oil} = \text{dry}) = \mathbb{N}(0, 0)$. If the hole is wet, then some oil is extracted: $\mathcal{L}(\text{Volume} | \text{Oil} = \text{wet}) = \mathbb{N}(6, 1)$. If the hole is soaking with oil, then a lot of oil is extracted: $\mathcal{L}(\text{Volume} | \text{Oil} = \text{soaking}) = \mathbb{N}(13.5, 4)$. The unit is a thousand barrels. The cost of drilling follows a Gaussian distribution $\mathcal{L}(\text{Cost} | \text{Drill} = \text{yes}) = \mathbb{N}(70, 100)$. We assume that the price of oil Price also follows a Gaussian distribution $\mathcal{L}(\text{Price}) = \mathbb{N}(20, 4)$.

Notice that the continuous utility functions U_2 and U_3 are not linear–quadratic in their continuous domain variables. □

4.2.3 Limited Memory Influence Diagrams

The framework of influence diagrams offers compact and intuitive models for reasoning about decision making under uncertainty. Two of the fundamental assumptions of the influence diagram representation are the no-forgetting assumption implying perfect recall of the past and the assumption of a total order on the decisions. The limited memory influence diagram framework (LIMID) (Lauritzen & Nilsson 2001) relaxes both of these fundamental assumptions.

Relaxing the no-forgetting and the total order (on decisions) assumptions largely increases the class of multistage decision problems that can be modeled. LIMIDs allow us to model more types of decision problems than the ordinary influence diagrams.

The graphical difference between the LIMID representation and the ordinary influence diagram representation is that the latter representation (as presented in this book) assumes some informational links to be implicitly present in the graph. This assumption is not made in the LIMID representation. For this reason, it is necessary to explicitly represent all information available to the decision maker at each decision.

The definition of a limited memory influence diagram is as follows.

Definition 4.6. A LIMID $\mathcal{N} = (\mathcal{X}, \mathcal{G}, \mathcal{P}, \mathcal{U})$ consists of:

- A DAG $\mathcal{G} = (V, E)$ with nodes V and directed links E encoding dependence relations and information precedence.
- A set of random variables, $\mathcal{X}_C$, and discrete decision variables, $\mathcal{X}_D$, such that $\mathcal{X} = \mathcal{X}_C \cup \mathcal{X}_D$ represented by nodes of $\mathcal{G}$.
- A set of conditional probability distributions, $\mathcal{P}$, containing one distribution, $P(X_v | X_{\text{pa}(v)})$, for each discrete random variable X_v.
- A set of utility functions, $\mathcal{U}$, containing one utility function, $u(X_{\text{pa}(v)})$, for each node v in the subset $V_U \subset V$ of utility nodes.

Fig. 4.14 A LIMID
representation of a decision
scenario with two unordered
decisions

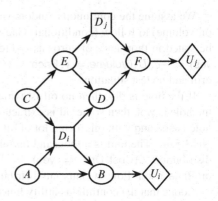

Using the LIMID representation, it is possible to model multistage decision problems with unordered sequences of decisions and decision problems in which perfect recall cannot be assumed or may not be appropriate. This makes the LIMID framework a good candidate for modeling large and complex domains using an appropriate assumption of forgetfulness of the decision maker. Notice that all decision problems that can be represented as an ordinary influence diagram can also be represented as a LIMID.

Example 4.12 (LIMID). Figure 4.14 shows an example of a LIMID representation $\mathcal{N} = (\mathcal{X}, \mathcal{G}, \mathcal{P}, \mathcal{U})$ of a decision scenario with two unordered decisions. Prior to decision D_i, observations on the values of A and C are made, while prior to decision D_j, an observation on the value of E is made. Notice that the observations on A and C made prior to decision D_i are not available at decision D_j and vice versa for the observation on E. □

Example 4.13 (Breeding Pigs (Lauritzen & Nilsson 2001)). A farmer is growing pigs for a period of four months and subsequently selling them. During this period, the pigs may or may not develop a certain disease. If a pig has the disease at the time, it must be sold for slaughtering; its expected market price is € 40. If it is disease free, its expected market price as a breeding animal is € 135.

Once a month, a veterinarian inspects each pig and makes a test for presence of the disease. If a pig is ill, the test will indicate this with probability 0.80, and if the pig is healthy, the test will indicate this with probability 0.90. At each monthly visit, the doctor may or may not treat a pig for the disease by injecting a certain drug. The cost of an injection is € 13.

A pig has the disease in the first month with probability 0.10. A healthy pig develops the disease in the following month with probability 0.20 without injection, whereas a healthy and treated pig develops the disease with probability 0.10, so the injection has some preventive effect. An untreated pig that is unhealthy will remain so in the following month with probability 0.90, whereas the similar probability is 0.50 for an unhealthy pig that is treated. Thus, spontaneous cure is possible, but treatment is beneficial on average.

Fig. 4.15   Three
test-and-treat cycles are
performed prior to selling a
pig

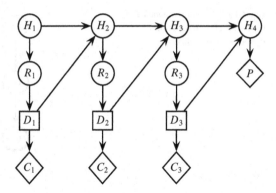

The qualitative structure of the LIMID representation of this decision problem is shown in Fig. 4.15. Notice that we make the assumption that the test result R_i is only available for decision D_i. This implies that the test result is not taken into account for future decisions as it is either forgotten or ignored.   □

The above example could be modeled as a standard influence diagram (assuming perfect recall), but if more test-and-treat cycles must be performed, the state space size of the past renders decision making intractable. Therefore, it is appropriate to make the decision on whether or not to treat based on the current test result (and not considering past test results and possible treatments)—in this case, individual records for the pigs need not be kept. In short, the example illustrates a situation where instead of keeping track of all past observations and decisions, some of these are deliberately ignored (in order to maintain tractability of the task of computing policies).

4.3   Object-Oriented Probabilistic Networks

As large and complex systems are often composed of collections of identical or similar components, models of such systems will naturally contain repetitive patterns. A complex system will typically be composed of a large number of similar or even identical components. This composition of the system should be reflected in models of the system to support model construction, maintenance, and reconfiguration. For instance, a diagnosis model for diagnosing car start problems could reflect the natural decomposition of a car into its engine, electrical system, fuel system, etc.

To support this approach to model development, the framework of object-oriented probabilistic networks has been developed, see, for example, (Koller & Pfeffer 1997, Laskey & Mahoney 1997, Neil, Fenton & Nielsen 2000). Object-orientation may be defined in the following way

Fig. 4.16 $\mathcal{M}$ is an instance of a network class $C_{\mathcal{M}}$ within another network class $C_{\mathcal{N}}$

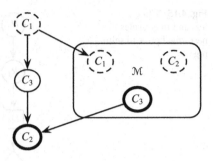

object-orientation = objects + inheritance,

where objects are instances of classes and inheritance defines a relationship between classes. Thus, we need to introduce the notion of objects and classes. In this section, we introduce the notion of *object-oriented probabilistic networks (OOPNs)*.

The basic OOPN mechanisms described below support a type of object-oriented specification of probabilistic networks, which makes it simple to reuse models, to encapsulate submodels (providing a means for hierarchical model specification), and to perform model construction in a top-down fashion, a bottom-up fashion, or a mixture of the two (allowing repeated changes of level of abstraction).

An object-oriented modeling paradigm provides support for working with different levels of abstraction in constructing network models. Repeated changes of focus are partly due to the fact that humans naturally think about systems in terms of hierarchies of abstractions and partly due to lack of ability to mentally capture all details of a complex system simultaneously. Specifying a model in a hierarchical fashion often makes the model less cluttered and thus provides a better means of communicating ideas among knowledge engineers, domain experts, and users.

In the OOPN paradigm we present, an *instance* or *object* has a set of variables and related functions (i.e., probability distributions, probability densities, utility functions, and precedence constraints). This implies that in addition to the usual types of nodes, the graph of an OOPN model may contain nodes representing instances of other networks encapsulated in the model. A node that does not represent an instance of a network class is said to represent a *basic* variable.

An instance represents an instantiation of a network class within another network class. A network class is a blueprint for an instance. As such, a *network class* is a named and self-contained description of a probabilistic network, characterized by its name, interface, and hidden part. As instances can be nested, an object-oriented network can be viewed as a hierarchical description of a problem domain. In this way, an instance $\mathcal{M}$ is the instantiation (or realization) of a network class $C_{\mathcal{M}}$ within another network class $C_{\mathcal{N}}$, see Fig. 4.16.

An instance connects to other variables via some of its (basic) variables. These variables are known as its *interface* variables. As we wish to support information hiding, the interface variables usually only constitute a subset of the variables in the network class.

Let us be more precise. A network class C is a DAG over three pairwise disjoint sets of nodes $\mathcal{I}(C)$, $\mathcal{H}(C)$, and $\mathcal{O}(C)$, where $\mathcal{I}(C)$ are the input nodes, $\mathcal{H}(C)$ are the hidden nodes, and $\mathcal{O}(C)$ are the output nodes of C. The set $\mathcal{I}(C) \cup \mathcal{O}(C)$ is the interface of C. Interface nodes may represent either decision or random variables, whereas hidden nodes may be instances of network classes, decision variables, random variables, and utility functions.

Definition 4.7. An OOPN network class $C = (\mathcal{N}, \mathcal{I}, \mathcal{O})$ consists of:

- A probabilistic network $\mathcal{N}$ over variables $\mathcal{X}$ with DAG $\mathcal{G}$
- A set of basic variables $\mathcal{I} \subseteq \mathcal{X}$ specified as input variables and a set of basic variables $\mathcal{O} \subseteq \mathcal{X}$ specified as output variables such that $\mathcal{I} \cap \mathcal{O} = \emptyset$ and $\mathcal{H} = \mathcal{X} \setminus (\mathcal{I} \cup \mathcal{O})$

In the graphical representation of an OOPN instances are represented as rectangles with arc-shaped corners, whereas input variables are represented as dashed ovals, and output variables are represented as bold ovals. If the interface variables of a network instance are not shown, then the instance is collapsed. Otherwise, it is expanded.

Since an OOPN implements information hiding through encapsulation, we need to be clear on scope rules. First, we define the notations of simple and qualified names. If X is a variable of a network instance $\mathcal{N}$, then X is the *simple name* of the variable, whereas $\mathcal{N}.X$ is the *qualified name* (also known as the long name) of the variable. The scope $\mathbb{S}(X)$ of a variable X (i.e., a basic variable or an instance) is defined as the part of a model in which the declaration of X can be referred to by its simple name.

The (internal) scope $\mathbb{S}(C)$ of a network class C is the set of variables and instances which can be referred to by their simple names inside C. For instance, the internal scope of the network $C_{\mathcal{N}}$ in Fig. 4.16 on the facing page is $\mathbb{S}(C_{\mathcal{N}}) = \{C_1, C_3, C_2, \mathcal{M}\}$. The scope of an instance $\mathcal{M}$ of a network class $C_{\mathcal{M}}$, that is, class$(\mathcal{M}) = C_{\mathcal{M}}$, is defined in a similar manner.

The interface variables $\mathcal{I}(C) \cup \mathcal{O}(C)$ of C are used to enlarge the visibility of basic variables in the instantiations of C. The visibility of a variable X can be enlarged by specifying it as either an input or an output variable of its class.

An input variable X of an instance $\mathcal{M}$ is a placeholder for a variable (the parent of X) in the encapsulating class of $\mathcal{M}$. Therefore, an input variable has at most one parent. An output variable X of an instance $\mathcal{M}$, on the other hand, enlarges the visibility of X to include the encapsulating network class of $\mathcal{M}$.

Notice that the scope of a variable is distinct from visibility of the variable. In Fig. 4.16, the scope of output variable C_3 is $\mathcal{M}$, whereas its visibility is enlarged to include $\mathcal{N}$ by defining it as an output variable of $\mathcal{M}$.

An input variable I of an instance $\mathcal{M}$ of network class C is *bound* if it has a parent X in the network class encapsulating $\mathcal{M}$. Each input random variable I of a class C is assigned a default prior probability distribution $P(I)$, which becomes the probability distribution of the variable I in all instances of C where I is an unbound input variable. A link into a node representing an input variable may be referred to as a *binding link*.

Let $\mathcal{M}$ be an instance of network class C. Each input variable $I \in \mathcal{I}(C)$ has no parent in C, no children outside C, and the corresponding variable of M has at most one parent in the encapsulating class of $\mathcal{M}$. Each output variable $O \in \mathcal{O}(C)$ may only have parents in $\mathcal{I}(C) \cup \mathcal{H}(C)$. The children and parents of $H \in \mathcal{H}(C)$ are subsets of the variables of C.

Example 4.14 (Object-Oriented Probabilistic Network). Figure 4.16 shows an instance $\mathcal{M}$ of a network class $C_{\mathcal{M}}$ instantiated within another network class $C_{\mathcal{N}}$. Network class $C_{\mathcal{N}}$ has input variable C_1, hidden variables C_3 and $\mathcal{M}$, and output variable C_2. The network class $C_{\mathcal{M}}$ has input variables C_1 and C_2, output variable C_3, and unknown hidden variables. The input variable C_1 of instance $\mathcal{M}$ is bound to C_1 of C_N, whereas C_2 is unbound.

Since $C_1 \in \mathcal{I}(C_{\mathcal{N}})$ is bound to $C_1 \in \mathcal{I}(\mathcal{M})$, the visibility of $C_1 \in \mathcal{I}(C_{\mathcal{N}})$ is extended to include the internal scope of $\mathcal{M}$. Hence, when we refer to $C_1 \in \mathcal{I}(C_{\mathcal{M}})$ inside $C_{\mathcal{M}}$, we are in fact referring to $C_1 \in \mathcal{I}(C_{\mathcal{N}})$ as $C_1 \in \mathcal{I}(C_{\mathcal{M}})$ in instance M is a placeholder for $C_1 \in \mathcal{I}(C_{\mathcal{N}})$ (i.e., you may think of $C_1 \in \mathcal{I}(C_{\mathcal{M}})$ as the formal parameter of $C_{\mathcal{M}}$ and $C_1 \in \mathcal{I}(C_{\mathcal{N}})$ as the actual parameter of $\mathcal{M}$).   □

Since an input variable $I \in \mathcal{I}(\mathcal{M})$ of an instance $\mathcal{M}$ is a placeholder for a variable Y in the internal scope of the encapsulating instance of $\mathcal{M}$, type checking becomes important when the variable Y is bound to I. The variable I enlarges the visibility of Y to include the internal scope of $\mathcal{M}$, and it should therefore be equivalent to Y. We define two variables Y and X to be equivalent as follows:

Definition 4.8. Two variables X and Y are *equivalent* if and only if they are of the same kind, category, and subtype with the same state labels in the case of discrete variables.

This approach to type checking is referred as *strong type checking*.

If a model contains a lot of repetitive structure, its construction may be tiresome, and the resulting model may even be rather cluttered. Both issues are solved when using object-oriented models. Another key feature of object-oriented models is modularity. Modularity allows knowledge engineers to work on different parts of the model independently once an appropriate interface has been defined. The following example will illustrate this point.

Example 4.15 (Apple Jack's Garden). Let us assume that Apple Jack from Example 4.1 on page 71 has a garden of three apple trees (including his finest apple tree). He may want to reason about the sickness of each tree given observations on whether or not some of the trees in the garden are losing their leaves.

Figure 4.17 shows the apple tree network class. The prior of each tree being sick will be the same, while the dryness of a tree is caused by a drought. The drought is an input variable of the apple tree network class. If there is a drought, this will impact the dryness of all trees. The prior on drought is $P(\mathsf{Drought}) = (0.9, 0.1)$, while the conditional distribution of Dry conditional on $\mathsf{Drought}$ is shown in Table 4.9.

Fig. 4.17 The apple tree
network class

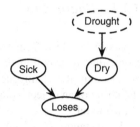

Table 4.9 The conditional
probability
distribution $P(\text{Drought}\,|\,\text{Dry})$

	Dry	
Drought	no	yes
no	0.85	0.15
yes	0.35	0.65

Fig. 4.18 The apple garden
network consisting of three
instantiations of the apple tree
network

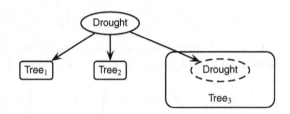

Figure 4.18 shows the network class of the apple garden. The input variable Drought of each of the instances of the apple tree network class is bound to the Drought variable in the apple garden network class. This enlarges the visibility of the Drought variable (in the apple garden network class) to the internal scope defined by each instance.

The two instances Tree$_1$ and Tree$_2$ are collapsed (i.e., not showing the interface variables), while the instance Tree$_3$ is expanded (i.e., not collapsed) illustrating the interface of the network class.

The Drought variable could be an input variable of the apple garden network class as well as it is determined by other complex factors. For the sake of simplicity of the example, we have made it a hidden variable of the apple garden network class. □

As mentioned above, a default prior distribution $P(X)$ is assigned to each input variable $X \in \mathcal{I}(C)$ of the class $C = (\mathcal{N}, \mathcal{O}, \mathcal{I})$. Assigning a default potential to each input variable X implies that any network class is a valid probabilistic network model.

4.3.1 Chain Rule

It should be clear from the above discussion that each OOPN encodes either a probability distribution or an expected utility function. For simplicity, we will discuss only the chain rule for object-oriented (discrete) Bayesian networks. The chain rule of an object-oriented Bayesian network reflects the hierarchical structure of the model.

An instance $\mathcal{M}$ of network class C encapsulates a conditional probability distribution over its random variables given its unbound input nodes. For further simplicity, let $C = (\mathcal{N}, \mathcal{I}, \mathcal{O})$ be a network class over basic discrete random variables only (i.e., no instances, no decisions, and no utilities) with $\mathcal{N} = (\mathcal{X}, \mathcal{G}, \mathcal{P})$ where $X \in \mathcal{X}$ is the only input variable, that is, $X \in \mathcal{I}$ and $|\mathcal{I}| = 1$. Since X has a default prior distribution, $\mathcal{N}$ is a valid model representing the joint probability distribution

$$P(\mathcal{X}) = P(X) \prod_{Y_v \neq X} P(Y_v \mid X_{\mathrm{pa}(v)}).$$

In general, an instance $\mathcal{M}$ is a representation of the conditional probability distribution $P(\mathcal{O} \mid \mathcal{I}')$ where $\mathcal{I}' \subseteq \mathcal{I}$ is the subset of bound input variables of $\mathcal{M}$

$$P(\mathcal{O} \mid \mathcal{I}') = \prod_{X \in \mathcal{I} \setminus \mathcal{I}'} P(X) \prod_{Y_v \notin \mathcal{I}} P(Y_v \mid X_{\mathrm{pa}(v)}).$$

4.3.2 Unfolded OOPNs

An object-oriented network $\mathcal{N}$ has an equivalent *flat* or *unfolded* network model representation $\mathcal{M}$. The unfolded network model of an object-oriented network $\mathcal{N}$ is obtained by recursively unfolding the instance nodes of $\mathcal{N}$. The unfolded network representation of a network class is important as it is the structure used for inference.

The joint distribution of an object-oriented Bayesian network model is equivalent to the joint distribution of its unfolded network model

$$P(\mathcal{X}) = \prod_{X_v \in \mathcal{X}_\mathcal{M}} P(X_v \mid X_{\mathrm{pa}(v)}),$$

where $\mathcal{M} = (\mathcal{X}, \mathcal{G}, \mathcal{P})$ is the unfolded network.

4.3.3 Instance Trees

An object-oriented model is a hierarchical model representation. The *instance tree* T of an object-oriented model $\mathcal{N}$ is a tree over the set of instances of classes in $\mathcal{N}$. Two

Fig. 4.19 An instance tree

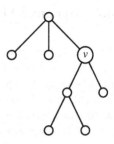

nodes v_i and v_j in T (with v_i closer to the root of T than v_j) are connected by an undirected link if and only if the instance represented by v_i contains the instance represented by v_j. The root of an instance tree is the top-level network class not instantiated in any other network class within the model. Notice that an instance tree is unique.

In addition to the notion of default potentials, there is the notion of the *default instance*. Let C be a network class with instance tree T. Each non-root node v of T represents an instance of a class C_v, whereas the root node r of T represents an instance of the unique class C_r, which has not been instantiated in any class. This instance is referred to as the default instance of C_r.

Example 4.16 (Instance Tree). Figure 4.19 shows the instance tree of a network class $\mathcal{N}$ where the root is the default instance of $\mathcal{N}$.

Each node v of T represents an instance $\mathcal{M}$, and the children of v in T represent instances in $\mathcal{M}$.   □

4.3.4   Inheritance

Another important concept of the OOPN framework is inheritance. For simplicity, we define inheritance as the ability of an instance to take its interface definition from another instance. Let C_1 be a network class with input variables $I(C_1)$ and output variables $O(C_1)$, that is, $C_1 = (\mathcal{N}_1, \mathcal{I}_1, \mathcal{O}_1)$. A network class $C_2 = (\mathcal{N}_2, \mathcal{I}_2, \mathcal{O}_2)$ may be specified as a subclass of C_1 if and only if $\mathcal{I}_1 \subseteq \mathcal{I}_2$ and $\mathcal{O}_1 \subseteq \mathcal{O}_2$. Hence, subclasses may enlarge the interface.

Inheritance is not to the knowledge of the authors implemented in any widely available software supporting OOPN.

4.4 Dynamic Models

The graph of a probabilistic network is restricted to be a finite acyclic directed graph, see Sect. 2.1. This seems to imply that probabilistic networks as such do not support models with feedback loops or models of dynamic systems changing over time. This is not the case. A common approach to representing and solving dynamic models or models with feedback loops is to unroll the dynamic model for the desired number of time steps and treat the resulting network as a static network. Similarly, a feedback loop can be unrolled and represented using a desired number of time steps. The unrolled static network is then solved using a standard algorithm applying evidence at the appropriate time steps.

As an example of a dynamic model, consider the problem of monitoring the state of a dynamic process over a specific period of time. Assume the network of Fig. 4.20 is an appropriate model of the causal relations between variables representing the system at any point in time. The structure of this network is static in the sense that it represents the state of the system at a certain point in time. In the process of monitoring the state of the system over a specific period of time, we will make observations on a subset of the variables in the network and make inference about the remaining unobserved variables. In addition to reasoning about the current state of the system, we may want to reason about the state of the system at previous and future points in time. For this usage, the network in Fig. 4.20 is inadequate. Furthermore, the state of the system at the current point in time will impact the state of the system in the future and be impacted by the state of the system in the past.

What is needed is a time-sliced model covering the period of time over which the system should be monitored. Figure 4.21 indicates a time-sliced model constructed based on the static network shown in Fig. 4.20. Each time-slice consists of the structure shown in Fig. 4.20, while the development of the system is specified by links between variables of different time-slices.

The *temporal links* of a time-slice t_i are the set of links from variables of time-slice t_{i-1} into variables of time-slice t_i. The temporal links of time-slice t_i define the conditional distribution of the variables of time-slice t_i given the variables of time-

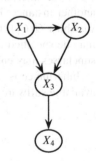

Fig. 4.20 The structure of a static network model

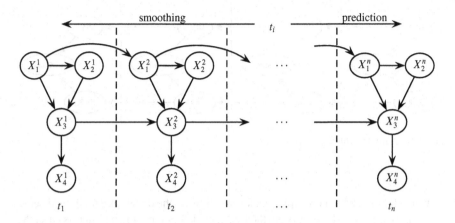

Fig. 4.21 The structure of a dynamic model with n time-slices

slice t_{i-1}. The temporal links connect variables of adjacent time-slices. For instance, the temporal links of time-slice t_2 in Fig. 4.21 is the set $\{(X_1^1, X_1^2), (X_3^1, X_3^2)\}$.

The *interface* of a time-slice is the set of variables with parents in the previous time-slice. For instance, the interface of time-slice t_2 in Fig. 4.21 is the set $\{X_1^2, X_3^2\}$.

Three additional concepts are often used in relation to dynamic models. Let i be the current time step, then *smoothing* is the process of querying about the state of the system at a previous time step $j < i$ given evidence about the system at time i, *filtering* is the process of querying about the state of the system at the current time step, and *prediction* is the process of querying about the state of the system at a future time step $j > i$.

A dynamic Bayesian network is *stationary* when the transition probability distributions are invariant between time steps. A dynamic Bayesian network is first-order Markovian when the variables at time step $i + 1$ are d-separated from the variables at time step $i - 1$ given the variables at time step i. When a system is stationary and Markovian, the state of the system at time $i + 1$ only depends on its state at time i, and the probabilistic dependence relations are the same for all i. The Markovian property implies that arcs between time-slices only go from one time-slice to the subsequent time-slice.

A dynamic Bayesian network is referred to as either a *dynamic Bayesian network* (DBN) or a *time-sliced Bayesian network* (TBN). See Kjærulff (1995) for more details on dynamic Bayesian networks.

Example 4.17 (Apple Jack's Finest Tree). Consider the Apple Jack network in Fig. 4.1 of Example 4.1 on page 71. The network is used for reasoning about the cause of Apple Jack's finest apple tree losing its leaves. The network is static and models the dependence relations between two diseases and a symptom at four specific points in time where Apple Jack is observing his tree.

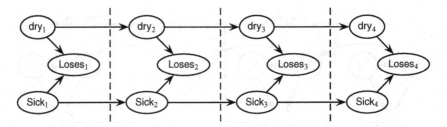

Fig. 4.22 A model with four time-slices

Consider the case where Apple Jack is monitoring the development of the disease over a period of time by observing the tree each day in the morning. In this case, the level of dryness of the tree on a specific day will depend on the level of dryness on the previous day and impact the level of dryness on the next day, similarly for the level of sickness. The levels of dryness and sickness on the next day are independent of the levels of dryness and sickness on the previous day given the levels of dryness and sickness on the current day. This can be captured by a dynamic model.

Figure 4.22 shows a dynamic model with four time-slices. Each time step models the state of the apple tree at a specific point in time (the dashed lines illustrate the separation of the model into time-slices). The conditional probability distributions $P(\text{Dry}_i | \text{Dry}_{i-1})$ and $P(\text{Sick}_i | \text{Sick}_{i-1})$ are the transition probability distributions. The interface between time-slices $i-1$ and i consists of Dry_i and Sick_i.

Assume that it is the second day when Apple Jack is observing his tree. The observations on Loses of the first and second day are entered as evidence on the corresponding variables. Filtering is the task of computing the probability of the tree being sick on the second day, smoothing is the task of computing the probability of sickness on the first day, and prediction is the task of computing the probability of the tree being sick on the third or fourth day.   □

Dynamic models are not restricted to be Bayesian networks. Influence diagrams and LIMIDs can also be represented as dynamic models.

4.4.1 Time-Sliced Networks Represented as OOPNs

Time-sliced networks are often represented using object-oriented networks as the following example illustrates.

Example 4.18 (Breeding Pigs). Example 4.13 shows a LIMID representation of a decision problem related to breeding pigs, see Fig. 4.15 on page 95. The decision problem is in fact modeled as a time-sliced model where the structure of each time-slice representing a test-and-treat cycle is shown in Fig. 4.23.

Fig. 4.23 The test-and-treat cycle of the breeding pigs network in Fig. 4.15

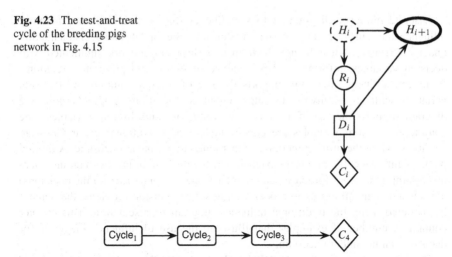

Fig. 4.24 The breeding pigs network as a time-sliced OOPN

Three instances of the network class in Fig. 4.23 are constructed to create the network in Fig. 4.24. The use of object-oriented modeling has simplified the network construction.

The network in Fig. 4.24 is equivalent to the network in Fig. 4.15 on page 95. □

Kjærulff (1995) has described a computational system for dynamic time-sliced Bayesian networks. The system implemented is referred to as dHugin. Boyen & Koller (1998) have described an approximate inference algorithm for solving dynamic Bayesian networks with bounds on the approximation error.

4.5 Summary

In this chapter we have introduced probabilistic networks for belief update and decision making under uncertainty. A probabilistic network represents and processes probabilistic knowledge. The qualitative component of a probabilistic network encodes a set of (conditional) dependence and independence statements among a set of random variables, informational precedence, and preference relations. The quantitative component specifies the strengths of dependence relations using probability theory and preference relations using utility theory.

We have introduced discrete Bayesian network models and CLG Bayesian network models for belief update. A discrete Bayesian network supports the use of discrete random variables, whereas a CLG Bayesian network supports the use of a mixture of continuous and discrete random variables. The continuous variables are constrained to be conditional linear Gaussian variables. This chapter contains a number of examples that illustrate the use of Bayesian networks for belief update.

Discrete influence diagrams, CLQG influence diagrams, and limited memory influence diagrams were introduced as models for belief update and decision making under uncertainty. An influence diagram is a Bayesian network augmented with decision variables, informational precedence relations, and preference relations. A discrete influence diagram supports the use of discrete random and decision variables with an additively decomposing utility function. A CLQG influence diagram supports the use of a mixture of continuous and discrete variables. The continuous random variables are constrained to be conditional linear Gaussian variables, while the utility function is constrained to be linear–quadratic. A limited memory influence diagram is an extension of the discrete influence diagram where the assumptions of no-forgetting and regularity (i.e., a total order on the decisions) are relaxed. This allows us to model a large set of decision problems that cannot be modeled using the traditional influence diagram representation. This chapter contains a number of examples that illustrate the use of influence diagrams for decision making under uncertainty.

Finally, we have introduced OOPNs. The basic OOPN mechanisms introduced support a type of object-oriented specification of probabilistic networks, which makes it simple to reuse models, to encapsulate submodels, and to perform model construction at different levels of abstraction. This chapter contains a number of examples that illustrate the use of the basic OOPN mechanisms in the model development process. OOPNs are well suited for constructing time-sliced networks. Time-sliced networks are used to represent dynamic models.

In Chap. 5, we discuss techniques for solving probabilistic networks.

Exercises

Exercise 4.1. Peter and Eric are chefs at Restaurant Bayes. Peter works 6 days a week, while Eric works one day a week. In 90% of the cases, Peter's food is high quality, while Eric's food is high quality in 50% of the cases. One evening Restaurant Bayes serves an awful meal.

Is it fair to conclude that Eric prepared the food that evening?

Exercise 4.2. One in a thousand people has a prevalence for a particular heart disease. There is a test to detect this disease. The test is 100% accurate for people who have the disease and is 95% accurate for those who do not (this means that 5% of people who do not have the disease will be wrongly diagnosed as having it).

(1) If a randomly selected person tests positive, what is the probability that the person actually has the heart disease?

Exercise 4.3. Assume a math class is offered once every semester, while an AI class is offered twice. The number of students taking a class depends on the subject. On average, 120 students take AI ($\sigma^2 = 500$), while 180 students take math ($\sigma^2 = 1,000$). Assume that on average 25% pass the AI exam ($\sigma^2 = 400$) while 50% pass the math exam ($\sigma^2 = 500$).

(a) What is the average number of students passing either a math or AI exam?
(b) What is the average number of students passing a math exam?
(c) What is the average number of students taking a math class when 80 students pass the exam?

Exercise 4.4. Frank goes to the doctor because he believes that he has got the flu. At this particular time of the year, the doctor estimates that one out of 1, 000 people suffers from the flu. The first thing the doctor checks is whether Frank appears to have the standard symptoms of the flu; if Frank suffers from the flu, then he will exhibit these symptoms with probability 0.9, but if he does not have the flu, he may still have these symptoms with probability 0.05. After checking whether or not Frank has the symptoms, the doctor can decide to have a test performed which may reveal more information about whether or not Frank suffers from the flu; the cost of performing the test is € 40. The test can either give a positive or a negative result, and the frequency of false-positives and false-negatives is 0.05 and 0.1, respectively. After observing the test result (if any), the doctor can decide to administer a drug that with probability 0.6 may shorten the sickness period if Frank suffers from the flu (if he has not got the flu, then the drug has no effect). The cost of administering the drug is € 100, and if the sickness period is shortened, the doctor estimates that this is worth € 1, 000.

(a) Construct an influence diagram for the doctor from the description above.
(b) Specify the probability distributions and the utility functions for the influence diagram.

Exercise 4.5. Assume that Frank is thinking about buying a used car for € 20, 000, and the market price for similar cars with no defects is € 23, 000. The car may, however, have defects which can be repaired at the cost of € 5, 000; the probability that the car has defects is 0.3. Frank has the option of asking a mechanic to perform (exactly) one out of two different tests on the car. $Test_1$ has three possible outcomes, namely, no defects, defects, and inconclusive. For $Test_2$ there are only two possible outcomes (no defects and defects). If Frank chooses to have a test performed on the car, the mechanic will report the result back to Frank who then decides whether or not to buy the car; the cost of $Test_1$ is € 300, and the cost of $Test_2$ is € 1, 000.

(a) Construct an influence diagram for Frank's decision problem.
(b) Calculate the expected utility and the optimal strategy for the influence diagram; calculate the required probabilities from the joint probability table (over the variables $Test_1$, $Test_2$, and StateOfCar) specified below.

		$Test_1$		
		no defects	defects	inconclusive
$Test_2$	no defects	(0.448, 0.00375)	(0.028, 0.05625)	(0.084, 0.015)
	defects	(0.112, 0.01125)	(0.007, 0.16875)	(0.021, 0.045)

Exercise 4.6. An environmental agency visits a site where a chemical production facility has previously been situated. Based on the agency's knowledge about the

facility, they estimate that there is a 0.6 risk that chemicals from the facility have contaminated the soil. If the soil is contaminated (and nothing is done about it), all people in the surrounding area will have to undergo a medical examination due to the possible exposure; there are 1,000 people in the area, and the cost of examining/treating one person is $100. To avoid exposure, the agency can decide to remove the top layer of the soil which, in case the ground is contaminated, will completely remove the risk of exposure; the cost of removing the soil is $30,000. Before making the decision of whether or not to remove the top layer of soil, the agency can perform a test which will give a positive result (with probability 0.9) if the ground is contaminated; if the ground is not contaminated, the test will give a positive result with probability 0.01. The cost of performing the test is $1,000.

(a) Construct an influence diagram for the environmental agency from the description above.
(b) Specify the probability distributions and the utility functions for the influence diagram.

Exercise 4.7. A company has observed that one of their software systems is unstable, and they have identified a component which they suspect is the cause of the instability. The company estimates that the prior probability for the component being faulty is 0.01, and if the component is faulty, then it causes the system to become unstable with probability 0.99; if the component is not faulty, then the system may still be unstable (due to some other unspecified element) with probability 0.001.

To try to solve the problem, the company must first decide whether to *patch* the component at a cost € 10,000 : if the component is faulty, then the patch will solve the fault with probability 0.95 (there may be several things wrong, not all of which may be covered by the patch), but if the component is not faulty, then the patch will have no effect. The company also knows that in the near future the vendor of the component will make another patch available at the cost of € 20,000; the two patches focus on different parts of the component. This new patch will solve the problem with probability 0.99, and (as for the first patch) if the component is not faulty, then the patch will have no effect. Thus, after deciding on the first patch, the company observes whether or not the patch solved the problem (i.e., is the system still unstable?) and it then has to decide on the second patch. The company estimates that (after the final decision has been made) the value of having a fully functioning component is worth € 100,000.

(a) Construct an influence diagram for the company from the description above.
(b) Specify the probability distributions and the utility functions for the influence diagram.

Exercise 4.8. Consider a stud farm with ten horses where Cecily has unknown mare and sire, John has mare Irene and sire Henry, Henry has mare Dorothy and sire Fred, Irene has mare Gwenn and sire Eric, Gwenn has mare Ann and unknown sire, Eric has mare Cecily and sire Brian, Fred has mare Ann and unknown sire, Brian

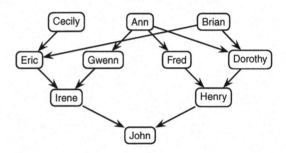

Fig. 4.25 The stud farm pedigree

has unknown mare and sire, Dorothy has mare Ann and sire Brian, and Ann has unknown mare and sire, see Fig. 4.25.

A sick horse has genotype aa, a carrier of the disease has genotype aA, and a noncarrier has genotype AA. $P(aa, aA, AA) = (0.04, 0.32, 0.64)$.

(a) Construct an object-oriented network representation of the stud farm problem.
(b) What is the probability of each horse being sick/a carrier/a noncarrier once we learn that John is sick?

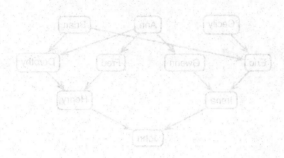

Chapter 5
Solving Probabilistic Networks

We build knowledge bases in order to formulate our knowledge about a certain problem domain in a structured way. The purpose of the knowledge base is to support our reasoning about events and decisions in a domain with inherent uncertainty. The fundamental idea of solving a probabilistic network is to exploit the structure of the knowledge base to reason efficiently about the events and decisions of the domain taking the inherent uncertainty into account.

An expert system consists of a knowledge base and an inference engine. The inference engine is used to solve queries against the knowledge base. In the case of probabilistic networks, we have a clear distinction between the knowledge base and the inference engine. The knowledge base is the Bayesian network or influence diagram, whereas the inference engine consists of a set of generic methods that applies the knowledge formulated in the knowledge base on task-specific data sets, known as evidence, to compute solutions to queries against the knowledge base. The knowledge base alone is of limited use if it cannot be applied to update our belief about the state of the world or to identify (optimal) decisions in the light of new knowledge.

As we saw in the previous chapter, the knowledge bases we consider are probabilistic networks. A probabilistic network may be an efficient representation of a joint probability distribution or a joint expected utility function. In the former case the model is a Bayesian network, while in the latter case it is an influence diagram.

In this chapter we consider the process of solving probabilistic networks. As the exact nature of solving a query against a probabilistic network depends on the type of model, the solution process of Bayesian networks and influence diagrams is considered separately in the following sections.

Section 5.1 considers probabilistic inference in Bayesian networks as the task of computing posterior beliefs in the light of evidence. A number of different approaches to inference are considered. We consider variable elimination, query-based inference, arc reversal, and message passing in junction trees. The inference process in discrete Bayesian networks is treated in detail, while the inference process in CLG Bayesian networks and CLQG influence diagrams is outlined.

U.B. Kjærulff and A.L. Madsen, *Bayesian Networks and Influence Diagrams: A Guide to Construction and Analysis*, ISS 22, DOI 10.1007/978-1-4614-5104-4_5,
© Springer Science+Business Media New York 2013

In Sect. 5.2, we consider the task of solving decision models. Solving a decision model amounts to computing (maximum) expected utilities. We derive a generic method for solving influence diagrams and LIMIDs.

Parts of this chapter have appeared in Madsen, Jensen, Kjærulff & Lang (2005).

5.1 Probabilistic Inference

We build Bayesian network models in order to support efficient belief update in a given domain. Belief update is the task of computing our updated beliefs in (unobserved) events given observations on other events, that is, evidence.

5.1.1 Inference in Discrete Bayesian Networks

One particular type of probabilistic inference task in Bayesian networks is the task of computing the posterior marginal of an unobserved variable Y given a (possibly empty) set of evidence ε, that is, $P(Y \mid \varepsilon)$. Let $\mathcal{N} = (\mathcal{X}, \mathcal{G}, \mathcal{P})$ be a Bayesian network over the set of discrete random variables $\mathcal{X} = \{X_1, \ldots, X_n\}$, and assume that $\varepsilon = \emptyset$. Exploiting the chain rule for Bayesian networks [see, e.g., (4.1) on page 71], for variable $Y \in \mathcal{X}$, we may compute

$$
\begin{aligned}
P(Y) &= \sum_{X \in \mathcal{X} \setminus \{Y\}} P(\mathcal{X}) \\
&= \sum_{X \in \mathcal{X} \setminus \{Y\}} \prod_{X_v \in \mathcal{X}} P(X_v \mid X_{\mathrm{pa}(v)}).
\end{aligned}
\tag{5.1}
$$

This is the prior marginal distribution $P(Y)$ of Y. The prior marginal of all variables may be computed by repetition for each variable.

Example 5.1 (Prior Probability (Apple Jack)). Given the example of Apple Jack (Example 4.1 on page 71), we may consider the task of computing the prior marginal distribution $P(L)$ over the events that the tree does lose its leaves and that the tree does not lose its leaves. The distribution $P(L)$ may be computed as

$$
P(L) = \sum_S \sum_D P(S) P(L \mid S, D) P(D).
$$

Using the quantification of the model specified as part of Example 4.1, we arrive at the prior distribution $P(L) = (0.82, 0.18)$. Hence, a priori, there is an 18% probability that the tree will lose its leaves. □

The above approach does not incorporate evidence into the inference task. In addition, it is a very inefficient approach for nontrivial Bayesian networks because the joint distribution $P(\mathcal{X})$ over $\mathcal{X}$ is constructed as an intermediate step and because a lot of calculations are repeated.

As we will see, it is possible to develop a more efficient approach to probabilistic inference by exploiting the independence relations induced by the structure of the DAG and the evidence and by minimizing the repetition of calculations. Having said that, let us turn to the general case of computing the posterior marginal $P(X \mid \varepsilon)$ of a variable, X, given evidence ε.

Let $\varepsilon = \{\varepsilon_1, \ldots, \varepsilon_m\}$ be a nonempty set of evidence over variables $\mathcal{X}(\varepsilon)$. For a (non-observed) variable $X_{v_j} \in \mathcal{X}$ of $\mathcal{N}$, the task is to compute the posterior probability distribution $P(X_{v_j} \mid \varepsilon)$. This can be done by exploiting the chain rule factorization of the joint probability distribution induced by $\mathcal{N}$:

$$P(X_{v_j} \mid \varepsilon) = \eta(P(\varepsilon \mid X_{v_j}) P(X_{v_j}))$$

$$= \frac{P(\varepsilon \mid X_{v_j}) P(X_{v_j})}{P(\varepsilon)} = \frac{P(X_{v_j}, \varepsilon)}{P(\varepsilon)}$$

$$\propto P(X_{v_j}, \varepsilon)$$

$$= \sum_{Y \in \mathcal{X} \setminus \{X_{v_j}\}} P(\mathcal{X}, \varepsilon)$$

$$= \sum_{Y \in \mathcal{X} \setminus \{X_{v_j}\}} \prod_{X_{v_i} \in \mathcal{X}} P(X_{v_i} \mid X_{\mathrm{pa}(v_i)}) \mathcal{E}_\varepsilon$$

$$= \sum_{Y \in \mathcal{X} \setminus \{X_{v_j}\}} \prod_{X_{v_i} \in \mathcal{X}} P(X_{v_i} \mid X_{\mathrm{pa}(v_i)}) \prod_{X \in \mathcal{X}(\varepsilon)} \mathcal{E}_X$$

for each $X_{v_j} \notin \mathcal{X}(\varepsilon)$, where $\mathcal{E}_X$ is the evidence function for $X \in \mathcal{X}(\varepsilon)$ and v_i is the node representing X_{v_i}. Notice that

$$L(X_{v_j} \mid \varepsilon) = P(\varepsilon \mid X_{v_j}) = \sum_{Y \in \mathcal{X} \setminus \{X_{v_j}\} \, i \neq j} \prod P(X_{v_i} \mid X_{\mathrm{pa}(v_i)}) \prod_{X \in \mathcal{X}(\varepsilon)} \mathcal{E}_X \qquad (5.2)$$

is the likelihood function of X_{v_j} given ε. Since $P(X_{v_j})$ may be obtained by inference over the empty set of evidence, we can—using Bayes' rule—compute

$$P(X_{v_j} \mid \varepsilon) \propto L(X_{v_j} \mid \varepsilon) P(X_{v_j}).$$

The proportionality factor is the normalization constant $\alpha = \eta(P(\mathcal{X}, \varepsilon)) = P(\varepsilon)$, which is easily computed from $P(\mathcal{X}, \varepsilon)$ by summation over $\mathcal{X}$ as $\alpha = \eta(P(\mathcal{X}, \varepsilon)) = \sum_{\mathcal{X}} P(\mathcal{X}, \varepsilon)$; see 3.3 on page 47.

Example 5.2 (Posterior Probability (Apple Jack)). One evening when Apple Jack is taking his usual after-dinner walk in the garden, he observes his finest apple tree to be losing its leaves. Given that he knows that this may be an indication of the tree being sick, he starts wondering whether or not the tree is sick.

Apple Jack is interested in the probability of the tree being sick given the observation on the tree losing its leaves

$$
P(S \mid \varepsilon) = \frac{P(S, \varepsilon)}{P(\varepsilon)}
$$

$$
= \frac{\sum_S \sum_D P(S) P(L \mid S, D) P(D) \mathcal{E}_L}{P(\varepsilon)}
$$

$$
\propto (0.0927, 0.0905),
$$

where $\mathcal{E}_L = (0, 1)$ is the evidence function reflecting the tree losing its leaves. The normalization constant is $\alpha = P(\varepsilon) = P(S = \text{no} \mid \varepsilon) + P(S = \text{yes} \mid \varepsilon) = 0.0927 + 0.0905 = 0.1832$. This produces the posterior distribution $P(S \mid \varepsilon) = (0.506, 0.494)$ over the tree losing its leaves. Hence, there is an increased probability that the tree is sick when it has been observed to lose its leaves. The prior distribution on the tree being sick is $P(S) = (0.9, 0.1)$. □

In general, probabilistic inference is an NP-hard task (Cooper 1990). Even approximate probabilistic inference is NP-hard (Dagum & Luby 1993). For certain classes of Bayesian network models, the complexity of probabilistic inference is polynomial or even linear in the number of variables in the network. The complexity is polynomial when the graph of the Bayesian network is a poly-tree (Kim & Pearl 1983, Pearl 1988) (a directed graph $\mathcal{G}$ is called a *poly-tree*, if its underlying undirected graph is singly connected), while it is linear when the graph of the Bayesian network is a tree.

The most critical problem related to the efficiency of the inference process is that of finding the optimal order in which to perform the computations. The inference task is, in principle, solved by performing a sequence of multiplications and additions.

Query-Based Inference

One approach to inference is to consider the inference task as the task of computing the posterior distribution of a set of variables. This is referred to as query-based inference. We define the notion of a query, Q, against a Bayesian network model $\mathcal{N}$ as follows.

Definition 5.1 (Query). Let $\mathcal{N} = (\mathcal{X}, \mathcal{G}, \mathcal{P})$ be a Bayesian network model. A query Q is a three-tuple $Q = (\mathcal{N}, \mathcal{T}, \varepsilon)$ where $\mathcal{T} \subseteq \mathcal{X}$ is the target set and ε is the evidence set.

The solution of a query, Q, is the posterior distribution over the target, that is, $P(\mathcal{T}|\varepsilon)$. A variable X is a target variable if $X \in \mathcal{T}$. Notice that computing all posterior marginals of a Bayesian network $\mathcal{N} = (\mathcal{X}, \mathcal{G}, \mathcal{P})$ corresponds to solving $|\mathcal{X}|$ queries, that is, $Q = (\mathcal{N}, \{X\}, \varepsilon)$ for each $X \in \mathcal{X}$.

Prior to solving the query Q, the graph $\mathcal{G}$ of $\mathcal{N}$ may be pruned to include only variables relevant for the query. One class of variables which may be pruned from the graph without any computation is the class of barren variables; see Sect. 3.3.4 on page 53 for an example. Here, we give a formal definition of a barren variable.

Definition 5.2 (Barren Variable). Let $\mathcal{N} = (\mathcal{X}, \mathcal{G}, \mathcal{P})$ be a Bayesian network and let $Q = (\mathcal{N}, \mathcal{T} \subseteq \mathcal{X}, \varepsilon)$ be a query against $\mathcal{N}$. A variable X is a *barren variable* with respect to Q, if $X \notin \mathcal{T}$, $X \notin \varepsilon$, and all descendants, if any, de(X), of X are barren.

When a variable X is classified as a barren variable, it is always relative to a target and given a set of evidence. A barren variable does not add any information to the inference process. It is computationally irrelevant to Q.

Once all barren variables with respect to Q have been pruned from the graph $\mathcal{G}$, the inference task can be solved by variable elimination as described in the previous section.

In addition to the concept of a barren variable, there is the concept of a *nuisance variable*.

Definition 5.3 (Nuisance Variable). Let $\mathcal{N} = (\mathcal{X}, \mathcal{G}, \mathcal{P})$ be a Bayesian network and let $Q = (\mathcal{N}, \mathcal{T} \subseteq \mathcal{X}, \varepsilon)$ be a query against $\mathcal{N}$. A non-barren variable X is a *nuisance variable* with respect to Q, if $X \notin \mathcal{T}$, $X \notin \varepsilon$, and X is not on a path between any pair of variables $Y \in \mathcal{T}$ and $Z \in \varepsilon$.

Notice that a nuisance variable is computationally relevant for a query Q, but it is not on a path between any pair of evidence and query variables. Given a query and a set of evidence variables, the contribution from a nuisance variable does not depend on the observed values of the evidence variables. Hence, if a query is to be solved with respect to multiple instantiations over the evidence variables, then the nuisance variables (and barren variables) may be eliminated in a preprocessing step to obtain the *relevant network* (Lin & Druzdzel 1997). The relevant network consists of target variables, evidence variables, and variables on paths between target and evidence variables only.

Example 5.3 (Barren Variables and Nuisance Variables). Let us return to the chest clinic example (Example 4.2 on page 73) and consider the task of computing the probability of each disease given the observations that the patient is a smoker and has a positive X-ray result. That is, we need to compute $P(Y|\varepsilon)$ for $Y \in \{T, L, B\}$ and $\varepsilon = \{S = \text{yes}, X = \text{yes}\}$.

The variables $\{A, T\}$ are nuisance variables with respect to posteriors for B and L. The variable D is a barren variable with respect to the posteriors for B, T, and L, whereas B is a barren variable with respect to the posteriors for T and L. Figure 5.1 shows the relevant networks for (a) computing $P(T|\varepsilon)$ and $P(L|\varepsilon)$ and for (b) computing $P(B|\varepsilon)$. $\qquad\square$

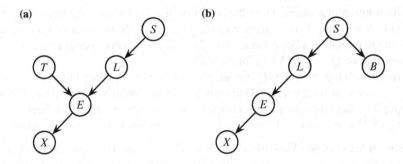

Fig. 5.1 The relevant networks for computing (a) $P(T \mid \varepsilon)$ and $P(L \mid \varepsilon)$ and (b) $P(B \mid \varepsilon)$

The approach to inference outlined above may be referred to as a *direct approach*. Arc reversal is a specific type of direct approach to inference (Olmsted 1983, Shachter 1986).

Arc Reversal

In Sect. 3.4.1.1 on page 56, we illustrated how application of Bayes' rule can be given a graphical interpretation as arc reversal. We mentioned that Olmsted (1983) and Shachter (1986) have exploited this view of inference in their arc reversal algorithms for inference in probabilistic networks. Here, we consider the process in more detail.

Let $\mathcal{G}$ be the DAG of a Bayesian network $\mathcal{N} = (\mathcal{X}, \mathcal{G}, \mathcal{P})$ and assume a query $Q = (\mathcal{N}, \{Z\}, \emptyset)$ against $\mathcal{N}$. The inference task is to compute $P(Z)$ by eliminating all variables $\mathcal{X} \setminus \{Z\}$.

The inference process on $\mathcal{G}$ has a natural graphical interpretation as a sequence of arc reversals and barren variable eliminations. The fundamental idea is to adjust the structure of $\mathcal{G}$ such that all variables except Z are pruned as barren variables while maintaining the underlying properties of the joint probability distributions over the remaining variables. The structure of $\mathcal{G}$ is adjusted through a sequence of arc reversal operations.

Assume X_w is the next variable to be eliminated as a barren variable. Let X_w have parents $X_{\mathrm{pa}(w)} = X_i \cup X_j$ and X_v have parents $X_{\mathrm{pa}(v)} = \{X_w\} \cup X_j \cup X_k$ where $X_i \cap X_j = X_i \cap X_k = X_j \cap X_k = \emptyset$ such that $X_i = X_{\mathrm{pa}(w)} \setminus X_{\mathrm{pa}(v)}$ are the parents specific for X_w, $X_j = X_{\mathrm{pa}(w)} \cap X_{\mathrm{pa}(v)}$ are the common parents, and $X_k = X_{\mathrm{pa}(v)} \setminus X_{\mathrm{fa}(w)}$ are the parents specific for X_v.

The reversal of arc (w, v) proceeds by setting $X_{\mathrm{pa}(w)} = X_i \cup X_j \cup X_k \cup \{X_v\}$ and $X_{\mathrm{pa}(v)} = X_i \cup X_j \cup X_k$ as well as performing the computations specified below; see Fig. 5.2 for a graphical representation

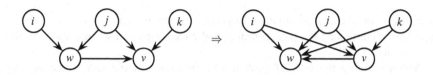

Fig. 5.2 An illustration of reversal of the arc (w, v)

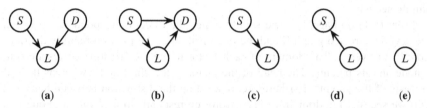

    (a)   (b)   (c)   (d)   (e)

Fig. 5.3 Computing $P(L)$ by arc reversal

$$P(X_v \mid X_i, X_j, X_k) = \sum_{X_w} P(X_w \mid X_i, X_j) P(X_v \mid X_w, X_j, X_k) \qquad (5.3)$$

$$P(X_w \mid X_v, X_i, X_j, X_k) = \frac{P(X_w \mid X_i, X_j) P(X_v \mid X_w, X_j, X_k)}{P(X_v \mid X_i, X_j, X_k)}. \qquad (5.4)$$

The operation of reversing an arc changes the structure of $\mathcal{G}$ without changing the underlying joint probability distribution over $\mathcal{X}$ induced by $\mathcal{N}$.

Once the arc (w, v) has been reversed, the variable X_w is a barren variable relative to the other variables (given the empty set of evidence) and can be pruned from $\mathcal{G}$ without further computations.

The basic idea of the inference process known as arc reversal is to perform a sequence of arc reversals and barren variable eliminations on the DAG $\mathcal{G}$ until a desired marginal or conditional is obtained. In this process a valid Bayesian network structure is maintained throughout the inference process.

Example 5.4 (Arc Reversal). We may compute the prior probability distribution $P(L)$ in the Apple Jack example (see Example 4.1 on page 71) using a sequence of arc reversals and barren variable eliminations as indicated in Fig. 5.3.

Notice that the arc reversal method does not have worse complexity than variable elimination.   ☐

Arc reversal is not a local computation algorithm in the sense that when reversing an arc (w, v), it is necessary to test for existence of a directed path from w to v not containing (w, v). If such a path exists, then the arc (w, v) cannot be reversed until one or more other arcs have been reversed as reversing (w, v) would otherwise create a directed path.

Graphical Representation of Inference

We may define the task of solving a Bayesian network model $\mathcal{N} = (\mathcal{X}, \mathcal{G}, \mathcal{P})$ as the problem of computing the posterior marginal $P(X \mid \varepsilon)$ given a set of evidence ε for all variables $X \in \mathcal{X}$.

When defining the task of probabilistic inference as the task of computing the posterior marginals $P(X \mid \varepsilon)$ for all X given evidence ε, the most common approach is to use a secondary computational structure. Performing inference in a secondary computational structure aims at reusing calculations solving all queries simultaneously.

From (5.1) on page 112, we should notice the direct correspondence between the acyclic, directed graph $\mathcal{G}$ and the factorization of the joint probability distribution $P(\mathcal{X})$ over $\mathcal{X}$. The domain of each factor in the factorization corresponds to a node and its parents. The head of the factor is the child node, whereas the tail consists of the parents. Furthermore, if we drop the distinction between head and tail, we see that the domain of each factor corresponds to a *clique* (a clique is a maximal complete subgraph) of $\mathcal{G}^m$—the moralization of $\mathcal{G}$. This is exploited to build a secondary structure for performing inference.

Assume we are in the process of computing $P(X_i)$. Let Y be the first random variable to eliminate. The elimination process proceeds by local computation in order to maintain efficiency (i.e., we exploit the distributive law to maintain the factorization of the joint probability distribution—see Sect. 3.3.3 on page 51). The set of probability potentials $\mathcal{P}$ can be divided into two disjoint subsets with respect to Y. Let $\mathcal{P}_Y \subseteq \mathcal{P}$ be the subset of probability potentials including Y in the domain

$$\mathcal{P}_Y = \{P \in \mathcal{P} \mid Y \in \mathrm{dom}(P)\},$$

where $\mathrm{dom}(P)$ denotes the domain of P (i.e., the set of variables over which it is defined). Then $\mathcal{P} \setminus \mathcal{P}_Y$ is the set of probability potentials not including Y in their domain. Let ϕ_Y be the probability potential obtained by eliminating Y (by summation) from the combination of all probability potentials in $\mathcal{P}_Y$. Using ϕ_Y as well as a generalized version of the distributive law, we may rewrite 5.1 on page 112 as

$$
\begin{aligned}
P(X_i) &= \sum_{X \in \mathcal{X} \setminus \{X_i\}} \prod_{X_v \in \mathcal{X}} P(X_v \mid X_{\mathrm{pa}(v)}) \\
&= \sum_{X \in \mathcal{X} \setminus \{X_i\}} \prod_{\phi \in \mathcal{P} \setminus \mathcal{P}_Y} \phi \prod_{\phi' \in \mathcal{P}_Y} \phi' \\
&= \sum_{X \in \mathcal{X} \setminus \{X_i, Y\}} \prod_{\phi \in \mathcal{P} \setminus \mathcal{P}_Y} \phi \sum_Y \prod_{\phi' \in \mathcal{P}_Y} \phi' \\
&= \sum_{X \in \mathcal{X} \setminus \{X_i, Y\}} \phi_Y \prod_{\phi \in \mathcal{P} \setminus \mathcal{P}_Y} \phi.
\end{aligned}
\tag{5.5}
$$

Fig. 5.4 A graphical illustration of the process of eliminating Y from $\phi(X_1, X_2, Y)$ and X_1 from $\phi(X_1, X_2, X_3, X_4)$, where the ovals represent the domain of a potential before elimination and rectangles represent the domain of a potential after elimination

Fig. 5.5 The Burglary or Earthquake network

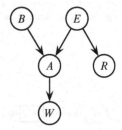

Equation 5.5 on the preceding page specifies a decomposition of the joint probability distribution over $\mathcal{X} \setminus \{Y\}$. The decomposition has the form of (5.1). The decomposition is the product over the elements of $\mathcal{P} \setminus \mathcal{P}_Y \cup \{\phi_Y\}$. In addition, we have performed the elimination over Y by local computations only involving potentials of which Y is a domain variable. We say that the set

$$\mathcal{P} \setminus \mathcal{P}_Y \cup \{\phi_Y\}$$

is a reduction of $\mathcal{P}$ where Y has been eliminated. The elimination of the next variable may proceed in the same manner on $\mathcal{P} \setminus \mathcal{P}_Y \cup \{\phi_Y\}$. The order in which variables are eliminated is the *elimination order*. An example of this process may be depicted graphically as shown in Fig. 5.4 where we assume $\mathrm{dom}(\phi_Y) = (X_1, X_2)$. The arrows in the figure are used to illustrate the flow of computations.

The elimination of Y from $\phi(X_1, X_2, Y)$ creates a potential over $\phi(X_1, X_2)$ which is included in the elimination of the next variable X_1 to be eliminated. In this way the process continues until the desired marginals are obtained. Let us consider an even more concrete example.

Example 5.5 (Burglary or Earthquake, page 25). Consider the Bayesian network in Fig. 2.3 on page 26, which is repeated in Fig. 5.5.

The prior marginal on A may be computed by elimination of $\{B, E, R, W\}$ as follows:

$$P(A) = \sum_E P(E) \sum_B P(B) P(A \mid B, E) \sum_R P(R \mid E) \sum_W P(W \mid A). \qquad (5.6)$$

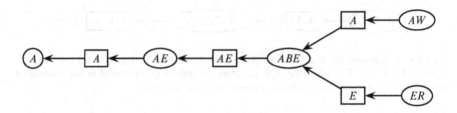

Fig. 5.6 A graphical illustration of the process of computing $P(A)$ in (5.6), where the ovals represent the domain of a potential before elimination and rectangles represent the domain of a potential after elimination

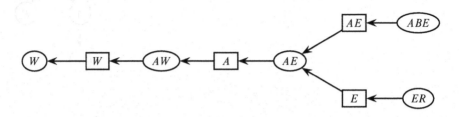

Fig. 5.7 A graphical illustration of the process of computing $P(W)$ in (5.7), where the ovals represent the domain of a potential before elimination and rectangles represent the domain of a potential after elimination

Figure 5.6 shows a graphical representation of the computations and potentials created the during process of computing $P(A)$.

Similarly, the prior marginal distribution over W may be computed by elimination of $\{A, B, E, R\}$ as follows:

$$P(W) = \sum_A P(W \mid A) \sum_E P(E) \sum_B P(B) P(A \mid B, E) \sum_R P(R \mid E). \quad (5.7)$$

Figure 5.7 shows a graphical representation of the computations and potentials created during the process of computing of $P(W)$.

Notice the similarity between the potentials created in the process of computing $P(A)$ and $P(W)$. There is a significant overlap between the potentials created and therefore the calculations performed. This is no coincidence. □

Junction Trees

The task of probabilistic inference may be solved efficiently by local procedures operating on a secondary computational structure known as the *junction tree* (also known as a join tree and a Markov tree) representation of a Bayesian network (Jensen & Jensen 1994, Jensen et al. 1994).

Fig. 5.8 A junction tree representation $\mathcal{T}$ for the chest clinic network

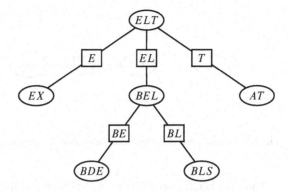

The junction tree representation is efficient when solving the inference task for multiple sets of different evidence and target variables. A junction tree representation $\mathcal{T}$ of a Bayesian network $\mathcal{N} = (\mathcal{X}, \mathcal{G}, \mathcal{P})$ is a pair $\mathcal{T} = (\mathcal{C}, \mathcal{S})$ where $\mathcal{C}$ is the set of cliques and $\mathcal{S}$ is the set of separators. The cliques $\mathcal{C}$ are the nodes of $\mathcal{T}$, whereas the separators $\mathcal{S}$ annotate the links of the tree. Each clique $C \in \mathcal{C}$ represents a maximal complete subset of pairwise connected variables of $\mathcal{X}$, that is, $C \subseteq \mathcal{X}$, of an undirected graph.[1] The link between two neighboring cliques C_i and C_j is annotated with the intersection $S = C_i \cap C_j$, where $S \in \mathcal{S}$.

Example 5.6 (Chest Clinic). Figure 4.2 on page 74 shows the DAG $\mathcal{G}$ of the chest clinic network $\mathcal{N} = (\mathcal{X}, \mathcal{G}, \mathcal{P})$; see Example 4.2 on page 73.

Figure 5.8 shows a junction tree representation $\mathcal{T} = (\mathcal{C}, \mathcal{S})$ of the chest clinic network. The junction tree consists of cliques

$$\mathcal{C} = \{\{A, T\}, \{B, D, E\}, \{B, E, L\}, \{B, L, S\}, \{E, L, T\}, \{E, X\}\}$$

and separators

$$\mathcal{S} = \{\{B, E\}, \{B, L\}, \{E\}, \{E, L\}, \{T\}\}.$$

The structure of $\mathcal{T}$ is determined from the structure of $\mathcal{G}$. □

The process of creating a junction tree representation of a DAG is beyond the scope of this book. Instead, we refer the interested reader to the literature; see, for example, Cowell, Dawid, Lauritzen & Spiegelhalter (1999).

The junction tree serves as an excellent control structure for organizing the computations performed during probabilistic inference. Messages are passed between cliques of the junction tree in two sweeps such that a single message is passed between each pair of neighboring cliques in each sweep. This process is referred to as a *propagation of information*.

---

[1] The undirected graph is constructed from the moral graph $\mathcal{G}^m$ of $\mathcal{G}$ by adding undirected edges until the graph is triangulated. A graph is triangulated if every cycle of length greater than three has a chord.

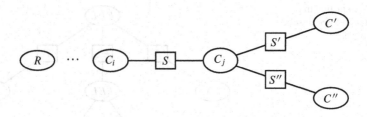

Fig. 5.9 When C_j has absorbed information from its other neighbors, C_i can absorb from C_j

Once the junction tree $\mathcal{T} = (\mathcal{C}, \mathcal{S})$ has been constructed, a probability potential is associated with each clique $C \in \mathcal{C}$ and each separator $S \in \mathcal{S}$ between two adjacent cliques C_i and C_j where $S = C_i \cap C_j$; see Fig. 5.9.

Inference involves the following steps:

1. Each item of evidence must be incorporated into the junction tree potentials. For each item of evidence, an evidence function is multiplied onto an appropriate clique potential.
2. Some clique $R \in \mathcal{C}$ of $\mathcal{T}$ is selected. This clique is referred to as the *root* of the propagation.
3. Then messages are passed toward the selected root. The messages are passed through the separators of the junction tree (i.e., along the links of the tree). These messages cause the potentials of the receiving cliques and separators to be updated. This phase is known as COLLECTINFORMATION.
4. Now messages are passed in the opposite direction (i.e., from the root toward the leaves of the junction tree). This phase is known as DISTRIBUTEINFORMATION.
5. At this point, the junction tree is said to be in equilibrium: the probability $P(X \mid \varepsilon)$ can be computed from any clique or separator containing X—the result will be independent of the chosen clique or separator.

Prior to the initial round of message passing, for each variable $X_v \in \mathcal{X}$, we assign the conditional probability distribution $P(X_v \mid X_{\mathrm{pa}(v)})$ to a clique C such that $X_{\mathrm{fa}(v)} \subseteq C$. Once all conditional probability distributions have been assigned to cliques, the distributions assigned to each clique are combined to form the initial clique potential.

Example 5.7 (Association of CPTs to Cliques). Consider again the junction tree of the chest clinic network shown in Fig. 5.8. Each conditional probability distribution $P \in \mathcal{P}$ is associated with a clique of $\mathcal{T}$ such that $\mathrm{dom}(P) \subseteq C$ for $C \in \mathcal{C}$. Notice that the association of distributions with cliques is unique in this example.   □

The basic inference algorithm is as follows. Each separator holds a single potential over the separator variables, which initially is a unity potential. During propagation of information, the separator and clique potentials are updated. Consider two adjacent cliques C_i and C_j as shown in Fig. 5.9. When a message is

Fig. 5.10 A junction tree
representation $\mathcal{T}$ of the
Bayesian network depicted in
Fig. 5.5 on page 119

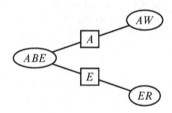

passed from C_j to C_i either during COLLECTINFORMATION or DISTRIBUTEIN-
FORMATION, C_i absorbs information from C_j. Absorption of information involves
performing the following calculations:

1. Calculate the updated separator potential:

$$\phi_S^* = \sum_{C_j \setminus S} \phi_{C_j}.$$

2. Update the clique potential of C_i:

$$\phi_{C_i} := \phi_{C_i} \frac{\phi_S^*}{\phi_S}.$$

3. Associate the updated potential with the separator:

$$\phi_S = \phi_S^*.$$

After a full round of message passing, the potential associated with any clique
(separator) is the joint probability distribution (up to the same normalization
constant) of the variables in the clique (separator) and the evidence. This algorithm
is known as the *HUGIN algorithm*. Details on the inference process can be found
in the literature (Lauritzen & Spiegelhalter 1988, Andersen, Olesen, Jensen &
Jensen 1989, Jensen et al. 1990, Dawid 1992, Jensen et al. 1994, Lauritzen &
Jensen 2001).

Example 5.8 (Cluster Trees Vs. Junction Trees). Figure 5.10 shows a junction tree
representation $\mathcal{T} = (\mathcal{C}, \mathcal{S})$ of the Bayesian network depicted in Fig. 5.5 on page 119
with cliques:

$$\mathcal{C} = \{\{A, B, E\}, \{E, R\}, \{A, W\}\}$$

and separators:

$$\mathcal{S} = \{\{E\}, \{A\}\}.$$

Notice the similarity between Fig. 5.10 and Figs. 5.6 and 5.7. The nodes
of Figs. 5.6 and 5.7 are clusters (i.e., subsets of variables), whereas the nodes of
Fig. 5.10 are cliques (i.e., maximal subsets of pairwise connected variables) of
undirected graphs.

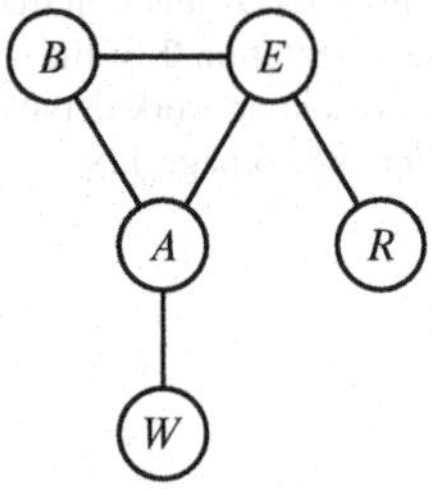

Fig. 5.11 The undirected
graph corresponding to
Figs. 5.6, 5.7, and 5.10

Fig. 5.12 Message passing
in $\mathcal{T}$

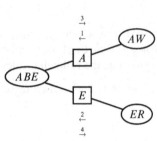

The undirected graph corresponding to a junction tree is obtained by adding undirected edges between each pair of variables contained in the same clique or cluster. Figure 5.11 is the undirected graph corresponding to Figs. 5.6, 5.7, and 5.10.

Figure 5.12 shows how messages are passed over $\mathcal{T}$ relative to the root ABE.

Underlying any approach to inference is the junction tree representation, although its presence may be implicit. Figure 5.6 shows the cluster tree representation underlying the computation of $P(A)$, whereas Fig. 5.7 shows the cluster tree representation underlying the computation of $P(W)$. Figures 5.6 and 5.7 are not junction trees, but cluster trees. The cliques of a junction tree are maximal complete subsets of pairwise connected variables, whereas clusters are not necessarily maximal.   □

The quality of the junction tree $\mathcal{T} = (\mathcal{C}, \mathcal{S})$ determines the efficiency of inference. A common score or criterion to use when considering the optimality of a junction tree is the maximum state space size over all cliques in $\mathcal{T}$, that is, $\max_{C \in \mathcal{C}} \|C\|$. Another similar score is the sum over all cliques in $\mathcal{T}$, that is, $\sum_{C \in \mathcal{C}} \|C\|$.

If the maximum state space size of a clique C is very large, for example, $\|C\| = 1,000,000,000$, then exact inference may become intractable on standard desktop computers. A very large state space size of a clique may be due to a poor triangulation of the underlying DAG of the Bayesian network, for instance. Thus, the first option should be to search for a better triangulation reducing (significantly) the state space size of the largest clique. If the triangulation, on the other hand, is known to be optimal (or near optimal), then options to change the structure of the graph should be considered, for instance, using the modeling techniques described in Chap. 7. An alternative to changing the structure is to use approximate algorithms for performing inference in the model.

All types of probabilistic networks considered in this book may be solved by message passing in a junction tree representation. However, we will restrict ourselves from a detailed treatment of this topic for all models presented as it is beyond the scope of this book.

The approach to inference outlined above may be referred to as an *indirect approach*.

5.1.2 Inference in CLG Bayesian Networks

Let $\mathcal{N} = (\mathcal{X}, \mathcal{G}, \mathcal{P}, \mathcal{F})$ be a CLG Bayesian network with continuous random variables, $\mathcal{X}_\Gamma$, and discrete random variables, $\mathcal{X}_\Delta$, such that $\mathcal{X} = \mathcal{X}_\Gamma \cup \mathcal{X}_\Delta$. To solve the probabilistic inference task on $\mathcal{N}$ is to compute the marginal for each $X \in \mathcal{X}$. Since $\mathcal{N}$ is a CLG Bayesian network, the task of performing inference becomes more subtle than in the case of a pure discrete Bayesian network.

The prior distribution, $P(X)$, of a discrete variable $X \in \mathcal{X}_\Delta$ is equal to the distribution of X in the discrete network $\mathcal{N}' = (\mathcal{X}_\Delta, \mathcal{P})$ obtained by removing all continuous variables from the model (all continuous variables are barren variables with respect to the joint over the discrete variables). The prior density of a continuous variable Y, on the other hand, will, in general, be a mixture of Gaussian distributions where the mixing factors are joint probabilities over configurations of discrete variables $I \subseteq \mathcal{X}_\Delta$. For each configuration i of I with nonzero probability, that is, $p(i) > 0$, the joint distribution of I and X has the form

$$P(I = i) * \mathbb{N}(\mu(i), \sigma^2(i)).$$

This implies that the marginal of $X \in \mathcal{X}_\Gamma$ is

$$\mathcal{L}(X) = \sum_{i : P(I=i)>0} P(i) * \mathbb{N}(\mu(i), \sigma^2(i)).$$

For each configuration i of I with $P(i) = 0$, the mean $\mu(i)$ and variance $\sigma^2(i)$ may be random numbers. Hence, the marginal density function for a continuous variable $X \in \mathcal{X}_\Gamma$ is, in general, a mixture of Gaussian distributions

$$f(x) = \sum_{i=0}^{n} \alpha_i f_i(x),$$

where each component f_i is a one-dimensional Gaussian density function in X and each coefficient α_i is the probability of a configuration of discrete variables. This implies that the marginal density function of $X \in \mathcal{X}_\Gamma$ is not necessarily a CLG distribution with the indicated mean μ and variance σ^2. That is, the result of a marginalization of a CLG distribution over continuous variables is a CLG distribution, whereas the result of a marginalization of a CLG distribution over discrete variables, in general, is not. The marginalization is performed using the operations defined by Lauritzen & Jensen (2001). The result of the first type of marginal is referred to as a *strong marginal*, whereas the latter is referred to as a *weak marginal*. The marginal is strong as we compute the mean μ and the variance σ^2, and we know the distribution is a CLG distribution; whereas in the case of a weak marginal, only μ and σ are computed; the exact density is usually not computed as part of belief update. It is computed separately.

Probabilistic inference is the task of updating our belief about the state of the world in light of evidence. Evidence on discrete variables, be it hard or soft evidence, is treated as in the case of discrete Bayesian networks. Evidence on a continuous variable, on the other hand, is restricted to be hard evidence, that is, instantiations.

In the general case where evidence ε is available, the marginal for a discrete variable $X \in \mathcal{X}_\Delta$ is a probability distribution $P(X \mid \varepsilon)$ conditional on the evidence ε, whereas the marginal for a continuous variable $X \in \mathcal{X}_\Gamma$ is a density function $f(x \mid \varepsilon)$ conditional on ε with a mean μ and a variance σ^2.

Example 5.9 (Density Function). Example 4.3 on page 77 shows an example of a simple CLG Bayesian network. Computing the prior probability density in X_3 amounts to eliminating the variables X_1 and X_2. With the quantification specified in Example 4.3, this produces the following mixture:

$$\mathcal{L}(X_3) = 0.75 * \mathbb{N}(-5, 5.1) + 0.25 * \mathbb{N}(5, 5.2)$$

with mean $\mu = -2.5$ and variance $\sigma^2 = 23.88$. Notice that the density for X_3 is not the density for the Gaussian distribution with mean $\mu = -2.5$ and variance $\sigma^2 = 23.88$. The density function is shown in Fig. 5.13 on the next page.

The prior probability density for X_2 and the prior probability distribution for X_1 are trivial to compute as $\{X_2, X_3\}$ are barren with respect to the prior for X_1 and similarly $\{X_1, X_3\}$ are barren with respect to the prior for X_2. $\square$

The above examples illustrates that the class of CLG distributions is not closed under the operation of discrete variable elimination. The marginal distribution $\mathbb{N}(\mu, \sigma^2)$ may, however, be used as an approximation of the *true* marginal. This marginal is the closest non-mixture to the true marginal in terms of the Kullback–Leibler distance (Lauritzen 1996).

Example 5.10 (Density Function vs. Weak Marginal). Again, let us consider the CLG Bayesian network $\mathcal{N}$ from Example 4.3 on page 77. Figure 5.13 on the next page shows the density function for X_3. Figure 5.14 shows both the density function $f(X_3)$ and the weak marginal $g(X_3)$ for X_3. It is obvious that the weak marginal is only an approximation of the exact density function. $\square$

Since the CLG distribution is not closed under the operation of discrete variable elimination and since the operation of discrete variable elimination is not defined when continuous variables are in the domain of the potential to be marginalized, it is required that continuous variables are eliminated before discrete variables. For this reason, when marginalizing over both continuous and discrete variables,

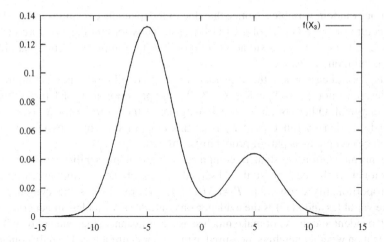

Fig. 5.13 The density function for X_3

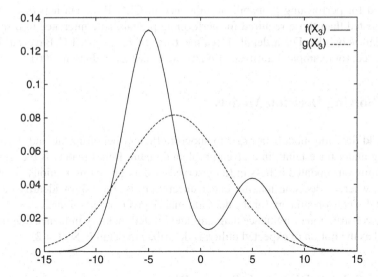

Fig. 5.14 The density function $f(X_3)$ for X_3 and its weak marginal $g(X_3)$

we first marginalize over the continuous variables and then over the discrete variables (Lauritzen 1992b).

This implies that the (exact) solution method for inference in CLG Bayesian networks induces the partial order $\mathfrak{X}_\Delta \prec \mathfrak{X}_\Gamma$ on the elimination order. Hence, the continuous variables $\mathfrak{X}_\Gamma$ should be eliminated before the discrete variables $\mathfrak{X}_\Delta$. A variable elimination order, which is restricted to induce a certain (partial) order, is referred to as a *strong* elimination order. Hence, we use a strong elimination order to solve a CLG Bayesian network by variable elimination. For this reason, inference in a CLG Bayesian network may be more resource intensive than inference in a corresponding Bayesian network with the same structure, but consisting only of continuous random variables. Notice that due to independence relations induced by the structure of $\mathcal{G} = (V, E)$ of a CLG Bayesian network and the structure of the evidence ε, it may in some situations be possible to eliminate discrete variables before continuous variables.

In the special case where the ancestors of $v \in V$ are all representing continuous variables (i.e., $\mathrm{an}(v) \subseteq V_\Gamma$) for $X_v \in \mathcal{X}$, the posterior marginal for X_v is a strong marginal. Otherwise, it is a weak marginal. If the posterior for X_v is a weak marginal, the density function of X_v is an unknown mixture of Gaussians, which needs to be computed as part of probabilistic inference.

The normalization constant α computed as part of probabilistic inference is proportional to the density at the observed values of the continuous variables. The proportionality constant is $P(\varepsilon(\Delta) | \varepsilon(\Gamma))$, where $\varepsilon(\Delta)$ is the evidence on discrete variables and $\varepsilon(\Gamma)$ is the evidence on continuous variables. In general, α is scale dependent and does not make much sense. For instance, the value of α will be dependent on whether height is measured in meters or centimeters. If ε only contains discrete variables, then α is the probability of ε.

The presence of both continuous and discrete variables makes the operations required for performing probabilistic inference in CLG Bayesian networks more complicated than those required for performing probabilistic inference in discrete Bayesian networks. For a detailed treatment on inference in CLG Bayesian networks, see, for example, Lauritzen (1992b) and Lauritzen & Jensen (2001).

5.2 Solving Decision Models

We build decision models in order to support efficient belief update and decision making under uncertainty in a given problem domain. Belief update is the task of computing our updated beliefs in (unobserved) events given observations on other events, whereas decision making under uncertainty is the task of identifying the (optimal) decision strategy for the decision-maker-given observations.

The construction of influence diagrams and limited memory influence diagrams, as well as the notion of expected utility, is described in detail in Sect. 4.2.

5.2.1 Solving Discrete Influence Diagrams

Inference in an (perfect recall) influence diagram $\mathcal{N} = (\mathcal{X}, \mathcal{G}, \mathcal{P}, \mathcal{U})$ is to determine an optimal strategy $\hat{\Delta} = \{\hat{\delta}_1, \dots, \hat{\delta}_n\}$ for the decision maker and compute the expected utility of adhering to $\hat{\Delta}$.

The influence diagram is a compact representation of a joint expected utility function due to the chain rule

$$\mathrm{EU}(\mathcal{X}) = \prod_{X_v \in \mathcal{X}_C} P(X_v | X_{\mathrm{pa}(v)}) \sum_{w \in V_U} u(X_{\mathrm{pa}(w)}).$$

Applying the $\sum$-max-$\sum$-rule (Jensen 1996) on the joint expected utility function, we may solve $\mathcal{N}$ by eliminating variables in the reverse order of the information precedence order $\prec$. That is, the precedence relation $\prec$ induces a partial order on the elimination of variables in $\mathcal{X}$. This implies that we use a strong variable elimination order to solve an influence diagram by variable elimination.

Starting with the last decision D_n, the $\sum$-max-$\sum$-rule says that we should average over the unknown random variables $\mathcal{I}_n$, maximize over the decision D_n, average over the random variables $\mathcal{I}_{n-1}$ known to the decision maker at D_n (but not known to the analyst), maximize over D_{n-1}, and so on. The principle is to average over the unknown random variables, maximize over the decision variable, and finally average over the observed random variables.

The intuition behind the application of the $\sum$-max-$\sum$-rule in reverse order of decisions is as follows. When we consider the last decision D_n, its past is fully observed, and the only unobserved variables are the variables never observed or observed after D_n, that is, $\mathcal{I}_n$. Hence, after averaging over $\mathcal{I}_n$, we can select a maximizing argument of the utility function $u(\mathcal{I}(D_n), D_n)$ for each configuration of $\mathcal{I}(D_n)$ as an optimal decision at D_n. Notice that we select a maximizing argument for each configuration of the past. In principle, this eliminates $\mathcal{I}_n$ and D_n from the decision problem, and we may consider D_{n-1} as the last decision. This implies that when we solve for D_{n-1}, we assume the decision maker to act optimally for D_n.

Notice that the variables are observed at the time of decision, but not (necessarily) at the time of analysis. Whether a random variable is known or unknown is defined from the point of view of the decision maker, and not the analyst. In this way we may solve $\mathcal{N}$ by computing the expected utility $\mathrm{EU}(\hat{\Delta})$ of the optimal strategy Δ as

$$\mathrm{EU}(\hat{\Delta}) = \sum_{\mathcal{I}_0} \max_{D_1} \sum_{\mathcal{I}_1} \max_{D_2} \cdots \sum_{\mathcal{I}_{n-1}} \max_{D_n} \sum_{\mathcal{I}_n} \mathrm{EU}(\mathcal{X})$$

$$= \sum_{\mathcal{I}_0} \max_{D_1} \sum_{\mathcal{I}_1} \max_{D_2} \cdots \sum_{\mathcal{I}_{n-1}} \max_{D_n}$$

$$\prod_{X_v \in \mathcal{X}_C} P(X_v \mid X_{\mathrm{pa}(v)}) \sum_{U_w \in \mathcal{X}_U} u(X_{\mathrm{pa}(w)}). \qquad (5.8)$$

As part of the process, prior to the elimination of each decision D, we record the maximizing arguments of D over the utility potential $\psi(D, \mathcal{I}(D))$ from which D is eliminated for each configuration of $\mathcal{I}(D)$. From $\psi(D, \mathcal{I}(D))$, we define the (probabilistic) policy function $\delta(D \mid \mathcal{I}(D))$ for D as

$$\delta(d \mid \mathcal{I}(D) = i) = \begin{cases} 1 & \text{if } d = \arg\max_{d'} \psi(d', i), \\ 0 & \text{otherwise,} \end{cases}$$

where we assume the maximizing argument $\arg\max_{d'} \psi(d', i)$ to be unique. If it is not unique, then any maximizing argument may be chosen.[2]

Example 5.11 (Oil Wildcatter). To solve the decision problem of the oil wildcatter of Example 4.5 on page 82 is to identify the optimal decision policies for the

---

[2] As we shall see in Sect. 5.2.4, the policy function need not be deterministic.

Table 5.1 The joint expected
utility function $EU(D, S, T)$

D	S	T	
no	cl	no	0
yes	cl	no	7
no	op	no	0
yes	op	no	7
no	di	no	0
yes	di	no	7
no	cl	yes	−2.4
yes	cl	yes	18.6
no	op	yes	−3.5
yes	op	yes	8
no	di	yes	−4.1
yes	di	yes	−16.6

Table 5.2 The expected
utility function $EU(T)$

T	
no	21
yes	22.5

test and drill decisions. From the joint expected utility function, $EU(\mathcal{X})$, over variables $\mathcal{X}$ of $\mathcal{N} = (\mathcal{X}, \mathcal{G}, \mathcal{P}, \mathcal{U})$, we may compute the expected utility, $EU(\hat{\Delta})$, of the optimal strategy, $\hat{\Delta} = \left\{\hat{\delta}_D(S, T), \hat{\delta}_T()\right\}$, and in the process determine the optimal strategy as

$$EU(\hat{\Delta}) = \max_T \sum_S \max_D \sum_O P(O) P(S \mid O, T)(U_1(T) + U_2(D, O)).$$

Table 5.1 shows the expected utility function over D, S, T from which the decision policy $\hat{\delta}_D(S, T)$ is identified as the maximizing argument of D for each configuration of S and T. The oil wildcatter should drill for oil unless he performed the test and obtained a diffuse pattern.

Table 5.2 shows the expected utility function over T from which the decision policy $\hat{\delta}_T()$ is identified as the maximizing argument of T. Hence, the test should always be performed.

The decision policies $\hat{\delta}_T()$ and $\hat{\delta}_D(S, T)$ are already known from Example 4.7 on page 84. The expected utility for the decision problem is 22.5.   □

Solving an influence diagram by performing the variable eliminations according to (5.8) will be highly inefficient even for simple influence diagrams. Instead, we will—as in the case of Bayesian networks—apply a generalized version of the distributive law to increase computational efficiency.

For notational convenience, the generalized marginalization operator M was introduced by Jensen et al. (1994). The marginalization operator works differently for marginalization of random variables and decision variables:

$$\underset{X}{\mathsf{M}}\,\rho \triangleq \sum_X \rho \quad \text{and} \quad \underset{D}{\mathsf{M}}\,\rho \triangleq \max_D \rho,$$

where X is a random variable while D is a decision variable. We will use the generalized marginalization operator to explain the process of solving an influence diagram; see Madsen & Jensen (1999) for details.

Using a generalized version of the distributive law, the solution of an influence diagram may proceed as follows. Let Y be the first random variable to eliminate. The set of utility potentials $\mathcal{U}$ can be divided into two disjoint subsets with respect to Y. Let $\mathcal{U}_Y \subseteq \mathcal{U}$ be the subset of utility potentials including Y in the domain

$$\mathcal{U}_Y = \{u \in \mathcal{U} \,|\, Y \in \mathrm{dom}(u)\}.$$

Then $\mathcal{U} \setminus \mathcal{U}_Y$ is the set of utility potentials not including Y in the domain. Similarly, let $\mathcal{P}_Y \subseteq \mathcal{P}$ be the subset of probability distributions including Y in the domain

$$\mathcal{P}_Y = \{P \in \mathcal{P} \,|\, Y \in \mathrm{dom}(P)\}.$$

Then $\mathcal{P} \setminus \mathcal{P}_Y$ is the set of probability potentials not including Y in the domain. The elimination process proceeds by local computation in order to maintain efficiency (i.e., we exploit the distributive law to maintain the factorization of the joint expected utility function). Let ϕ_Y be the probability potential obtained by eliminating Y from the combination of all probability potentials in $\mathcal{P}_Y$ and let ψ_Y be the utility potential obtained by eliminating Y from the combination of all probability and utility potentials in $\mathcal{P}_Y \cup \mathcal{U}_Y$ such that

$$\phi_Y = \underset{Y}{\mathsf{M}} \prod_{\phi \in \mathcal{P}_Y} \phi,$$

$$\psi_Y = \underset{Y}{\mathsf{M}} \phi_Y \sum_{\psi \in \mathcal{U}_Y} \psi. \tag{5.9}$$

The two potentials ϕ_Y and ψ_Y will be used to enforce the factorization of the joint expected utility function over $\mathcal{X} \setminus \{Y\}$. The factorization may be achieved by rewriting (5.8) using ϕ_Y and ψ_Y as well as applying the distributive law

$$\mathrm{EU}(\hat{\Delta}) = \underset{X \in \mathcal{X}}{\mathsf{M}} \left(\prod_{\phi \in \mathcal{P}} \phi \sum_{\psi \in \mathcal{U}} \psi \right)$$

$$= \underset{X \in \mathcal{X}}{\mathsf{M}} \left[\left(\prod_{\phi \in \mathcal{P} \setminus \mathcal{P}_Y} \phi \prod_{\phi' \in \mathcal{P}_Y} \phi' \right) \left(\sum_{\psi \in \mathcal{U} \setminus \mathcal{U}_Y} \psi + \sum_{\psi' \in \mathcal{U}_Y} \psi' \right) \right]$$

$$= \underset{X \in \mathcal{X} \setminus \{Y\}}{\mathsf{M}} \left[\left(\prod_{\phi \in \mathcal{P} \setminus \mathcal{P}_Y} \phi \right) \underset{Y}{\mathsf{M}} \left(\prod_{\phi' \in \mathcal{P}_Y} \phi' \right) \left(\sum_{\psi \in \mathcal{U} \setminus \mathcal{U}_Y} \psi + \sum_{\psi' \in \mathcal{U}_Y} \psi' \right) \right]$$

$$= \bigwedge_{X \in \mathcal{X} \setminus \{Y\}} \left[\left(\prod_{\phi \in \mathcal{P} \setminus \mathcal{P}_Y} \phi \right) \left(\left(\sum_{\psi \in \mathcal{U} \setminus \mathcal{U}_Y} \psi \right) \phi_Y + \psi_Y \right) \right] \qquad (5.10)$$

$$= \bigwedge_{X \in \mathcal{X} \setminus \{Y\}} \left[\left(\prod_{\phi \in \mathcal{P} \setminus \mathcal{P}_Y} \phi \right) \phi_Y \left(\sum_{\psi \in \mathcal{U} \setminus \mathcal{U}_Y} \psi + \frac{\psi_Y}{\phi_Y} \right) \right]. \qquad (5.11)$$

Equation (5.11) specifies a decomposition of the joint expected utility function over $\mathcal{X} \setminus \{Y\}$, and decomposition has the form of (5.8). The decomposition is the product of the summation over the elements of $\mathcal{U} \setminus \mathcal{U}_Y \cup \{\frac{\psi_Y}{\phi_Y}\}$ and the product over the elements of $\mathcal{P} \setminus \mathcal{P}_Y \cup \{\phi_Y\}$. In addition, we have performed the elimination of Y by local computations only involving potentials with Y as a domain variable. We say that the sets

$$\mathcal{P} \setminus \mathcal{P}_Y \cup \{\phi_Y\} \quad \text{and} \quad \mathcal{U} \setminus \mathcal{U}_Y \cup \left\{ \frac{\psi_Y}{\phi_Y} \right\}$$

are a value preserving reduction of $\mathcal{P}$ and $\mathcal{U}$ where Y has been eliminated. The elimination of the next variable may proceed in the same manner on $\mathcal{U} \setminus \mathcal{U}_Y \cup \{\frac{\psi_Y}{\phi_Y}\}$ and $\mathcal{P} \setminus \mathcal{P}_Y \cup \{\phi_Y\}$.

The division operation in (5.11) is introduced because the combination of probability potentials and utility potentials is nonassociative. Thus, either the division should be performed or the probability potentials have to be distributed over the terms of the utility function as in (5.10).

Example 5.12 (Oil Wildcatter). Utilizing the local computation approach explained above, we may solve the oil wildcatter problem as follows:

$$\mathrm{EU}(\hat{\Delta}) = \max_T \sum_S \max_D \sum_O P(O) P(S \mid O, T)(C(T) + U(D, O))$$

$$= \max_T \left(C(T) + \sum_S P(S) \max_D \sum_O \frac{P(O) P(S \mid O, T)}{P(S)} U(D, O) \right).$$

The division by $P(S)$ is necessary in order to obtain the correct conditional expected utility for D. This division does not effect the policy.

The benefit of the local computation approach is more profound on large and more complex influence diagrams.   □

5.2.2 Solving CLQG Influence Diagrams

Inference in a CLQG influence diagram $\mathcal{N} = (\mathcal{X}, \mathcal{G}, \mathcal{P}, \mathcal{F}, \mathcal{U})$ is similar to inference in a discrete influence diagram. The task is to determine an optimal strategy, $\hat{\Delta} = \{\hat{\delta}_1, \ldots, \hat{\delta}_n\}$, for the decision maker and compute the expected utility of adhering to $\hat{\Delta}$.

Fig. 5.15 Optimization of price given marketing budget size

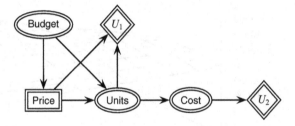

The influence diagram is a compact representation of a joint expected utility function due to the chain rule

$$\text{EU}(X_\Delta = i, X_\Gamma) = \prod_{v \in V_\Delta} P(i_v \,|\, i_{\text{pa}(v)}) * \prod_{w \in V_\Gamma} p(y_w \,|\, X_{\text{pa}(w)}) * \sum_{z \in V_U} u(X_{\text{pa}(z)}).$$

The solution process for CLQG influence diagrams follows the same approach as the solution process for discrete influence diagrams. The solution process proceeds by applying an extension of the $\sum$-max-$\sum$-rule (Madsen & Jensen 2005).The extension is that we need to eliminate the continuous random variables X_Γ by integration as opposed to summation. We refer the interested reader to the literature for details on the solution process (Kenley 1986, Shachter & Kenley 1989, Poland 1994, Madsen & Jensen 2005).

The optimal strategy $\hat{\Delta} = \{\hat{\delta}_1, \ldots, \hat{\delta}_n\}$ will consist of decision policies for both discrete and continuous decision variables. The decision policy for a discrete decision variable $D_i \in X_\Delta \cap X_D$ is a mapping from the configuration of its past $\mathfrak{I}(D_i)$ to $\text{dom}(D_i)$, whereas the decision policy for a continuous decision variable $D_j \in X_\Gamma \cap X_D$ is a multilinear function in its continuous past $\mathfrak{I}(D_i) \cap X_\Gamma$ conditional on its discrete past $\mathfrak{I}(D_i) \cap X_\Delta$.

Example 5.13 (Marketing Budget (Madsen & Jensen 2005)). Consider a company manager has to decide on a unit price, Price, to charge for a certain item he/she wants to sell. The number of items sold, Units, is a function of the price and marketing budget, Budget, whereas the cost of production, Cost, is a function of the number of items sold. This scenario can be modeled using the CLQG influence diagram shown in Fig. 5.15. Prior to making the decision on price, he/she will be allocated a marketing budget.

The decision problem may be quantified as follows where the unit of utility is thousands of euros. The distributions of items sold and production cost are

$$\mathcal{L}(\text{Units}\,|\,\text{Budget} = b, \text{Price} = p) = \mathbb{N}(20 + 0.2 * b - 0.1 * p, 25)$$

$$\mathcal{L}(\text{Cost}\,|\,\text{Units} = u) = \mathbb{N}(400 + 10 * u, 2500)$$

The distribution of marketing budget is

$$\mathcal{L}(\text{Budget}) = \mathbb{N}(100, 400).$$

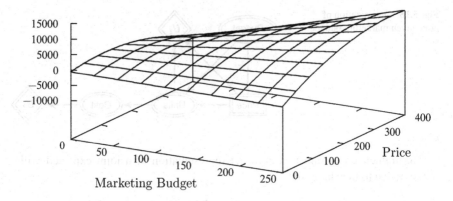

Fig. 5.16 Expected utility as a function of price and marketing budget

The cost function is

$$U_2(\text{Cost} = c) = -c$$

and the revenue function is

$$U_1(\text{Price} = p, \text{Units} = u) = u * p.$$

Figure 5.16 shows the expected utility function as a function of M and P. The optimal decision policy $\delta_P(m)$ for P is a linear function in M: $\delta_P(m) = 105 + m$.

□

5.2.3   Relevance Reasoning

As mentioned in the previous section, a policy δ for D is a mapping from past observations and decisions to the possible decision options at D. When modeling a large and complex decision problem involving a sequence of decisions, the past observations and decisions of a decision may involve a large set of variables. At the same time, it may be that only a small subset of these are essential for the decision. Informally speaking, an observation (or decision) is essential (also known as *requisite*) for a decision, if the outcome of the observation may impact the choice of decision option.

Assume we are able to identify and remove non-requisite parents of each decision. This makes a policy for decision D a function from the requisite past RP(D) to the decision options such that $\delta : \text{RP}(D) \rightarrow \text{dom}(D)$. It is not a trivial task to determine the requisite past of a decision D, that is, the variables observed prior to D, whose values have an impact on the choice of decision option for D (Shachter 1998, Lauritzen & Nilsson 2001, Nielsen 2001).

Definition 5.4 (Requisite Observation). Let $\mathcal{N} = (\mathcal{X}, \mathcal{G} = (V, E), \mathcal{P}, \mathcal{U})$ be an influence diagram. The observation on variable $Y_v \in \mathcal{I}(D_i)$ is requisite for decision D_i in $\mathcal{N}$ if and only if $v \not\perp_\mathcal{G} V_U \cap \text{de}(v_i) \mid (V_{\mathcal{I}(D_i)} \setminus \{v\})$, where v_i is the node representing D_i.

The solution algorithm will identify some of the non-requisite parents for each decision, but there is no guarantee that all non-requisite parents will be identified and ignored. The implicit identification of non-requisite parents is due to conditional independence properties of the graph.

Similar to the concept of a requisite observation is the concept of a *relevant variable*. The set of variables relevant for a decision, D, is the set of variables observed and the set of decisions made after decision D, which may impact the expected utility of D.

Definition 5.5 (Relevant Variable). Let $\mathcal{N} = (\mathcal{X}, \mathcal{G} = (V, E), \mathcal{P}, \mathcal{U})$ be an influence diagram. A variable $Y_v \in \mathcal{F}(D_i)$ is relevant for decision D_i if and only if $v \not\perp_\mathcal{G} V_U \cap \text{de}(v_i) \mid (V_{\mathcal{I}(D_i)} \setminus \{v\})$, where v_i is the node representing D_i.

Using the concepts of relevant variables and requisite observations, it is possible to decompose the structure of an influence diagram $\mathcal{N} = (\mathcal{X}, \mathcal{G} = (V, E), \mathcal{P}, \mathcal{U})$ into a submodels consisting only of requisite parents and relevant variables for each decision in $\mathcal{N}$.

Example 5.14 (Decomposition of Influence Diagrams (Nielsen 2001)). Let us consider the influence diagram shown in Fig. 4.10 on page 89. Traversing the decision variables in reverse order, we may for each decision variable construct the submodel consisting of relevant variables and requisite parents only.

We consider the decisions in reverse order starting with D_4. The reasoning proceeds by searching for non-requisite parents of D_4. By inspection of the diagram, it becomes clear that G blocks the flow of information from observations made prior to D_4 to the only utility descendant U_3 of D_4. Hence, all other parents are non-requisite. Similarly, we identify the set of relevant variables. Figure 5.17 on the following page shows the DAG induced by the subset of requisite observations and relevant variables for D_4.

Similarly, Figs. 5.18 and 5.19 show the DAGs induced by the subsets of requisite observations and relevant variables for D_3 and D_2, respectively.

The DAG induced by the subset of requisite observations and relevant variables for D_1 is equal to the DAG shown in Fig. 4.10 on page 89. $\square$

Decomposing an influence diagram into its submodels of requisite observations and relevant variables for each decision is very useful for model validation.

Fig. 5.17 The DAG induced
by the subset of requisite
observations and relevant
variables for D_4

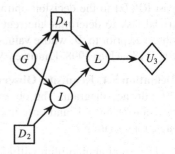

Fig. 5.18 The DAG induced
by the subset of requisite
observations and relevant
variables for D_3

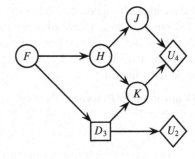

Fig. 5.19 The DAG induced
by the subset of requisite
observations and relevant
variables for D_2

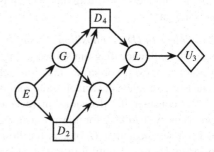

5.2.4   Solving LIMIDs

The LIMID representation relaxes the two fundamental assumptions of the influence
diagram representation. The assumptions are the total order on decisions and
the perfect recall of past decisions and observations. These two assumptions are
fundamental to the solution algorithm for influence diagrams described above. Due
to the relaxation of these two assumptions, the solution process of LIMIDs becomes
more complex than the solution process of influence diagrams.

Let $\mathcal{N} = (\mathcal{X}, \mathcal{G}, \mathcal{P}, \mathcal{U})$ be a LIMID representation of a decision problem. The *Single Policy Updating* (SPU) algorithm is an iterative procedure for identifying (locally) optimal decision policies for the decisions of $\mathcal{N}$. The basic idea is to start an iterative process from some initial strategy where the policy at each decision is updated while keeping the remaining policies fixed until convergence. The starting point can be the uniform strategy where all options are equally likely to be chosen by the decision maker.

As mentioned in Chap. 4, a decision policy δ_{D_i} is a mapping from the decision past of D_i to the state space $\mathrm{dom}(D_i)$ of D_i such that $\delta_{D_i} : \mathcal{I}(D_i) \to \mathrm{dom}(D_i)$. This implies that we may use the probabilistic policy function $\delta'_i(D_i \mid \mathcal{I}(D_i))$ of $\delta_{D_i}(\mathcal{I}(D_i))$ introduced in Sect. 5.2.1

$$
\delta'_i(d_i \mid \mathcal{I}(D_i) = j) = \begin{cases} 1 & \text{if } d_i = \delta_{D_i}(j), \\ 0 & \text{otherwise.} \end{cases}
$$

This encoding will play a central role in the process of solving a LIMID.

Let $\mathcal{N} = (\mathcal{X}, \mathcal{G}, \mathcal{P}, \mathcal{U})$ be a LIMID model with chance and decision variables $\mathcal{X}_C$ and $\mathcal{X}_D$, respectively. A strategy $\Delta = \{\delta_D : D \in \mathcal{X}_D\}$ for $\mathcal{N}$ induces a joint probability distribution $P_\Delta(\mathcal{X})$ over $\mathcal{X}$ as it specifies a probability distribution for each decision variable:

$$
P_\Delta(\mathcal{X}) = \prod_{X_v \in \mathcal{X}_C} P(X_v \mid X_{\mathrm{pa}(v)}) \prod_{D_i \in \mathcal{X}_D} \delta'_i. \tag{5.12}
$$

The aim of solving $\mathcal{N}$ is to identify a strategy, Δ, maximizing the expected utility

$$
\mathrm{EU}(\Delta) = \sum_{X \in \mathcal{X}} P_\Delta(\mathcal{X}) U(\mathcal{X}) = \prod_{X_v \in \mathcal{X}_C} P(X_v \mid X_{\mathrm{pa}(v)}) \prod_{D_i \in \mathcal{X}_D} \delta'_i \sum_{u \in \mathcal{U}} u.
$$

The SPU algorithm starts with some initial strategy and iteratively updates a single policy until convergence has occurred. Convergence has occurred when no single policy modification can increase the expected utility of the strategy. As mentioned above, a common initial strategy is the uniform strategy $\overline{\Delta} = \{\overline{\delta_1}', \ldots, \overline{\delta_n}'\}$ consisting of uniform policies $\overline{\delta_1}', \ldots, \overline{\delta_n}'$ where $\overline{\delta_i}'(d) = \frac{1}{\|D_i\|}$ for each $d \in \mathrm{dom}(D_i)$ and each $D_i \in \mathcal{X}_D$.

Assume Δ is the current strategy and D_i is the next decision to be considered for a policy update, then SPU proceeds by performing the following steps:

Retract   Retract the policy δ'_i from Δ to obtain $\Delta_{-i} = \Delta \setminus \{\delta'_i\}$ (i.e., Δ_{-i} is a strategy for all decisions except D_i).

Update   Identify a new policy $\hat{\delta}'_i$ for D_i by computing

$$
\hat{\delta}'_i = \arg\max_{\delta'_i} EU(\Delta_{-i} \cup \{\delta'_i\}).
$$

Replace   Set $\Delta = \Delta_{-i} \cup \{\hat{\delta}'_i\}$.

SPU may consider the decisions in an arbitrary order. However, if the graph $\mathcal{G}$ specifies a partial order on a subset of decisions $D_{i_1} \prec \cdots \prec D_{i_j} \prec \cdots \prec D_{i_m}$, then these decisions are processed in reverse order, cf. the solution process of ordinary influence diagrams.

Example 5.15 (Solving LIMID: Breeding Pigs). To solve the breeding pigs decision problem of Example 4.13 on page 94 is to identify a strategy consisting of one policy for each decision on whether or not to treat each pig for the disease. Using the SPU algorithm described above, we may solve the decision problem by iteratively updating each single policy until convergence has occurred.

The uniform strategy will serve as the initial strategy. Hence, we assign a uniform policy $\overline{\delta}_i$ to each decision D_i. As there is a total temporal order on the decisions, we consider them in reverse temporal order.

The SPU algorithm updates the policy of each decision iteratively until convergence. Once convergence has occurred, we have obtained the strategy $\Delta = \{\delta_{D_1}, \delta_{D_2}, \delta_{D_3}\}$, where

$$\delta_{D_1}(R_1) = \begin{cases} \text{no} & R_1 = \text{unhealthy} \\ \text{no} & R_1 = \text{healthy} \end{cases}$$

$$\delta_{D_2}(R_2) = \begin{cases} \text{yes} & R_2 = \text{unhealthy} \\ \text{no} & R_2 = \text{healthy} \end{cases}$$

$$\delta_{D_3}(R_3) = \begin{cases} \text{yes} & R_3 = \text{unhealthy} \\ \text{no} & R_3 = \text{healthy} \end{cases}$$

For R_2 and R_3, the strategy is to treat a pig when the test indicates that the pig is unhealthy. For R_1, no treatment should be performed. Notice that each policy is only a function of the most recent test result. This implies that previous results and decisions are ignored.   □

Probability of Future Decisions

Equation (5.12) specifies a factorization of the joint probability distribution P_Δ over $\mathcal{X}$ encoded by a strategy Δ. This factorization may be interpreted as a Bayesian network model. With this interpretation, we are able to compute the probability of future events under the assumption that the decision maker adheres to the strategy Δ. This property also holds for ordinary influence diagrams.

Example 5.16 (Probability of Future Decisions). In Example 5.15, we identified a strategy $\Delta = \{\delta_{D_1}, \delta_{D_2}, \delta_{D_3}\}$ for the breeding pigs problem. Having identified a strategy, the farmer may be interested in knowing the probability of a pig being healthy when it is sold for slaughtering. This probability may be computed using (5.12).

The probability of a pig being healthy under strategy Δ is $P_\Delta(H_4 = \text{true}) = 67.58$, whereas the probability of a pig being healthy under the uniform strategy $\overline{\Delta}$ is $P_{\overline{\Delta}}(H_4 = \text{true}) = 70.55$. The uniform strategy has a lower expected utility though. □

Minimal LIMIDs

LIMIDs relax the assumption of perfect recall of the decision maker. This implies that the structure of a LIMID defines what information is available to the decision maker at each decision. In addition to specifying what information is available to the decision maker, we may perform an analysis of which information is relevant to each decision.

It is not always obvious which informational links to include in a LIMID with graph $\mathcal{G} = (V, E)$. Sometimes a link $(v, w) \in E$ from $X_v \in \mathcal{X}_C$ to $D_w \in \mathcal{X}_D$ may be removed from the graph $\mathcal{G}$ without affecting the policies and the expected utility of the computed policies. When this is the case, we say that the link (v, w) (and the parent X_v given by the link) is non-requisite for D_w.

Removing all non-requisite informational links from a LIMID $\mathcal{N} = (\mathcal{X}, \mathcal{G} = (V, E), \mathcal{P}, \mathcal{U})$ produces the minimal reduction $\mathcal{N}^{\min} = (\mathcal{X}, \mathcal{G} = (V, E^*), \mathcal{P}, \mathcal{U})$ of $\mathcal{N}$. Any LIMID $\mathcal{N}$ has a unique minimal reduction $\mathcal{N}^{\min}$ obtained by iterative removal of informational links from non-requisite parents into decisions.

Since removing a non-requisite parent X from decision D_i may make another previously requisite parent $Y \in X_{\text{pa}(v_i)}$ a non-requisite parent, it is necessary to iteratively process the parent set of each decision until no non-requisite parents are identified. If $\mathcal{N}$ is an ordinary influence diagram, it is sufficient to perform a single pass over the decisions starting with the last decision first. The reason is that we have a total order on the decisions and all decisions are extremal (see Definition 5.6 below).

Optimal Strategies

In order to characterize the conditions under which SPU is guaranteed to find an optimal solution, we need to define the notion of an *extremal decision*.

Definition 5.6 (Extremal Decision). Let $\mathcal{N} = (\mathcal{X}, \mathcal{G}, \mathcal{P}, \mathcal{U})$ be a LIMID. A decision variable D_i is extremal if and only if

$$(V_U \cap \text{de}(D_i)) \perp_{\mathcal{G}} \bigcup_{j \neq i} \text{fa}(D_j) \,|\, \text{fa}(D_i).$$

That is, a decision variable is extremal if all other decisions and their parents are d-separated from the utility descendants of D_i given the family of D_i.

A LIMID is *soluble* if all decisions are extremal. If D_i is extremal in $\mathcal{N}$, then it has an optimal policy. If all policies in Δ are optimal, then Δ is an optimal strategy.

Example 5.17 (Unsoluble LIMID: Breeding Pigs). The breeding pigs network in Fig. 4.15 on page 95 is not soluble as all decisions are non-extremal. This implies that the local optimal strategy identified is not necessarily a globally optimal strategy.

Similarly, Fig. 4.14 of Example 4.12 on page 94 shows an example of a non-soluble LIMID $\mathcal{N} = (\mathcal{X}, \mathcal{G} = (V, E), \mathcal{P}, \mathcal{U})$. On the other hand, the LIMID $\mathcal{N} = (\mathcal{X}, \mathcal{G} = (V, E \setminus \{(D_i, D)\}), \mathcal{P}, \mathcal{U})$ is soluble as both D_i and D_j are extremal.   □

Notice that since any ordinary influence diagram may be represented as a limited memory influence diagram, the SPU solution process may be used to solve influence diagrams; see, for example, Madsen & Nilsson (2001). Any ordinary influence diagram is a special case of a limited memory influence diagram. The LIMID representation of an ordinary influence diagram will produce an optimal strategy.

See Lauritzen & Nilsson (2001) for more details on the solution process.

5.3   Solving OOPNs

For the purpose of inference, an object-oriented model is unfolded. The unfolded network is subsequently transformed into the computational structure used for inference. This implies that to solve an object-oriented model is equivalent to solving its unfolded network. Hence, from the point of view of inference, there is no difference between an object-oriented network and a flat network.

5.4   Summary

In this chapter we have considered the process of solving probabilistic networks. As the exact nature of solving a query against a probabilistic network depends on the type of model, the solution processes of Bayesian networks and influence diagrams have been considered separately.

We build Bayesian network models in order to support efficient belief update in a given domain. Belief update is the task of computing our updated beliefs in (unobserved) events given observations on other events, that is, evidence.

We have considered the task of computing the posterior marginal of each unobserved variable, Y, given a (possibly empty) set of evidence ε, that is, $P(Y \mid \varepsilon)$. We have focused on the solution process that computes the posterior marginal for all unobserved variables using a two-phase message passing process on a junction tree structure.

We build decision models in order to support efficient belief update and decision making under uncertainty in a given problem domain. Belief update is the task of computing our updated beliefs in (unobserved) events given observations on other events, whereas decision making under uncertainty is the task of identifying the (optimal) decision strategy for the decision maker given observations.

We have derived a method for solving influence diagrams by variable elimination. In the process of eliminating variables, we are able to identify the decision policy for each decision. The resulting set of policies is the optimal strategy for the influence diagram.

The LIMID representation relaxes the two fundamental assumptions of the influence diagram representation. The assumptions are the total order on decisions and the perfect recall of past decisions and observations. These two assumptions are fundamental to the solution algorithm for influence diagrams described above. Due to the relaxation of these two assumptions, the solution process of LIMIDs becomes more complex than the solution process of influence diagrams.

We have described how the single policy updating algorithm iteratively identifies a set of locally optimal decision policies. A decision policy is globally optimal when the decision is extremal.

Finally, an OOPN is solved by solving its equivalent unfolded network.

Exercises

Exercise 5.1. You are confronted with three doors, A, B, and C. Behind exactly one of the doors, there is a big prize. The money is yours if you choose the correct door. After you have made your first choice of door but still not opened it, an official opens another one with nothing behind it, and you are allowed to alter your choice.

(a) Construct a model for reasoning about the location of the prize.
(b) Compute by hand the probability distribution over the location of the prize given you select door A and the official opens door B.

Exercise 5.2. Consider the Asia network shown in Fig. 5.20 (see Example 4.2 on page 73 for more details).

(a) Determine the set of barren variables for queries:

$$Q_1 = (\mathcal{N}, \{\text{Bronchitis}\}, \{\}),$$
$$Q_2 = (\mathcal{N}, \{\text{Bronchitis}\}, \{\text{Dyspnoea} = \text{yes}\}),$$
$$Q_3 = (\mathcal{N}, \{\text{Bronchitis}\}, \{\text{Dyspnoea} = \text{no}, \text{X\_ray} = \text{yes}\}),$$

where $Q_i = (\mathcal{N}, \mathcal{T}, \varepsilon)$ with $\mathcal{N}$ denoting the model, $\mathcal{T}$ the target, and ε the evidence set.
(b) Determine the set of nuisance variables for the same queries.

Exercise 5.3. Consider again the Asia network shown in Fig. 5.20.

Fig. 5.20 A graph
specifying the
independence and
dependence relations
of the Asia example

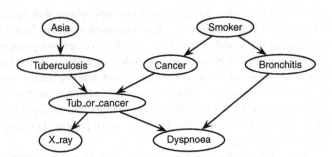

(a) Identify the domain of potentials created by the elimination sequence $\sigma =$
   (Tub_or_cancer, Asia, Smoker, Bronchitis, Cancer, X_ray, Dyspnoea, Tuberculosis).
(b) Construct a junction tree representation using the elimination order σ.
(c) Compare the answer to (a) with Fig. 5.8 on page 121.

Exercise 5.4. Consider the influence diagram in Fig. 4.10 on page 89.

(a) Which variables are relevant for each decision node?
(b) Which observed variables are requisite for each decision node?
(c) Identify the partial order of the chance nodes relative to the decision nodes.
(d) Identify the domains of each decision policy.

Exercise 5.5. Interpret the graph in Fig. 4.10 on page 89 as a LIMID.

(a) Identify the domains of decision policies and compare the results with the
   policies identified in Exercise 5.4(d).
(b) Are any of the decision nodes extreme?

Exercise 5.6. Consider the decision problem in Exercise 5.1.

(a) Calculate the EU for the first choice/decision.
(b) Explain the results of your calculations.
(c) Construct an influence diagram for this problem, and check if the results are
   consistent with your calculations.
(d) What is the optimal policy?

Part II
Model Construction

Chapter 6
Eliciting the Model

This chapter, focusing on fundamental issues of manual construction of probabilistic networks, and Chap. 7, focusing on techniques and tricks to attack commonly recurring modeling problems, jointly comprise the complete account on manual model construction offered in this book.

A probabilistic network can be constructed manually, (semi-)automatically from data, or through a combination of a manual and a data-driven process. In this chapter we will focus exclusively on the manual approach. See Chap. 8 for approaches involving data.

Faced with a new problem, one first has to figure out whether or not probabilistic networks are the right choice of "tool" for solving the problem. Depending on the nature of the problem, probabilistic networks might or might not be a natural choice of modeling methodology. In Sect. 6.1, we consider criteria to be fulfilled for probabilistic networks to be a suitable modeling methodology.

A probabilistic network consists of two components: structure and parameters [i.e., conditional probabilities and utilities (statements of preference)]. The structure of a probabilistic network is often referred to as the qualitative part of the network, whereas the parameters are often referred to as its quantitative part. As the parameters of a model are determined by its structure, the model elicitation process always proceeds in two consecutive stages: First, the variables and the causal, functional or informational relations among the variables are identified, providing the qualitative part of the model. Second, once the model structure has been determined through an iterative process involving testing of variables and conditional independences, and verification of the directionality of the links, the values of the parameters are elicited.

Manual construction of probabilistic networks can be a labor-intensive task, requiring close communication with problem-domain experts. Two key problems need to be addressed in the process of establishing the model structure: identification of the relevant variables and identification of the links between the variables.

The notion of variables (be they discrete chance or decision variables or continuous chance or decision variables) plays a key role in probabilistic networks.

U.B. Kjærulff and A.L. Madsen, *Bayesian Networks and Influence Diagrams: A Guide to Construction and Analysis*, ISS 22, DOI 10.1007/978-1-4614-5104-4_6,
© Springer Science+Business Media New York 2013

At a first glance, it might seem like a relatively simple task to identify the variables of a probabilistic network, but experience shows that in some cases, it can be a difficult task. In Sect. 6.2, we carefully introduce the notion of variables, discuss the various types of variables and their roles, and provide a test that each variable should be able to pass.

In Sect. 6.3, we discuss the process of eliciting the structure of a probabilistic network for a problem domain, discussing both a basic approach, utilizing variable classification and typical causal relations among these, and a more elaborate approach based on identification of archetypical semantical substructures.

Although the use of structured approaches to elicitation of model structure might drastically reduce the risk of misplacing and/or reversing links, model verification ought to be performed before elicitation of model parameters is initiated. In Sect. 6.4, we discuss the importance of inspecting the model structure to verify that the dependence and independence properties encoded in the structure are reasonable.

Having the structure of the probabilistic network in place, the parameters [conditional probabilities and the utilities (if any)] of the network are identified. Quite often, this is the most labor-intensive task, as the number of parameters can be counted in hundreds or even thousands, possibly each requiring consultation of a domain expert. In Sect. 6.5, we consider ways in which the elicitation of the numbers (parameters) can be eased.

In Sect. 6.6 we make some concluding remarks about the art of constructing probabilistic networks by hand and point out the importance of being aware of the limitations of models and the conditions and assumptions under which they are supposed to work. We also stress that an object-oriented approach, which facilitates a modular model construction approach, should preferably be used for large models or models with repetitive structures (e.g., dynamic models). Finally, we point out that manual model construction is an iterative process that can be quite labor intensive.

6.1   When to Use Probabilistic Networks

Following up on the discussion in Sect. 1.5, let us discuss in greater detail the prerequisites to be fulfilled for probabilistic networks to be a suitable framework for modeling problem domain knowledge and supporting belief updating and decision making under uncertainty.

A probabilistic network is a compact and intuitive representation of causal relations among entities of a problem domain, where these entities are represented as discrete variables over either finite sets of mutually exclusive and exhaustive sets of possible values or as continuous variables defined over a space ranging from minus infinity to plus infinity.

There are many good reasons to choose probabilistic networks as the modeling framework, including the coherent and mathematically sound handling of uncertainty and normative decision making; the automated construction and adaptation of models based on data; the intuitive and compact representation of cause–effect relations and (conditional) dependence and independence relations; the efficient solution of queries given evidence; and the ability to support a whole range of analyses of the results produced, including conflict analysis, sensitivity analysis (with respect to both parameters and evidence), and value of information analysis. There are, however, some requirements to the nature of the problem that have to be fulfilled for probabilistic networks to be the right choice of paradigm for solving the problem.

6.1.1   Characteristics of Probabilistic Networks

To understand the power as well as the limitations of the framework of probabilistic networks, let us briefly discuss the main characteristics of probabilistic networks. Probabilistic networks are *normative*, meaning that they provide answers to queries that are mathematically coherent and in agreement with a set of fundamental principles (axioms) of probability calculus and decision theory. There are four ground characteristics that constitute the foundation of (normative) probabilistic models:

- *Graphical representation of causal relations among domain entities (variables).* The notion of causality is central in probabilistic networks, meaning that a directed link from one variable to another (usually) signifies a causal relation among the two. For example, in the chest clinic model (see Example 4.2 on page 73), the directed link from Smoker to Bronchitis indicates that Smoker is a (possible) cause of Bronchitis.
- *Strengths of probabilistic relations are represented by (conditional) probabilities.* Causal relations among variables are seldomly deterministic in the sense that if the cause is present, then the effect can be concluded by certainty. For example, $P(\text{Bronchitis} = \text{yes} \mid \text{Smoker} = \text{yes}) = 0.6$ indicates that among smokers entering the chest clinic, 60 % suffer from bronchitis.
- *Preferences are represented as utilities on a numerical scale.* All sorts of preferences that are relevant in a decision scenario must be expressed on a numerical scale. In a medical scenario, for example, some relevant factors might be medicine expenses and patient comfort.
- *Recommendations are based on the principle of maximal expected utility.* As the reasoning performed by a probabilistic network is normative, the outcome (e.g., most likely diagnosis or suggested decision) is guaranteed to provide a recommended course of action that maximizes the expected utility to the extent that the model is a "true" representation of problem domain.

In addition, it is important to realize the fact that real-world problem domains exist in an open world, whereas any model is based on a closed-world assumption, that is, that it is valid only in a particular context and thus only works correctly under certain assumptions and conditions.

6.1.2   Some Criteria for Using Probabilistic Networks

Given the characteristics of probabilistic networks, there are obviously some problems that can be modeled nicely with probabilistic networks and others that cannot.

For probabilistic networks to be a good choice of modeling paradigm, there should normally be an element of uncertainty associated with the problem definition, implying a desire to maximize some sort of expected utility.

There are a few problem domains where probabilistic networks might not be the ideal choice of modeling paradigm. For example, some problems concerning pattern recognition (e.g., recognition of fingerprints), where there are no well-understood mechanisms underlying the layout of the pattern, probabilistic networks most probably would not be the ideal choice. Also, if the cause–effect relations change over time (i.e., there is no fixed structure of the corresponding probabilistic network), other modeling paradigms might be considered.

So, we might set up the following criteria to be met for probabilistic networks to potentially be a good candidate technology for solving the problem at hand:

- *Well-defined variables.* The variables and events (i.e., possible values of the variables) of the problem domain need to be well defined. For example, in many medical problems, the set of relevant factors (e.g., headache, fever, fatigue, and abdominal pain) are well defined. On the other hand, the variables that determine whether or not a person likes a painting may not be well defined.
- *Highly structured problem domain with identifiable cause–effect relations.* Well-established and detailed knowledge should be available concerning structure (variables and (causal) links), conditional probabilities, and utilities (preferences). In general, the structure needs to be static (i.e., not changing over time), although re-estimation of structure (often through the usage of learning tools; see Chap. 8) can be performed. The values of the probability parameters of a probabilistic network might be drifting, in which case, adaptation techniques can be used to update the parameter values (see Chap. 8).
- *Uncertainty associated with the cause–effect relations.* If all cause–effect relations are deterministic (i.e., all conditional probabilities either take the value 0 or the value 1), more efficient technologies probably exist. In almost all real-world problem domains, there are, however, various kinds of uncertainty associated with cause–effect mechanisms, be it incomplete knowledge about the mechanisms, noisy observations (measurement error), or abstraction of information (e.g., discretization of real-valued observations).
- *Repetitive problem solving.* Often, for the (sometimes large) effort invested in constructing a probabilistic network to pay off, the problem solved should be of repetitive nature. A physician diagnosing respiratory diseases, an Internet company profiling its customers, and a bank deciding to grant loans to its customers are all examples of problems that need to be solved over and over again, where the involved variables and causal mechanisms are invariant over

time, and only the values observed for (some of) the variables differ. Although the repetitiveness criterion is characteristic of many real-world decision problems for which model-based decision support is well suited in terms of payoff of investment, there are important exceptions. Non-repetitive decision problems with high stakes comprise an important exception to this criterion. Examples include decisions on whether or not to establish an offshore oil rig, build a highway bridge, launch a Mars exploration mission, go to war, etc.

- *Maximization of expected utility.* For the probabilistic network framework to be a natural choice, the problem at hand should most probably involve an element of maximization of an expected utility, typically, to support decision making.

6.2 Identifying the Variables of a Model

The set of variables of a probabilistic network comprises the cornerstone of the model. Basically, there are two kinds of variables in probabilistic networks, namely, *chance* and *decision* variables. Chance variables model events of the problem domain, which are not under control of the decision maker, whereas decision variables represent precisely the decisions taken by the decision maker.

Variables can be *discrete* or *continuous*. Discrete variables take values from a finite set of possible values, whereas continuous variables take values from the set of real numbers. We refer the reader to Sect. 4.1.2 for details regarding networks with continuous variables.

6.2.1 Well-Defined Variables

A discrete chance variable of a probabilistic network must represent an exhaustive set of mutually exclusive events. That is, all possible values of the variable must be represented in its state space (exhaustiveness), and no pair of values from the set must exclude each other (mutual exclusiveness).

Example 6.1 (Set of States of a Variable). Let us consider the following sets of events:

1. {heads, tails},
2. {1, 2, 3}, and
3. {black or white, black, white}.

Assume that the first set is meant to describe the set of possible outcomes of a flip with a coin. Thus, assuming that the coin cannot end up in an upright position, the set constitutes an exhaustive set of mutually exclusive events describing the outcomes and is thus a positive example of a set of possible states of a variable of a probabilistic network.

Assume that the second set is meant to describe the outcomes of a roll with an ordinary die. The events of the set are mutually exclusive, but being non-exhaustive, the second set is a negative example of a set of possible states of a variable of a probabilistic network.

Assume that the third set is meant to describe the colors of the keys of a piano. The set of events is exhaustive, but not mutually exclusive, and is thus another negative example of a set of possible states of a variable of a probabilistic network.

□

In addition to representing an exhaustive set of mutually exclusive events, a variable of a probabilistic network typically must represent a *unique* set of events. This means that, usually, a state of a variable should not be mutually exclusive with a state of a single other variable. In other words, the state of a variable should not be given deterministically by the state of a single other variable. If there are states of two variables, say A and B, that are mutually exclusive, it most probably means that the two variables should be merged into one variable having $\{A, B\}$ as part of its set of states. We shall refer to a test of a variable for fulfillment of this uniqueness property as the *uniqueness test*.

Although we usually require variables to pass the uniqueness test, we do allow (the state of) a variable to be deterministically given by the states of two (or more) other variables. Consider, for example, the chest clinic example (see Example 4.2), where Tub or Cancer depends deterministically on variables Tuberculosis and Cancer through a logical OR relation. Constraint variables (see Chap. 7) also depend deterministically on its parent variables. Such "artificial" variables can be handy in many modeling situations, for example, reducing the number of conditional probabilities needed to be specified or enforcing constraints on the combinations of states among a subset of the variables.

Example 6.2 (Variables of a Probabilistic Network). Consider some candidate variables of a probabilistic network:

1. High temperature
2. Low temperature
3. Error occurred
4. No error

all having state space {no, yes}. Assume that the variables pairwise refer to the state of the same phenomenon (e.g., temperature of cooling water of a power plant and state of operation of a particular device of the plant, respectively). Then there are obviously states of variables High temperature and Low temperature that are mutually exclusive (i.e., High temperature = no implies that Low temperature = yes and vice versa). Obviously, the same problem pertains to variables error occurred and no error.

New variables, say Temperature and Error, with sets of states {high, low} and {error occured, no error}, respectively, should be defined, and the variables listed above should be eliminated.   □

Finally, a candidate variable of a problem domain needs to be clearly defined so that everyone involved in the construction or application of the model (i.e., knowledge engineers, domain experts, and decision makers) knows the exact semantics of each variable of the model. For example, the variable It will rain tomorrow might not be clearly defined, as there could be open questions regarding the location of the rain, the amount of rain, the time of observation of rain, etc.

To test if a candidate variable, say V, is indeed a proper and useful variable of the problem domain, it should be able to pass the *clarity test*:

1. The state space of V must consist of an exhaustive set of mutually exclusive values that V can take.
2. Usually, V should represent a unique set of events (i.e., there should be no other candidate variable of the problem domain a state of which is mutually exclusive with a state of V). If this principle is violated, the model most probably should include one or more "constraint" variables (see Sect. 7.1.4) to enforce mutual exclusivity.
3. V should be clearly defined, leaving no doubts about its semantics. In general, a variable is well defined if a clairvoyant can answer questions about the variable without asking further clarifying questions.

Identifying the variables of a problem domain is not always an easy task and requires some practicing. Defining variables corresponding to the (physical) objects of a problem domain is a common mistake made by most novices. Instead of focusing on objects of the problem domain, one needs to focus on the problem (possible diagnoses, classifications, predictions, decisions, etc., to be made) and the relevant pieces of information for solving the problem.

Example 6.3 (Clarity Test: Doors). Consider the task of constructing a probabilistic network for the following decision problem (cf. Exercise 5.1 on page 141):

> You are confronted with three doors, A, B, and C. Behind exactly one of the doors there is a big prize. The prize is yours if you choose the correct door. After you have made your first choice of door but still not opened it, an official opens another one with nothing behind it, and you are allowed to alter your choice. Now, the question is: Should you alter your choice?

A small probabilistic network can be constructed for solving the problem and which provides exact odds of winning given the two options. By experience, though, most novices construct models with one variable for each door, instead of variables modeling the information available and the problem to be solved.

By defining a variable for each door (with each variable having state space {Prize, No prize}, say), one violates the principle that variables should represent unique sets of events (unless constraint variables are included) and thus do not pass the clarity test.

Instead, one needs to take a different perspective, focusing on the problem and the available information:

1. *Problem:* Where is the prize? This gives rise to a variable, Prize location, with state space {A, B, C}, corresponding to doors A, B, and C, respectively.

2. *Information 1:* Which door did you choose originally? This gives rise to a variable, say First choice, with state space {A, B, C}.
3. *Information 2:* Which door were opened by the host? This gives rise to a variable, say Host choice, with state space {A, B, C}.

These variables pass the clarity test.   □

6.2.2   Types of Variables

In the process of identifying the variables, it can be useful to distinguish between different types of variables:

* *Problem variables:* These are the variables of interest, that is, those for which we want to compute their posterior probability given observations of values for information variables (see next item). Usually, the values of problem variables cannot be observed; otherwise, there would not be any point in constructing a probabilistic network in the first place. Problem variables (also sometimes called *hypothesis variables*) relate to diagnoses, classifications, predictions, decisions, etc., to be made.

* *Information variables:* These are the variables for which observations may be available and which can provide information relevant for solving the problem. Two subcategories of information variables can be distinguished:

   – *Background information:* Background information for solving a problem (represented by one or more problem variables) is information that was available before the occurrence of the problem and that has a causal influence on problem variables and symptom variables (see next item) and is thus usually among the "root" variables of a probabilistic network. For example, in a medical setting, relevant background information could include patient age, smoking habits, blood type, gender, etc.

   – *Symptom information:* Symptom information, on the other hand, can be observed as a consequence of the presence of the problem and hence will be available after the occurrence of the problem. In other words, problem variables have causal influences on its symptoms. Hence, symptom variables are usually descendants of background and problem variables. Again, in a medical setting, relevant symptom information could include various outcomes of clinical tests and patient interviews, for example, blood pressure, fever, headache, and weight.

* *Mediating variables:* These are unobservable variables for which posterior probabilities are not of immediate interest but which play important roles for achieving correct conditional independence and dependence properties and/or efficient inference. Mediating variables often have problem and background variables as parents and symptom variables as children.

Table 6.1 Typical causal dependence relations for different variable classes

Type	Causally influenced by
Background variables	None
Problem variables	Background variables
Mediating variables	Background and problem variables
Symptom variables	Background, problem, and mediating variables

Table 6.1 summarizes the typical causal dependence relations for the four different variable classes.

In Example 6.3, there are one problem variable (Prize location) and two information variables (First choice and Host choice), where First choice represents a piece of background information, as it was available before the problem occurred, and Host choice represents a piece of symptom information that became available only after the occurrence of the problem and as a consequence of it.

Example 6.4 (Classification). Assume that we wish to construct a probabilistic network for classifying scientific papers into the two classes of:

(1) Books referring to real-world applications of Bayesian networks
(2) Other books

We identify a problem variable, say Class, with two states, say bn appl books and other books. Assume that the classification is going to be based on detection of keywords, where keywords like "Bayesian network," "Bayes net," "application," "industrial," "decision support," etc., found in a book might indicate reference to real-world applications of Bayesian networks. Then we might define an information (symptom) variable for each keyword (or phrase). Each information variable could be binary, for example, Boolean, with states 0 ("false") and 1 ("true") indicating if the keyword is absent or present, respectively, in a particular book. In a more refined version, each information variable could represent the number of occurrences of the associated keyword, in which case the variable needs several states, say $\{[0; 1[, [1; 5[, [5; 15[, [15; \infty[\}$.   □

Example 6.5 provides a simple example in which the need for a mediating variable is crucial for achieving correct dependence and independence properties of a model (and, consequently, to get reliable answers from the model).

Example 6.5 (Insemination (Jensen 2001)). Six weeks after insemination of a cow, two tests can be performed to investigate the pregnancy state of the cow: blood test and urine test. We identify pregnancy state as the problem variable (PS) and the results of the blood and urine tests as information (symptom) variables (BT and UT, respectively), where PS has states {pregnant, not pregnant} and variables (BT and UT have states {positive, negative}). As the state of pregnancy has a causal influence on the outcome of the tests, we identify an initial model as shown in Fig. 6.1. Using d-separation, we find that this model assumes BT and UT to be

Fig. 6.1 A model for
determining pregnancy state

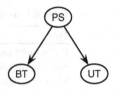

Fig. 6.2 A refined model for
determining the pregnancy
state, reflecting the fact that
both tests are indications of
the hormonal state, which in
turn is an indication of
pregnancy state

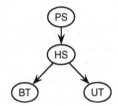

independent given information about the state of PS (i.e., $P(\text{PS} = \text{pregnant}) = 0$
or $P(\text{PS} = \text{pregnant}) = 1$). Assume now that a domain expert (e.g., a veterinarian)
informs us that this independence statement is false, that is, the expert expects the
outcome of one of the tests to be informative about the outcome of the other test even
if the pregnancy state of the cow is known for sure. As there are no natural causal
mechanisms linking BT and UT that could counter for the dependence between these
two variables, we need to solve the problem by introducing a fourth (mediating)
variable (HS) describing the hormonal state of the cow, which determines the
outcomes of both tests (i.e., HS has a causal influence on BT and UT). The resulting
model is shown in Fig. 6.2, where BT and UT are dependent (as they should be) even
if the state of pregnancy is known.   □

6.3   Eliciting the Structure

We shall consider two structured ways of eliciting the model structure. A basic
approach relies on the natural causal ordering that exists among the four categories
of variables that were discussed in Sect. 6.2.2. A more refined approach has been
developed by Neil et al. (2000) where model fragments are identified by recognizing
archetypical relations (known as *idioms*) among groups of variables.

6.3.1   A Basic Approach

Given an initial set of variables identified for a given problem domain, the next
step in the model construction process concerns the identification and verification
of (causal) links of the model. As discussed in Sect. 2.4, maintaining a causal

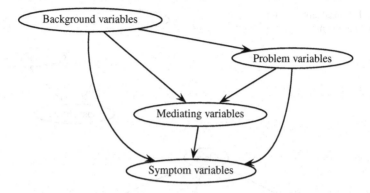

Fig. 6.3 Typical overall causal structure of a probabilistic network

perspective in the model construction process may prove valuable in terms of correctly representing the dependence and independence relations as well as in terms of ease of elicitation of the conditional probabilities of the model. Notice that maintaining a causal perspective is crucial when constructing influence diagrams (see Chap. 4).

As discussed in Sect. 6.2.2, there are four categories of variables of a probabilistic network: (1) background (information) variables, (2) problem variables, (3) mediating variables, and (4) symptom (information) variables. Also, as discussed above, background variables have a causal influence on problem variables and symptom variables, and problem variables have a causal influence on symptom variables. Mediating variables, if present, are most often causally influenced by problem variables and background variables, and they causally influence symptom variables. This gives us a typical overall causal structure of a probabilistic network as shown in Fig. 6.3.

Notice that the structure of the chest clinic example on page 73 fits nicely with this overall structure of a probabilistic network, where Asia and Smoker are background variables; Tuberculosis, Cancer, and Bronchitis are problem variables; Tub or cancer is a mediating variable; and X ray and Dyspnoea are symptom variables.

Example 6.6 (Doors, cont.). For the problem of Example 6.3 on page 151, we identified First choice as a background variable, Prize location as a problem variable, and Host choice as a symptom variable. Obviously, First choice has no influence on Prize location (i.e., no cheating by the host). Also, clearly, the choice of the host depends on your initial choice (First choice) as well as on the host's private information about the location of the prize (Prize location). Thus, following the overall structure in Fig. 6.3, we arrive at a structure for the problem as shown in Fig. 6.4.   □

Fig. 6.4 Causal structure for
the prize problem

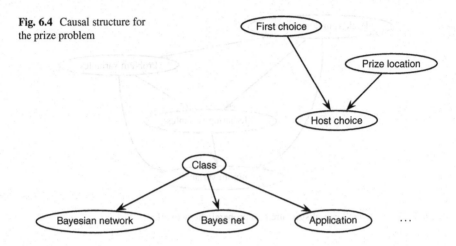

Fig. 6.5 Structure of the model for the classification problem of Example 6.4

Example 6.7 (Classification, cont.). In Example 6.4 on page 153, we have one
problem variable Class and a number of symptom variables, Bayesian network,
Bayes net, Application, Industrial, Decision support, etc. According to the overall
structure in Fig. 6.3, we get a network structure as shown in Fig. 6.5.   □

6.3.2 Idioms

Neil et al. (2000) have developed an approach to elicitation of model structure,
which is based on describing the semantics and syntax of five commonly occurring
substructures (called *idioms*), representing different modes of uncertain reasoning.
These five idioms are believed to cover the vast majority, if not all, substructures
that can occur in a Bayesian network. Each idiom can be considered an archetypical
set of relations among a set of variables. Thus, the use of idioms encourages the
knowledge engineer to think in terms of semantical relations among a (small) group
of variables rather than in terms of nodes and links. The modeling paradigm is thus
moved to a higher level of abstraction, leaving details about which links to include
and their directionality to be handled automatically through the predefined structures
of the idioms.

The five idioms are:

1. *Definitional/synthesis:* Models the combination of variables into a single variable, including deterministic or uncertain definition/function of a variable in
   terms of other variables.
2. *Cause–consequence:* Models cause–effect mechanisms (causal processes).
3. *Measurement:* Models the uncertainty associated with an observation or measurement.

Fig. 6.6 Sample instantiation
of the definitional/synthesis
idiom (Neil et al. 2000)

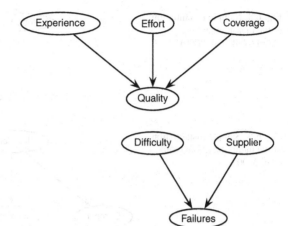

Fig. 6.7 Sample instantiation
of the cause–consequence
idiom (Neil et al. 2000)

4. *Induction:* Models inductive reasoning based on observations from similar
   entities to infer something about an unobserved entity.
5. *Reconciliation:* Models the reconciliation of results from competing statements
   that arise from different sources of information.

Let us consider some examples on the use of the idioms, all taken from
the problem domain of risk assessment in software development processes (Neil
et al. 2000).

Example 6.8 (Definitional/Synthesis Idiom (Neil et al. 2000)). The quality of software testing is defined in terms of the experience of the tester, the effort put into the
testing process, and the coverage of the test (i.e., how many modules of the software
system is tested). A submodel implementing this definition of testing quality as an
instance of the definitional/synthesis idiom is shown in Fig. 6.6.   □

Example 6.9 (Cause–Consequence Idiom (Neil et al. 2000)). Supplier quality and
problem difficulty have causal impacts on the number of failures in a software
product. The three variables, Difficulty, Supplier, and Failures, comprise a submodel
implemented as a join of two instantiations of the cause–consequence idiom, as
illustrated in Fig. 6.7.   □

One might argue that in Example 6.8, Experience, Effort, and Coverage are all
causes of Quality, and hence, these four variables comprise a submodel implemented as a join of three instantiations of the cause–consequence idiom. Also, in
Example 6.9, one might argue that variable Failures is defined in terms of (or is
a synthesis of) variables Difficulty and Supplier and hence should give rise to a
submodel implemented as an instance of the definitional/synthesis idiom. Which
of the two idioms is chosen, however, is immaterial, and depends on how the model
constructor perceives the relations among the variables.

Fig. 6.8 Sample instantiation
of the measurement
idiom (Neil et al. 2000)

Fig. 6.9 Sample instantiation
of the induction idiom (Neil
et al. 2000)

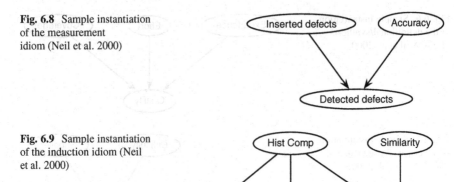

Example 6.10 (Measurement Idiom (Neil et al. 2000)). The number of defects in a
software system can only be estimated up to a certain accuracy. Still, however, the
true number of defects is the important variable in assessing the risk of employing
the system. Hence, based on the observed number of defects (Detected defects),
we need to estimate the true number (Inserted defects). In doing that, we need to
know the accuracy of the testing procedure applied. This is a classical example of
an instantiation of the measurement idiom, as illustrated in Fig. 6.8.   □

Example 6.11 (Induction Idiom (Neil et al. 2000)). Assume that information is
available on the competence of a software testing organization on two previous
occasions, where the organization tested noncritical software products. Given this
information and a measure on the similarity of these previous software products
with a safety-critical software product, we wish to estimate the competence of
the organization in testing the safety-critical product. In other words, we wish to
induce the competence from previous competencies on similar tasks. This induction
problem is implemented as an instantiation of the induction idiom, as illustrated in
Fig. 6.9.   □

Example 6.12 (Reconciliation Idiom (Neil et al. 2000)). Statements on fault tolerance of a software system can be derived either through a cause–consequence
relation, where the quality of the software development process has a causal
influence on the fault tolerance of the system or through a definitional relation
involving the contributions of various fault tolerance strategies such as error
checking and error recovery mechanisms. Thus, if two such competing submodels
provide statements about fault tolerance, we need to reconcile the two statements.
Figure 6.10 shows how this problem can be solved through an instantiation of
the reconciliation idiom, where Reconciliation is a binary variable with states on
and off that forces $P(\text{Fault tol } 1 \mid \varepsilon)$ and $P(\text{Fault tol } 2 \mid \varepsilon)$ to be identical whenever
Reconciliation $=$ on, that is,

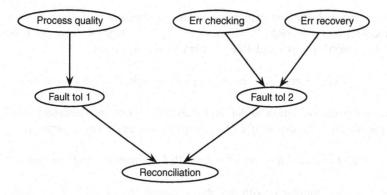

Fig. 6.10 Sample instantiation of the reconciliation idiom (Neil et al. 2000)

$$P(R = \text{on}\,|\,F_1, F_2) = \begin{cases} 1 \text{ whenever } F_1 = F_2 \\ 0 \text{ otherwise,} \end{cases}$$

where F_1, F_2, and R are abbreviations for, respectively, Fault tol 1, Fault tol 2, and Reconciliation. Assume that $\text{dom}(F_1) = \text{dom}(F_2) = (\text{low}, \text{high})$ and that before reconciliation has taken place (i.e., R has not been instantiated), we have

$$P(F_1 = \text{high}\,|\,\varepsilon) = 0.7$$
$$P(F_2 = \text{high}\,|\,\varepsilon) = 0.8.$$

Using Bayes' rule (3.11), we then get

$$P(F_1 = \text{high}\,|\,R = \text{on}, \varepsilon)$$
$$= \frac{P(F_1 = \text{high}, R = \text{on}\,|\,\varepsilon)}{P(R = \text{on}\,|\,\varepsilon)}$$
$$= \frac{P(R = \text{on}\,|\,F_1 = \text{high}, F_2 = \text{high})\,P(F_1 = \text{high}\,|\,\varepsilon)\,P(F_2 = \text{high}\,|\,\varepsilon)}{P(R = \text{on}\,|\,\varepsilon)}$$
$$+ \frac{P(R = \text{on}\,|\,F_1 = \text{high}, F_2 = \text{low})\,P(F_1 = \text{high}\,|\,\varepsilon)\,P(F_2 = \text{low}\,|\,\varepsilon)}{P(R = \text{on}\,|\,\varepsilon)}$$
$$= \frac{1 \cdot P(F_1 = \text{high}\,|\,\varepsilon)\,P(F_2 = \text{high}\,|\,\varepsilon) + 0 \cdot P(F_1 = \text{high}\,|\,\varepsilon)\,P(F_2 = \text{low}\,|\,\varepsilon)}{\sum_{F_1, F_2} P(R = \text{on}\,|\,F_1, F_2)\,P(F_1\,|\,\varepsilon)\,P(F_2\,|\,\varepsilon)}$$
$$= \frac{P(F_1 = \text{high}\,|\,\varepsilon)\,P(F_2 = \text{high}\,|\,\varepsilon)}{P(F_1 = \text{high}\,|\,\varepsilon)\,P(F_2 = \text{high}\,|\,\varepsilon) + P(F_1 = \text{low}\,|\,\varepsilon)\,P(F_2 = \text{low}\,|\,\varepsilon)}.$$

(Note that with the last expression being symmetrical in F_1 and F_2, we get (as expected) $P(F_1 = \text{high} | R = \text{no}, \varepsilon) = P(F_2 = \text{high} | R = \text{no}, \varepsilon)$.) Now, with $P(F_1 = \text{high} | \varepsilon) = 0.7$ and $P(F_2 = \text{high} | \varepsilon) = 0.8$, we get

$$P(F_1 = \text{high} | R = \text{no}, \varepsilon) = P(F_2 = \text{high} | R = \text{no}, \varepsilon) = 0.903.$$

That is, the reconciliation model reinforces both statements whenever the statements support each other, that is, if $P(F_1 = \text{high} | \varepsilon) > \frac{1}{2}$ and $P(F_2 = \text{high} | \varepsilon) > \frac{1}{2}$, then

$$P(F_i = \text{high} | R = \text{no}, \varepsilon) > \max\{P(F_1 = \text{high} | \varepsilon), P(F_2 = \text{high} | \varepsilon)\}$$

for $i = 1, 2$. Similarly, with the above sample values for $P(F_1 = \text{high} | \varepsilon)$ and $P(F_2 = \text{high} | \varepsilon)$,

$$P(F_i = \text{low} | R = \text{no}, \varepsilon) < \min\{P(F_1 = \text{low} | \varepsilon), P(F_2 = \text{low} | \varepsilon)\}$$

for $i = 1, 2$.   □

Some remarks concerning the feasibility of the reconciliation idiom are in order. As shown in Example 6.12 on page 158, the model reinforces statements supporting each other. That is, the posterior probability (i.e., after reconciliation) of a statement is greater (less) than the prior probability of the statement if the prior probability of each of the contributing statements is greater (less) than $\frac{1}{2}$. In applications like the one sketched in Example 6.12, this might make perfect sense, as two different sources of information form the bases of the two statements about fault tolerance, that is, one is based on an assessment of process quality and another on the extent to which errors have been checked for and the ability of the system to recover from errors. Thus, whenever two such independent statements about the fault tolerance of a system both point in the same direction, there is reason to believe that the combined statement is stronger than each individual statement.

Care should be taken, however, not to apply the reconciliation model in cases where the contributing/competing statements are based on the same source of information. Consider, for example, the statements from two, otherwise independent, astronomers about the risk of an asteroid hitting the Earth. If each of them states that the risk of collision is 10 %, it would definitely be a mistake to conclude that the risk then would be only 1.2 %, which would be the result of applying the reconciliation model in this case! In Sect. 7.2.2, we shall present a model for dealing with competing statements that are based on the same source of information.

The basic structures of the idioms are illustrated in Fig. 6.11. Notice that these basic structures can be combined into more complex structures. For example, parents in a definitional/synthesis idiom can be a child in another definitional/synthesis idiom etc., breaking down a definitional/synthesis idiom with many parents into a hierarchy through a parent divorcing process (see Chap. 7). Also, basic cause–consequence idioms are typically combined into more complex structures where causes have common effects and effects have common causes.

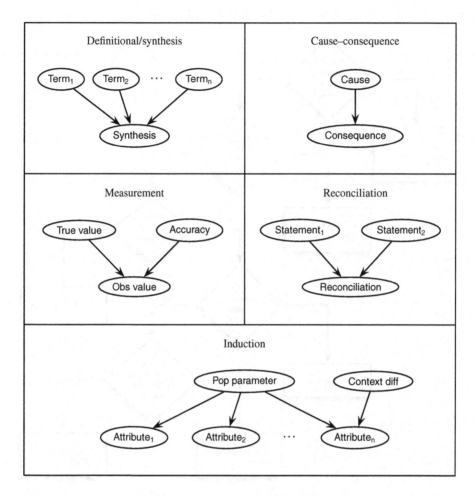

Fig. 6.11   The five basic kinds of idioms defined by Neil et al. (2000)

Also notice that depending on whether or not a root variable of an idiom is a root variable of the overall model structure, it may be categorized as either a background variable, a problem variable, or a mediating variable. Similarly, a non-root variable of an idiom may be a problem variable, a mediating variable, or a symptom variable of the overall model. For example, in the induction idiom of Fig. 6.9 on page 158, variables Hist comp and Similarity could typically be characterized as background variables, Comp 1 and Comp 2 as symptom variables, and Competence as a problem variable.

Probably the most frequent idiom used is the cause–consequence idiom. Thus, in determining the "right" idiom to use, it might be advisable to start considering

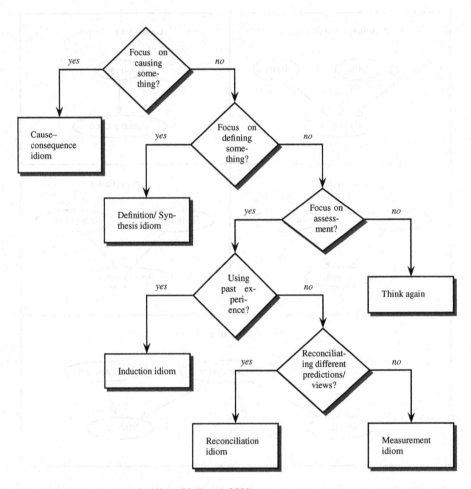

Fig. 6.12 Choosing the right idiom (Neil et al. 2000)

whether the relations among the subset of variables under consideration are best described using one or more cause–consequence relations. Also, the measurement, induction, and reconciliation idioms all deal with assessments of some sort. Therefore, another relevant question to ask is whether the relations among the variables under consideration describe some sort of assessment. As a guide to choosing the right idiom, one might consult the flowchart in Fig. 6.12.

6.3.3 An Example: Extended Chest Clinic Model

To get a better understanding of the structure elicitation process, let us consider a larger, more realistic version of the chest clinic problem introduced in Example 4.2 on page 73. Although this extended version of the model is closer to a realistic, useful model, it still is only a layman's rough interpretation of medical texts. However, whether or not the model is correct is irrelevant to the points discussed in this section.

We wish to construct a model for diagnosing four diseases related to the respiratory system: bronchitis, lung cancer, pneumonia, and tuberculosis. The diagnostic task can be formulated as computing posterior probabilities for these four diseases given observations about a particular patient.

Problem Variables

Therefore, our problem variables are variables whose posterior probabilities provide information about which of the four diseases the patient suffers from. As the diseases can coexist (i.e., they are not mutually exclusive), we need one problem variable per disease.

Since each problem variable is supposed to express presence (possibly in various forms or degrees) or absence of the disease it represents, the variable can assume a limited number of (mutually exclusive) discrete states. To decide on the set of possible states, we need to closely define our problem:

Bronchitis is an inflammation in the bronchi that can be classified as either acute or chronic. Since we need to be able to distinguish between the two for treatment purposes, the problem variable Bronchitis then has three possible states: No, Acute, and Chronic.

Lung cancer is characterized by uncontrolled cell growth in lung tissue. The disease can be categorized in terms of three main types: small cell lung carcinoma (SCLC), non-small cell lung carcinoma (NSCLC), and other (less common) lung cancers. Again, the distinction between the different types is important because of different treatment regimes. The problem variable Lung cancer then has four possible states: no, sclc, nsclc, and other.

Pneumonia is an inflammation in the lungs, which can be classified in several ways. In our model, we shall restrict the problem variable Pneumonia to be Boolean, as we shall assume that pneumonia in any case is treated using a broad-spectrum antibiotic.

Tuberculosis is an infectious disease, which usually attacks the lungs but can also attack other parts of the body. We shall assume that tuberculosis can be classified as primary pulmonary tuberculosis (initial infection form), tuberculosis pleuritis (variant of the former, where the disease is located between the lungs and the abdomen), miliary tuberculosis (the disease has spread to other parts of the body), and reactivated tuberculosis (return of a previous outbreak of the disease). Thus, we shall define a problem variable Tuberculosis, which can assume one of the following states: no, primary, pleuritis, miliary, and reactivated.

Having the problem variables defined, we next need to identify the information variables, that is, the background (information) variables, describing risk factors (causes) of the diseases, and the symptom (information) variables. A variable qualifies as a relevant information variable for the model if a value observed for the variable provides information about a patient that can potentially influence the diagnosis. That is, the value can affect the posterior probability distributions of the problem variables.

Background Variables

To identify background variables, let us consider the risk factors (causes) of the diseases:

Bronchitis: We distinguish between acute and chronic forms of the disease:

  Acute bronchitis   is most often caused by viruses but can also be caused by bacteria. As other inflammatory diseases, bronchitis is more likely to hit people with weak immune systems, such as old people and people suffering from HIV and other diseases weakening the immune system. The inflammation can be caused by irritation of the airways by tobacco smoke or other forms of polluted or dusty air. The disease often occurs during the course of a common cold or influenza. Lung cancer can cause repeated outbreak of bronchitis.

  Chronic bronchitis   is most commonly caused by long-term smoking, but also long-term inhalation of dust and fume can cause chronic bronchitis.

Lung cancer   is most often caused by carcinogens of tobacco smoke that lead to cumulative changes in the DNA of the lung tissue. Long-term exposure to air polluted by asbestos, radon, or chemicals are other potential causes of lung cancer.

Pneumonia   is most often caused by viruses and bacteria and is frequently seen in conjunction with a preceding common cold or influenza. Again, as for other inflammatory diseases, people with weak immune system are more likely to acquire pneumonia. Exposure to polluted air (including tobacco smoke) can also cause pneumonia. Lung cancer can cause repeated outbreak of pneumonia.

Tuberculosis,   being an infectious disease spread through the air when people who have the disease cough, sneeze, or spit, is acquired by health-care workers, people who live (or have lived) in countries or regions with high intensity of tuberculosis, people exposed to crowded living conditions, or people of low socioeconomic status who abuse alcohol (typically homeless). Weak immune system is another risk factor of tuberculosis.

Remembering that background variables represent information that was available before occurrence of the problem and that they causally influence the problem, the

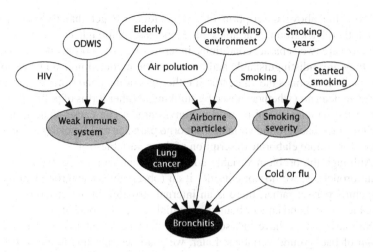

Fig. 6.13 Structure of potential causes of bronchitis

immediate background variables identified for the Bronchitis problem variable are then the following:

- *Acute bronchitis:* As the disease is caused by viruses or bacteria, it would seem natural to define background variables directly representing these causes. First, however, to detect viruses and bacteria, costly and time-consuming laboratory tests must be conducted, which are not cost-effective enough for a relatively mild disease like bronchitis. Second, even if presence of viruses or bacteria was to be detected indirectly through observations of some other immediately observable background variables, we would not be better off, as the causal mechanisms linking these other variables to the viruses and bacteria variables are not well known. Therefore, we shall exclude background variables representing viruses and bacteria altogether and instead rely on immediately observable background variables.

  Weak immune system appears to be an important piece of background information, although it is observable only indirectly through the observable background variables Elderly, HIV, and ODWIS (i.e., "other diseases weakening the immune system"). We could choose to model these three variables as direct causes of Bronchitis, but since they all cause bronchitis only indirectly by weakening the immune system, we choose to introduce Weak immune system as an *intermediate* background variable (see Fig. 6.13). Apart from matching better with our conceptual understanding of the causal mechanisms underlying the relationship between Elderly, HIV, and ODWIS on the one side and Bronchitis on the other, we also obtain a model in which the specification of the quantitative knowledge, specifically the conditional probability table of Bronchitis given its parents (i.e., causes), is significantly less complicated (cf. Occam's razor; see page 184).

From the above description of risk factors, we get that Cold or flu must be another background variables of Bronchitis. Although common cold and influenza are two separate conditions, in terms of causing bronchitis, we assume that they are indistinguishable and hence can be modeled in a single (Boolean) variable that assumes the value true if either one of the conditions are present.

As unclean air (whether caused by smoking or other forms of polluted or dusty air) in the patients living or working environment can cause acute bronchitis, we introduce background variables Airborne particles and Smoking severity. See below for a more elaborate description of these variables.

Although the problem variable Lung cancer can cause repeated outbreak of acute bronchitis, we do not consider it an (observable) background variable of bronchitis per se; rather, there is an internal structure in the group of problem variables, with Lung cancer being a potential cause of Bronchitis.

Similarly, as we have just seen, there might be an internal structure in the group of background variables. Later, we shall see that this might also be the case in the group of symptom variables.

• *Chronic bronchitis:* From the description of the risk factors of chronic bronchitis, we infer that the extent to which the patient smokes and has been smoking (i.e., smoking history) provides an important piece of background information. Thus, Smoking severity, introduced above, (obviously) depends on whether or not the patient is still smoking. We shall further assume that it depends on how many years the patient has been smoking and on the age at which the smoking started. Hence, we introduce the three causal variables Smoking, Smoking years, and Started Smoking of Smoking severity.

Being another possible cause of chronic bronchitis, long-term exposure to dust, fume, or polluted air should be modeled through background variables Dusty working environment (assuming that the exposure has been induced by the patient's working environment) and Air pollution. For simplicity, we shall assume that Air pollution and Dusty working environment can be synthesized into an intermediate background variable Airborne particles (introduced above), which will hence be a cause of both acute and chronic bronchitis and which, as Smoking severity, will not be directly observed. (Consequently, we do not distinguish between short-term and long-term exposure to airborne particles, which (of course) is an approximation that could lead to model inaccuracies.)

Figure 6.13 shows the resulting causal structure of Bronchitis and its risk factors (background variables), where the immediately observable "root causes" are shown as white ovals, the indirectly observable intermediate causes are shown as gray ovals, and the unobservable problem variables as black ovals.

NB: Some of the background variables (e.g., Smoking, Smoking years, *and* Started smoking) *obviously are not marginally independent (e.g., knowing that the patient started smoking at the age of 16 has a significant influence on the probability of the patient currently being a smoker). Thus, there ought to be links between some of the background variables to model such direct dependences. However, we shall ignore such dependences for all background variables, partly to simplify the exposition and partly because they are unimportant, since the values of all background variables are supposed to be known for any given patient prior to making diagnostic inference.*

Continuing along the same lines, we identify the immediate background variables for the Lung cancer problem variable as follows:

- *Lung cancer:* The above description of the risk factors tells us that Lung cancer and Bronchitis share Smoking severity as a risk factor.

  In addition, Lung cancer can be caused by airborne carcinogenic substances, including asbestos, chemicals (typically related to working conditions), and radon. Hence, we define the background variables Asbestos, Work chemicals, and Radon, representing exposure (possibly in various degrees) to asbestos, chemicals in the working environment, and the radioactive gas radon.

  Assuming that public polluted air, represented by the background variable Air pollution discussed above, can cause lung cancer, this variable will also be a shared as background variable for Bronchitis and Lung cancer. Analogous to the above discussion in the case of factors contributing to a weak immune system, the four variables Asbestos, Work chemicals, Radon, and Air pollution all contribute as risk factors of Lung cancer by releasing carcinogenic substances to air inhaled by the patient. In other words, there is a joint causal mechanism involved which can be described as inhalation of air containing carcinogenic substances. Thus, we introduce an intermediate background variable Carcinogenic air which joins the contributions from the four root-cause variables.

The resulting structure of Lung cancer and Bronchitis and their potential causes are shown in Fig. 6.14.

We observe that the causes of Pneumonia are a subset of the causes of Bronchitis, so the structure of the causes of Pneumonia is a substructure of the one shown in Fig. 6.13.

Continuing in the same fashion with Tuberculosis, we arrive at the structure shown in Fig. 6.15.

There are, however, a couple of aspects of the structure in Fig. 6.15 that are worth dwelling a bit on.

Although it is not a root cause, the variable Homeless is indeed immediately observable (indicated by the white coloring of the oval). Conditional independence rules (e.g., d-separation analysis) readily tell us that information about the states of variables Alcohol Aause and Low socioeconomic status is irrelevant once the state of Homeless is observed. Thus, other than communicating the causal relationships between the three variables, there is no reason to include Alcohol abuse and Low socioeconomic status in the model.

Another interesting aspect of the structure in Fig. 6.15 is the complex of intermediate background variables Exposed to TB, Bad living conditions, and TB environment which collectively reduce the maximum number of parent variables to two in the part of the model containing these variables and their ancestors. In the alternative model (see Fig. 6.16) without these three intermediate background variables, we would have the root causes of the complex be parents of the

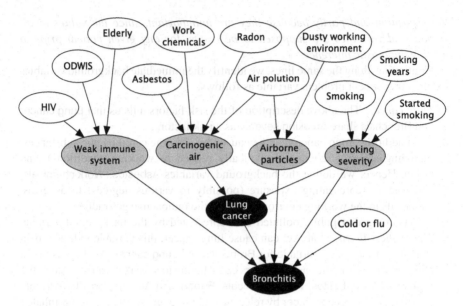

Fig. 6.14 Structure of potential causes of bronchitis and lung cancer

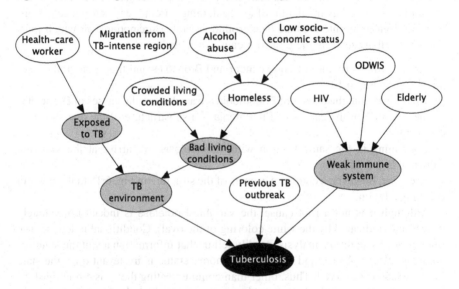

Fig. 6.15 Structure of potential causes of tuberculosis

problem variable. Assuming that |dom(Tuberculosis)| = 5 (as discussed above), |dom(TB environment)| = 3 (say, low, medium, or high), and that all other variables in our complex are Boolean, the size of the conditional probability table for Tuberculosis given its parent variables is going to be 5 * 3 * 2 * 2 = 60 in our

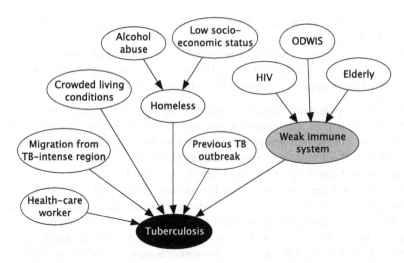

Fig. 6.16 Alternative structure of potential causes of tuberculosis

original model and $5 * 2^6 = 320$ in our alternative model, assuming that variables Previous TB outbreak and Weak immune system are both Boolean. Counting in also the conditional probability tables of the three intermediate background variables, we arrive at a total of $60 + 8 + 8 + 12 = 88$ numbers in the conditional probability tables of Tuberculosis and the three intermediate background variables.

So our three intermediate background variables play a very important role, not only in interpreting and communicating how the root causes affect the problem variable but also in significantly reducing the task of eliciting the probability parameters of the model.

It should be emphasized, however, that this trick of "divorcing" the parents of Tuberculosis is *not* universally applicable. Some requirements must be fulfilled for applying this trick without reducing the expressiveness of the model. We shall discuss these requirements in Sect. 7.1.1 on page 192.

Symptom Variables

To identify variables representing symptoms of the four diseases, let us consider the various effects that can be observed when a patient suffers from one or more of the diseases:

Bronchitis: As before, we distinguish between acute and chronic forms of the disease:

Acute bronchitis most often causes coughing (with or without production of sputum), shortness of breath, and wheezing. Other symptoms occasionally seen include chest pain, fever, sore throat, nasal congestion, and aching muscles.

Chronic bronchitis   can be characterized by the presence of a productive cough that lasts for 3 months or more, and it causes shortness of breath and wheezing. Chest pains may also occasionally be observed.

Lung cancer   potentially causes a variety of symptoms, including shortness of breath, chest pain, persistent fatigue, poor appetite, weight loss, hoarseness, wheezing, coughing, a new coughing pattern, worsening cough, blood in sputum, and repeated outbreak of bronchitis and/or pneumonia. Indications of lung cancer can also be found by performing a chest X-ray or a CT scan.

Pneumonia   is often associated with a productive cough, high fever, shortness of breath, and chest pain. The disease produces a distinct pattern on a chest X-ray.

Tuberculosis   causes chest pain, night sweats, fever, persistent fatigue, poor appetite, weight loss, and a productive, prolonged cough. Diagnosis relies (among other things) on chest X-rays and sputum examination. Chest X-rays can be used to distinguish between pulmonary tuberculosis and miliary (i.e., disseminated) tuberculosis.

Considering the symptoms of (acute and chronic) bronchitis, we observe a need to distinguish between "cough" and "old cough" (i.e., a cough that has lasted for at least 3 months), as the latter is characteristic of the chronic form of bronchitis. We could assume that "old cough" is a specific variant of coughing, and hence model all patterns of coughing in a single variable with "old cough" being one of several mutually exclusive variants (i.e., states) of coughing. However, for at least two reasons, we choose to model "old cough" as a separate (Boolean) variable. First, a cough can be dry, productive, productive with blood, etc., irrespective of the cough being "old" or not. Thus, the state space of the variable representing coughing would span the Cartesian product of, say, {old, ¬old} and {dry, productive (sputum), productive (sputum and blood), . . .}. Second, coughing is a symptom shared by all four diseases, but "old cough" is relevant only for (chronic) bronchitis.

The variables Coughing and Old cough are clearly related. For example, if Old cough is observed to be true, we would clearly expect $P(\text{Coughing} = \text{no}) = 0$. This cannot necessarily be guaranteed by a diverging connection through the problem variables, since the observation that the patient suffers from an old cough might not be sufficient for diagnosing chronic bronchitis with certainty, and hence, $P(\text{Coughing} = \text{no}) > 0$. Therefore, we need a direct connection between Coughing and Old cough. Since an old cough is defined as a cough that has lasted for a certain amount of time, we make the connection Coughing $\rightarrow$ Old cough (cf. the definitional/synthesis idiom) and specify $P(\text{Old cough} = \text{true} \,|\, \text{Bronchitis} = *, \text{Coughing} = \text{no}) = 0$, that is, no matter the state of Bronchitis, whenever Coughing $=$ no, $P(\text{Old cough} = \text{true}) = 0$ (and vice versa).

Apart from this (causal) relationship between "cough" and "old cough," assume that the symptoms of bronchitis and pneumonia are conditionally independent given bronchitis and pneumonia (i.e., given that we know whether or not a patient suffers

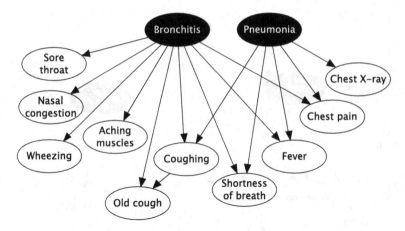

Fig. 6.17 Structure of potential symptoms of bronchitis and pneumonia

from bronchitis and pneumonia, observing one or more of their symptoms does not make us change our belief about the states of the yet unobserved symptoms).

Then we can model the causal relations between the problem variables Bronchitis and Pneumonia and their symptom variables as shown in Fig. 6.17.

Lung cancer has a number of symptoms in common with bronchitis and pneumonia, including shortness of breath, coughing, and chest pain. Moreover, lung cancer and pneumonia share findings from chest X-rays as symptoms.

Considering coughing, worsening of an existing cough and the appearance of a new coughing pattern are characteristic of lung cancer. Thus, in addition to the symptom variables Coughing and Old cough, we need two other symptom variables, Worsening cough and New cough, both being causally linked to Coughing and Lung cancer via directed links from the latter to the former (i.e., Worsening cough and New cough are linked to Coughing in the same way as Old cough, viz., by links emanating from Coughing).

The fact that lung cancer can cause repeated outbreak of bronchitis and pneumonia means that bronchitis and pneumonia are potential symptoms of lung cancer. To model this fact, we add causal links from Lung cancer to Bronchitis and Pneumonia. This is not enough, however. The model must be able to accommodate the pieces of information that a patient might have *repeatedly* suffered from bronchitis and/or pneumonia. Thus, we add variables Repeated Bronchitis and Repeated pneumonia with sets of parent variables {Bronchitis, Lung cancer} and {Pneumonia, Lung cancer}, respectively.

The structure of the submodel covering Bronchitis and Pneumonia and their symptom variables augmented with Lung cancer and the symptoms that Lung Cancer shares with Bronchitis and Pneumonia are shown in Fig. 6.18.

Continuing along similar lines to model the remaining symptoms of lung cancer and those of tuberculosis, we arrive at the complete structure shown in Fig. 6.19.

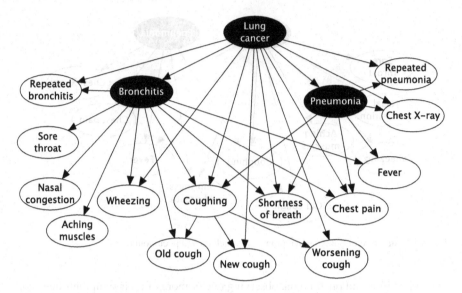

Fig. 6.18 Structure of potential symptoms shared by bronchitis, pneumonia, and lung cancer

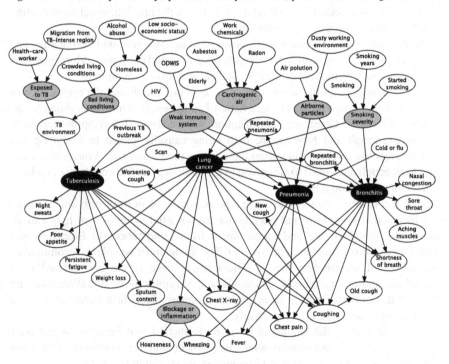

Fig. 6.19 Complete structure of the extended chest clinic model

Notice the intermediate symptom variable Blockage or inflammation, which plays the same role as variable HS in Example 6.5 on page 153 because hoarseness and wheezing are both symptoms of a blockage or an inflammation in the lungs caused by the lung cancer. Thus, without this intermediate variable, the (ground) symptom variables Hoarseness and Wheezing would be direct symptoms of Lung cancer and hence conditionally independent whenever the state of Lung cancer is known. This would be incorrect, as Hoarseness and Wheezing are conditionally independent only if the state of their true common cause, namely, Blockage or inflammation, is known. The gray coloring of the node in Fig. 6.19 on the facing page representing Blockage or inflammation indicates the (maybe not quite true) assumption that a blockage or an inflammation can be observed only indirectly through observation of Hoarseness and Wheezing.

Conclusion

This completes the (first iteration of) elicitation of a structure of the extended chest clinic model. The next phase of the model elicitation process consists of defining the domains of the variables not already determined as part of the structure elicitation process.

In general, we wish variable domains to be as small as possible to minimize the complexity of eliciting the probability parameters of the model and to minimize the complexity of inference. On the other hand, the number of states needs to be big enough to achieve sufficient expressive power to adequately represent the relationships between the domain variables and thereby give the model the necessary reasoning power and precision.

As a simple example, let us consider the task of defining the domain of variable Smoking severity, which is a potential cause of Bronchitis (in its chronic form) and Lung cancer. Assuming that lung cancer can be caused by even small doses of tobacco smoke and that chronic bronchitis only occurs in cases where the patient has been smoking intensely for many years, we can assume that Smoking severity needs to have states none, mild, medium, and heavy. To keep the presentation of the elicitation process for the extended chest clinic example as simple as possible, we have deliberately avoided discussing how the idioms approach could have been used. However, one of the exercises of this chapter addresses this issue.

6.3.4 The Generic Structure of Probabilistic Networks

From the elicitation of the structure of the extended chest clinic model, we have experienced that the typical overall structure of a probabilistic network shown in Fig. 6.3 on page 155 can be elaborated a bit, as mediating variables can appear also

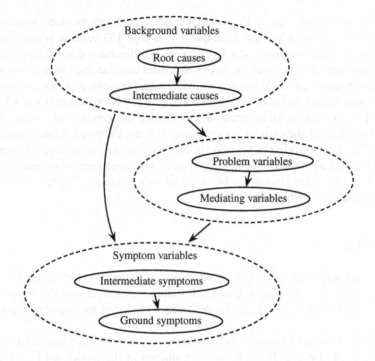

Fig. 6.20 Typical expanded overall causal structure of a probabilistic network

as intermediate causes and intermediate symptoms. Figure 6.20 shows an elaborated
(or expanded) version of the typical overall causal structure of a probabilistic
network.

A directed link between two dotted areas in Fig. 6.20 indicates that a directed
link from anyone of the two node categories in the "source area" to anyone of the
two node categories in the "destination area."

6.4   Model Verification

Proper use of idioms or identification and categorization of variables as background
variables, problem variables, symptom variables, and mediating variables and
adhering to the overall causal structure in Fig. 6.3 reduces the typical error of letting
links (arrows) point from symptom variables to problem variables. If, however, a
model contains links that point from effects to causes, inspection of the dependence
and independence properties represented by the model will often reveal a problem
caused by the reversed links.

Example 6.13 (Model Verification). Consider the following three Boolean variables
(i.e., having state space {false, true}):

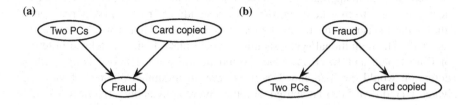

Fig. 6.21 Two competing models for Example 6.13

- Two PCs: Two or more PCs have been bought within a few days using the same credit card.
- Card copied: The credit card has been used at almost the same time at different locations.
- Fraud: The credit card has been subject to fraud.

Now, the question is if model A in Fig. 6.21a is correct or model B in Fig. 6.21b is correct. Experience shows that most novices tend to prefer model A, probably based on the notion that "input" leads to "output" (i.e., observations imply a hypothesis). That is, given the two pieces of information that the credit card has been used to buy two or more PCs within a few days and that the card has been used almost at the same time at different locations (the input), we can conclude that the card has been subject to fraud (the output). According to model A, however, information about Two PCs (Card copied) does not provide us with any information about Card copied (Two PCs) when no information is available about Fraud. This is obviously incorrect, as one piece of evidence confirming (or disconfirming) fraud would obviously make us believe more (or less) in observing the other symptom of fraud. Based on model A, we would thus get wrong posterior probabilities. Notice also that model A does not have the typical causal structure as displayed in Fig. 6.3, as we identify Fraud as the problem variable and Two PCs and Card copied as information (symptom) variables.

Model B, on the other hand, rightfully tells us that:

- Two PCs and Card copied are dependent when we have no hard evidence on Fraud: Observing Two PCs (Card copied) will increase our belief that we will also observe Card copied (Two PCs).
- Two PCs and Card copied are independent when the state of Fraud is known: If we know that we are considering a case of fraud, then observing Two PCs (Card copied) will not change our belief about whether or not we are going to observe Card copied (Two PCs).

As this example shows, close inspection of the dependence and independence relations of a model may reveal links pointing in the wrong direction. However, adherence to the overall causal structure of Fig. 6.3 would eliminate the possibility of arriving at model A.

Using the idiom approach to structure elicitation in this case, one would have to decide whether to use the definitional/synthesis idiom or the cause–consequence idiom (the measurement, reconciliation, and induction idioms would be readily rejected). The definitional/synthesis idiom would make Fraud be defined in terms of Card copied and Two PCs, which would be awkward. For one thing, such a definition would be open-ended, as there are an unlimited number of ways in which a credit card can be abused. Another, more convincing, argument why the definitional/synthesis idiom is the wrong choice is that it violates the overall causal structure of Fig. 6.3.   □

Despite its simplicity, Example 6.13 shows that it might be beneficial to combine the idiom approach and the basic approach to structure elicitation. In other words, keeping in mind the overall causal structure of Fig. 6.3 might be helpful when a choice of idiom has to be made.

As illustrated in Example 6.5 on page 153, model verification may reveal a need to introduce additional (mediating) variables. The mediating variable, HS, of Fig. 6.2 is a common cause of variables BT and UT. Identification of such common causes most often requires a close collaboration between a problem-domain expert and a knowledge engineer, as the domain expert often lacks the ability to read the dependence and independence properties displayed by a network structure and the knowledge engineer lacks insight into the causal mechanisms among the variables of problem domain. As illustrated in the pregnancy example, failure to identify such hidden common causes (i.e., causes for which we have neither any immediate interest in their probability distributions nor any (easy) way of observing their states) may lead to models that provide wrong answers. In the case of the pregnancy model, exclusion of the variable HS would make the model exaggerate the influence from BT and UT when both are observed, as they are both indicative of the same phenomenon, namely, a possible change in the hormonal state of the cow.

6.5 Eliciting the Numbers

Once the structure of the probabilistic network has been established—probably through an iterative process involving model verification and model revisions, including identification of new variables; deletion and redefinition of existing variables; and addition, deletion, and reversal of links—the next and usually the most difficult phase of constructing a probabilistic network concerns the elicitation of the quantitative information, including (conditional) probability distributions and utilities (jointly referred to as the "numbers" or the "parameters").

Due to the (most often) quite demanding effort involved in eliciting the numbers, it is important to carefully verify the structure of the model before proceeding to the quantitative part of the model construction. Otherwise, one runs the risk of having to reject or alter already elicited numbers, as the kind of numbers required is dictated by the model structure. Also, eliciting conditional probabilities with causal links reversed may be difficult and prone to errors. In practice, however, minor model structure adjustments are often made during the number elicitation process (e.g., to reduce the number of parameters).

The quantitative information of a probabilistic network (often referred to as the parameters of the model) is represented as real numbers in conditional probability tables (CPTs) and utility tables (UTs). CPTs represent (conditional) probability distributions with domains spanning the interval $[0; 1]$, whereas UTs represent utility functions with domains spanning $] - \infty; \infty[$.

The parameters of a probabilistic network can be retrieved from databases, elicited from (subjective) domain expert knowledge (e.g., from literature or interviews of domain experts), or established through a mathematical model (e.g., based on assumptions that a given probabilistic distribution can be approximated through a mixture of normal distributions). In this section, we shall focus only on the latter two approaches; see Chap. 8 for learning probability parameters from data.

6.5.1 Eliciting Subjective Conditional Probabilities

Part of the reason why elicitation of values of probability parameters can be rather demanding is that human reasoning is seldomly based on probabilities. In other words, a domain expert might find it awkward to express her/his domain-specific knowledge in terms of conditional probabilities and utility values. Thus, different indirect ways of eliciting the quantitative knowledge may be used.

The fact that small differences in the values of probability (or utility) parameters often make no difference in the recommendations provided by a model allows for parameter elicitation methods based on qualitative approaches. A qualitative approach often makes the domain expert more comfortable specifying her/his knowledge about strengths of causal relations and relative preferences associated with decision problems. An example of a qualitative approach for assessing subjective probabilities includes the usage of a so-called *probability wheel*. A probability wheel is a circle subdivided into n pie wedges, where n equals the number of possible states of the variable for which a (conditional) probability distribution is requested. The domain expert then estimates a probability for a particular state by sizing the corresponding pie wedge to match her/his best assessment of that probability.

Example 6.14 (Probability Wheel). A climate researcher asked to provide an estimate of the increase in the average global temperature over the next 100 years might use a probability wheel as shown in Fig. 6.22, working with a granularity of $< 2\,°C$, 2–$5\,°C$, and $> 5\,°C$. □

Another example of a qualitative approach is the use of verbal statements like "very unlikely" or "almost certain" that are then mapped to probabilities (see Fig. 6.23) (Renooij & Witteman 1999, van der Gaag, Renooij, Witteman & Taal 2002). The use of such a limited set of verbal statements often makes it quite a lot easier for the domain expert to provide assessments of (conditional) probabilities.

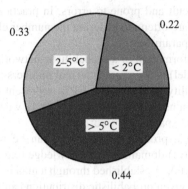

Fig. 6.22 Probability wheel with three pie wedges corresponding to three states of a variable representing the increase in the average global temperature over the next 100 years. The relative area occupied by a particular pie wedge represents the (conditional) probability for the associated state (indicated by the numbers next to the pie wedges)

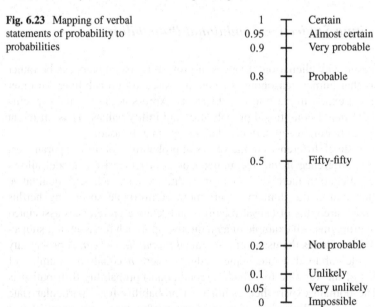

Fig. 6.23 Mapping of verbal statements of probability to probabilities

The reason why the scale in Fig. 6.23 is more fine-grained close to 0 and 1 is that small and large probabilities usually have greater influence on the outcome than those closer to 0.5.

A simple gamble-based approach can be used for assessing the value of a probability parameter. Assume that a domain expert is asked to assess the conditional probability that $X = x$ when $Y = y$, that is, $P(X = x \mid Y = y)$. An indirect way of making the expert assess this quantity would be to ask the expert to choose an n, where $0 \leq n \leq 100$, such that he/she finds the following two gambles equally attractive:

1. If X attains the value x when $Y = y$, you receive \$10.
2. If you draw a red ball from an urn with n red balls and $100 - n$ white balls, you receive \$10.

If all balls are red, he/she prefers Gamble 2, and if all balls are white, he/she prefers Gamble 1. The n for which he/she finds the two gambles equally attractive is her/his estimate of $100 * P(X = x|Y = y)$, that is, $P(X = x|Y = y) = n/100$. To reduce the parameter elicitation burden, it is advisable to perform the elicitation of probability parameters in a two-step process:

1. Quickly provide rough initial parameter estimates.
2. Perform sensitivity analysis (see Chap. 10) to identify parameter assessments that need to be made with care, as small variations can have a large impact on the posterior probability of problem variables given evidence.

The second step should probably be repeated, as the careful assessment of the critical parameters might reveal new critical parameters, etc.

See, for example, Renooij (2001) for a more in-depth discussion of the issues related to elicitation of probabilities for probabilistic networks.

6.5.2 Eliciting Subjective Utilities

Similar to the gamble-based approach for eliciting subjective conditional probabilities, we define a gamble-based approach for eliciting subjective utilities.

Assume that we have the ordering $(a_1, \ldots, a_n)$ of outcomes to which we need to assign subjective utilities, where the outcomes are ordered from worst to best. We first assign a utility of 0 to the worst outcome and a utility of 1 to the best outcome. So,

$$U(a_1) = 0$$
$$U(a_n) = 1.$$

Then we consider gambles (or lotteries) L_p in which we get the best outcome (a_n) with probability p and the worst outcome (a_1) with probability $1 - p$. The utility of an outcome a_i $(i = 2, \ldots, n - 1)$ is then defined to be the expected utility of L_p for which we are indifferent between playing the gamble and getting the outcome a_i for sure. Thus,

$$U(a_i) = \mathrm{EU}(L_p)$$
$$= p \cdot U(a_n) + (1 - p) \cdot U(a_1)$$
$$= p.$$

In other words, the probability, p, for which we are indifferent between the gamble and getting a_i for sure is our utility for a_i.

6.5.3   Specifying CPTs and UTs Through Expressions

Probability distributions and utility functions in probabilistic networks often follow (at least approximately) certain functional or distributional forms. In such cases, the CPTs/UTs can be described compactly through mathematical expressions. Apart from ease of specification, specification through expressions also makes maintenance and reuse much easier.

An expression may be built using standard statistical distributions (e.g., normal, binomial, beta, gamma), arithmetic operators, standard mathematical functions (e.g., logarithmic, exponential, trigonometric, and hyperbolic functions), logical operators (e.g., and, or, if-then-else), and relations (e.g., less than, equals).

The different operators used in an expression have different return types and different type requirements for arguments. Thus, in order to provide a rich language for specifying expressions, it is convenient to have a classification of the discrete chance and decision variables into different groups:

- *Labeled variables* have state spaces of arbitrary qualitative descriptors provided in the form of character strings. Labeled variables can be used in equality comparisons and to express deterministic relationships. For example, a labeled variable C_1 with states state 1 and state 2 can appear in an expression like if(C_1 = state 1, Distribution(0.2, 0.8), Distribution (0.4, 0.6)) for $P(C_2 | C_1)$, where C_2 is another discrete chance variable with two possible states and where if(p, a, b) is read as if predicate p is true, then a else b.
- *Boolean variables* represent the truth values "false" and "true" (in that order) and can be used in logical operators. For example, for a Boolean variable, B_0, being the logical OR of its (Boolean) parents, B_1, B_2, and B_3, $P(B_0 | B_1, B_2, B_3)$ can be specified simply as or(B_1, B_2, B_3).
- *Numbered variables* represent increasing sequences of numbers (integers or reals) and can be used in arithmetic operators, mathematical functions, etc. For example, a numbered variable with state space $\{1, 2, 3, 4, 5, 6\}$ can represent the outcome of the roll of a die.
- *Interval variables* represent disjoint intervals on the real line and can be used in the same way as numbered variables. In addition, they can be used when specifying the intervals over which a continuous quantity is to be discretized. For example, an interval variable with state space $\{[0; 2[, [2; 5[, [5; 10[\}$ may represent the increase in the average global temperature over the next 100 years (cf. Example 6.14 on page 177).

Table 6.2 shows some examples of possible states of the various subtypes of discrete chance variables.

Based on the semantics of discrete chance nodes provided by the above subtyping, an algorithm for automatic generation of CPTs and UTs can be implemented. The functionality of such a table generator will be dependent on the subtypes of the variables involved. Table 6.3 shows how the functionality of a table generator algorithm might be dependent on variable subtypes.

Table 6.2 Examples of possible states of the various subtypes of discrete chance variables

Subtype	Sample states
Labeled	red, blue, low
Boolean	false, true
Numbered	$-\infty, \ldots, -2.2, -1.5, 0, 1, 2, 3, \ldots, \infty$
Interval	$]-\infty; -10[, [-10; -5[, [-5; -1[$

Table 6.3 The functionality of a table generator algorithm is dependent on the subtype of the involved variables

Operation	Labeled	Boolean	Numbered	Interval	Utility
If-then-else	+	+	+	+	+
Arithmetic operators			+	+	+
Boolean operators		+			
Boolean comparison		+			
Boolean distributions		+			
Continuous distributions				+	
Discrete distributions			+		
Custom distribution	+				

Example 6.15 (Number of People (HUGIN 2006)). Assume that in some application, we have probability distributions over the number of males and females, where the distributions are defined over intervals $[0; 100[, [100; 500[, [500; 1000[$ and that we wish to compute the distribution over the total number of individuals given the two former distributions. (Note that this is an obvious sample usage of the definitional/synthesis idiom presented in Sect. 6.3.2 on page 157.) It is a simple but tedious task to specify $P(N_I \mid N_M, N_F)$, where N_I, N_M, and N_F stand for the number of individuals, the number of males, and the number of females, respectively. A much more expedient way of specifying this conditional probability distribution would be to let N_M and N_F be interpreted as interval variables with states $[0; 100[, [100; 500[,$ and $[500; 1000[$ and to let N_I be interpreted as an interval variable with states $[0; 200[, [200; 1000[,$ and $[1000; 2000[,$ for example, and then define $P(N_I \mid N_M, N_F)$ through the simple expression $N_I = N_M + N_F$. The alternative would require specification (including computation) of 27 probability parameters; see Table 6.4. □

Example 6.16 (Fair or Fake Die (HUGIN 2006)). Consider the problem of computing the probabilities of getting n sixes in n rolls with a fair die and a fake die, respectively. A random variable, X, denoting the number of sixes obtained in n rolls with a fair die is binomially distributed with parameters $(n, 1/6)$. Thus, the probability of getting k sixes in n rolls with a fair die is $P_{\text{fair}}(X = k)$, where P_{fair} is a Binomial$(n, 1/6)$. Assuming that for a fake die the probability of six pips facing up is $1/5$, the probability of getting k sixes in n rolls with a fake die is $P_{\text{fake}}(X = k)$, where P_{fake} is a Binomial$(n, 1/5)$.

A model of this problem is shown in Fig. 6.24, where #6's depends on #rolls and Fake die?. Now, if we let #rolls be interpreted as a numbered variable with state space $\{1, 2, 3, 4, 5\}$, let Fake die? be interpreted as a Boolean variable, and

Table 6.4 The CPT for $P(N_I \mid N_M, N_F)$ in Example 6.15 generated from the expression $N_I = N_M + N_F$

		N_I		
N_F	N_M	[0; 200[	[200; 1000[	[1,000; 2,000[
[0; 100[	[0; 100[	1	0	0
[0; 100[	[100; 500[	0.1248	0.8752	0
[0; 100[	[500; 1000[	0	0.8960	0.1040
[100; 500[	[0; 100[	0.1248	0.8752	0
[100; 500[	[100; 500[	0	1	0
[100; 500[	[500; 1000[	0	0.4	0.6
[500; 1000[	[0; 100[	0	0.8960	0.1040
[500; 1000[	[100; 500[	0	0.4	0.6
[500; 1000[	[500; 1000[	0	0	1

Fig. 6.24 A model for the fake die problem

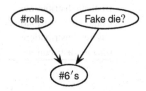

Table 6.5 The CPT for $P(\#6's \mid \#\text{rolls}, \text{Fake die?})$ in the fake die problem of Example 6.16 generated from the expression $\texttt{Binomial}(\#\text{rolls}, \texttt{if}(\text{Fake die?}, 1/5, 1/6))$

		$\#6's$					
Fake die?	#rolls	0	1	2	3	4	5
false	1	0.8333	0.1666	0	0	0	0
false	2	0.6944	0.2777	0.0277	0	0	0
false	3	0.5787	0.3472	0.0694	0.0046	0	0
false	4	0.4822	0.3858	0.1157	0.0154	0.0007	0
false	5	0.4018	0.4018	0.1607	0.0321	0.0032	0.0001
true	1	0.8	0.2	0	0	0	0
true	2	0.6400	0.3200	0.0400	0	0	0
true	3	0.5120	0.3840	0.0960	0.0080	0	0
true	4	0.4096	0.4096	0.1536	0.0256	0.0016	0
true	5	0.3276	0.4096	0.2048	0.0512	0.0064	0.0003

let $\#6's$ be interpreted as a numbered variable with state space $\{0, 1, 2, 3, 4, 5\}$, then $P(\#6's \mid \#\text{rolls}, \text{Fake die?})$ can be specified elegantly using the expression

$$P(\#6's \mid \#\text{rolls}, \text{Fake die?}) = \texttt{Binomial}(\#\text{rolls}, \texttt{if}(\text{Fake die?}, 1/5, 1/6)).$$

Filling in the probabilities by hand would require computation of 60 values of the binomial function with different parameters; see Table 6.5.   □

Example 6.17 (Discretizing a Variable (HUGIN 2006)). Assume that $P(C_1 \mid C_2)$ can be approximated by a normal distribution with mean given by C_2 and with

Table 6.6 The CPT for $P(C_1 | C_2)$ in the discretization problem of Example 6.17 generated from the expression Normal$(C_2, 1)$

C_2	C_1					
	$]-\infty; -5[$	$[-5; -2[$	$[-2; 0[$	$[0; 2[$	$[2; 5[$	$[5; \infty[$
$[-5; -1[$	0.0996	0.6297	0.2499	0.0207	$9.4E-5$	$3.8E-11$
$[-1; 0[$	$7.1E-6$	0.0748	0.6096	0.3075	0.0081	$5.3E-8$
$[0; 1[$	$5.3E-8$	0.0081	0.3075	0.6096	0.0748	$7.1E-6$
$[1; 5[$	$3.8E-11$	$9.4E-5$	0.0207	0.2499	0.6297	0.0996

variance 1, where C_2 is an interval variable with states $[-5; -1[$, $[-1; 0[$, $[0; 1[$, $[1; 5[$. If the discretization of C_1 given by the intervals $]-\infty; -5[$, $[-5; -2[$, $[-2; 0[$, $[0; 2[$, $[2; 5[$, $[5, \infty[$ is found to be suitable, then we can specify $P(C_1 | C_2)$ simply as Normal$(C_2, 1)$. The probability distribution $P(C_1 | C_2 = c_2)$ (i.e., the conditional distribution for C_1 given that the value of C_2 belongs to interval c_2) is generated by computing a large number of probability distributions for C_2, each distribution being obtained by instantiating C_1 to a value in the interval, c_2, under consideration. The average of these distributions (based on, e.g., the midpoints of 25 subintervals) is used as $P(C_1 | C_2 = c_2)$. Hence, for expressions involving several interval variables as parents, the generation of the CPT may be computationally intensive. Table 6.6 shows $P(C_1 | C_2)$.   □

6.6 Concluding Remarks

In this chapter we have tried to convey a set of good practices, routines, and hints that can be helpful for novices wanting to construct a probabilistic network model for a problem domain.

When constructing a model (probabilistic or not), it is crucial to realize that real-world problem domains are usually embedded in a complex reality involving interaction with numerous different aspects of the real world in a way that can never be fully captured in a model. Also, the internal causal mechanisms of a problem domain can almost always only be approximately described in a model. Thus, it is important to bear in mind that *all models are wrong*, but that some might be useful.

Based on this insight, it is important to clearly state the context of the model as well as the assumptions and conditions under which it is supposed to work. Real-world problem domains exist in an open world, whereas models for the problem domains are based on a (most often, erroneous) closed-world assumption.

The construction of a probabilistic network typically runs through four main phases:

Design   of the network structure, covering identification of variables and (causal) relations among the variables. In addition, verification of the network structure is an essential activity of the design phase, where the dependence and independence relations represented in the model are stated and verified.

Implementation   of the network, covering the process of populating the CPTs and UTs with (conditional) probabilities and utilities. This phase is often the most labor intensive of the four phases.

Test   of the network to check if it provides sensible outputs to a carefully selected set of inputs. In a medical setting, for example, testing may amount to entering patient data and comparing network recommendations (e.g., diagnoses or treatment plans) with recommendations provided by medical experts. If the test phase does not reveal any flaws of the network, construction of the network is considered successfully completed.

Analysis   of the network is performed to pinpoint problematic aspects of the network revealed in the test phase. Various tools may be brought into play, including conflict analysis (see Chap. 9), sensitivity analyses (see Chap. 10), and value of information analysis (see Chap. 11).

In the design phase, it is crucial to clearly define the problem that is going to be addressed by the model and to pay close attention to this problem definition when identifying the variables of the model. It is strongly recommended to keep the number of variables and (causal) relations among them to a minimum; only those variables and relations that are thought to have significant influences on the problem variable(s) should be included.

In his writings, William of Occam (or Ockham) (1284–1347) stressed the Aristotelian principle that entities must not be multiplied beyond what is necessary. This principle became known as *Occam's razor* or the *law of parsimony*; a problem should be stated in its basic and simplest terms. In science, the simplest theory that fits the facts of a problem is the one that should be selected. This rule is interpreted to mean that the simplest of two or more competing theories is preferable and that an explanation for unknown phenomena should first be attempted in terms of what is already known.[1]

One argument why one should go for simpler rather than complex solutions to a problem lies in the fact that simpler solutions impose less assumptions about the problem (e.g., about dependences and independences) and hence postulate fewer hypothetical solutions. The underlying idea is thus that simpler solutions are more likely to be "correct."

A key design principle applied in the construction of virtually any complex system is the principle of a modular top-down design in which the level of abstraction changes from the more abstract to the more concrete. To support a modular design approach, clear descriptions of the interface mechanisms of the modules must be provided. Also, given clear interface descriptions, cooperation among sub-teams, reuse modules (submodels), and support for bottom-up design are made possible. There are several reasons why an object-oriented modeling approach (see Sect. 4.3), which facilitates exactly a modular model construction approach that

---

[1]This paragraph is taken from http://www.2think.org/occams_razor.shtml.

allows for both top-down and bottom-up design, is recommended for constructing large models:

- Large and complex systems are often composed of collections of identical or similar components. Therefore, models of such systems will naturally contain repetitive submodels. Object orientation allows such components to be instantiated from a generic class. Both construction and maintenance become a whole lot easier in this way: Each component is generated simply through instantiation, and changes that apply to all instances should be made only in the class from which the components have been instantiated.
- Many complex real-world systems (e.g., mechanical and biological systems) are naturally described in terms of hierarchies of components (i.e., the system consists of components, which consist of subcomponents, etc.). Thus, often an object-oriented probabilistic network (OOPN) more naturally describes the structure of the system modeled.
- Object-oriented model construction supports both top-down and bottom-up modes of construction, which are often used, respectively, to maintain a good overview of the model by abstracting away irrelevant details and to focus on subcomponents with a well-defined interfaces to their surroundings. Thus, the OOPN framework provides support for working with different levels of abstraction in the model constructing process.
- Object-oriented model construction provides a natural means to reuse of existing submodels. That is, the methodology provides a means to maintain a library of submodels that can be instantiated in many different OOPNs.
- Specifying a model in a hierarchical fashion often makes the model less cluttered and thus provides a better means of communicating ideas among knowledge engineers, domain experts, and users.
- The composition of a model by a number of components with well-defined interfaces supports a collaborative model construction process, where different model constructors work on different parts of the model.

Finally, it is important to realize that construction of a probabilistic network is an iterative process in the sense that if model testing reveals flaws of the model, another cycle through the model construction phases mentioned above is necessary. In most practical model construction projects, many iterations are needed before a workable model is found. This iterative or spiral process, which is well known from almost all areas of engineering, is illustrated in Fig. 6.25.

6.7 Summary

Manual construction of a probabilistic network for a complex decision or diagnosis problem is usually a demanding task, involving different sources of expertise that provide model engineering skills as well as deep understanding of the problem domain. The model elicitation process requires careful problem definition, careful

Fig. 6.25 Model
development is an activity
that iteratively passes through
design, implementation, test,
and analysis phases until
model tests no longer uncover
undesired behavior of the
model

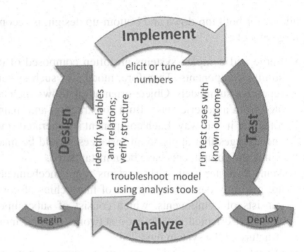

identification of the relevant variables and dependences/independences among the
variables, and elicitation of many (conditional) probabilities and utilities.

However, appealing a probabilistic network might seem in terms of compactness
of representation and in terms of serving as an intuitive means for communication
of problem-domain characteristics, there exist problems for which probabilistic
networks are not the ideal tool to use. In Sect. 6.1, we discussed some characteristics
of probabilistic networks and some criteria for using them. Briefly, and most impor-
tantly, the variables of the problem domain should be well defined, causal relations
among the variables should be identifiable, uncertainty should be associated with
the causal relations, and the problem should contain an element of decision making
with a desire to maximize the expected utility of a decision.

In Sect. 6.2, we discussed how to identify the right set of variables of a
probabilistic network and what it takes for a variable to be well defined. A simple
taxonomy of variables was introduced, which includes three basic types of variables
and their causal relations. Problem variables (or hypothesis variables) represent
the diagnoses, classifications, predictions, decisions, etc., to be made. Information
variables represent the available information (evidence) that can provide information relevant for solving the problem. Finally, mediating variables represent
unobservable entities of the problem domain for which posterior probabilities are
of no immediate interest but which play an important role for achieving the right
dependence and independence properties of the network or for making efficient
inference.

In Sect. 6.3, we first described a basic approach to structure elicitation, showing
how the variable taxonomy can be used in the attempt to elicit the model structure.
Next, we described how the notion of idioms can be helpful in identifying fractions
of a network structure, depending on the nature of the semantic relations that exist
among a small set of variables. Five idioms, thought to cover the vast majority of
commonly occurring semantic relationships, were presented. The five idioms can

be thought of as five archetypical modes of uncertain reasoning, and thus, using the idioms approach to elicitation of model structure, one is encouraged to think at a higher level of abstraction, leaving behind details about which links to include and their directionality.

Although the basic approach to structure elicitation can be quite feasible for some problems, for most (large) real-world problems, the use of idioms is preferable, as the idioms approach splits the problem into smaller and more manageable chunks.

In Sect. 6.4, we briefly touched upon the issue of model verification, an important activity immediately following the structure elicitation effort. In the model verification process, one checks if the dependence and independence statements imposed by the structure are consistent with the knowledge of problem-domain experts.

In Sect. 6.5, we presented some techniques that might be considered in the attempt to elicit the (subjective) values of the parameters (i.e., (conditional) probabilities and utilities) dictated by the structure of the model. Also, we presented a lower-level taxonomy for variables, distinguishing among them in terms of their types of domains. Knowledge about the types of domains of variables allows for automatic generation of CPTs and UTs through a language of mathematical operations, including if-then-else statements, arithmetic and Boolean operations, and a variety of discrete and continuous distributions. The ability to define CPTs and UTs in terms of compact mathematical expressions might greatly reduce the burden of eliciting the numbers (parameter values) of a probabilistic network.

We concluded the discussion on model elicitation in Sect. 6.6 by pointing out some typical main phases of the model construction process, and how these phases are repeated iteratively until model tests no longer uncover undesired behavior of the model. Also, we pointed to the fact that the best models are usually constructed through deliberate use of the law of parsimony (or Occam's razor). Finally, we touched upon the potential benefits of applying an objected-oriented modeling approach, which facilitates modular model construction with the freedom to use a top-down or a bottom-up approach. The use of an object-oriented approach is especially beneficial for construction of large models.

Exercises

Exercise 6.1. There are three condemned prisoners A, B, and C. The governor has announced that one of the three, chosen at random, has been pardoned, but does not say which. Prisoner A, realizing that he only has a $1/3$ chance of having been pardoned, reasons with the warden as follows: "Please tell me the name of one of the other prisoners B or C who will be executed. I already know that at least one of them will be executed so you will not be divulging any information." The warden then asks how he should choose between B or C in case both are to be executed. "In that case," A tells him, "simply flip a coin (when I'm not around) to choose randomly between the two." The warden agrees and later tells A that B will be executed. On hearing this news, A smiles and thinks to himself, "Now my chances of having been pardoned have increased from $1/3$ to $1/2$."

(a) Identify the variables of a Bayesian network model of the reasoning made by prisoner A.
(b) Specify the domains of the variables.
(c) Are your variables well defined? Why or why not?
(d) Characterize the variables in terms of the taxonomy presented in Sect. 6.2.2 on page 152 and specify the causal links of your model using the prototypical causal structure shown in Fig. 6.3.
(e) Specify the (conditional) probabilities of your model and check if your model agrees with the conclusion drawn by prisoner A.

Exercise 6.2. In Exercise 6.1, consider the suggestion to define three variables A, B, and C to represent the three prisoners. Are these variables well defined? If so, why? If not, why not?

Exercise 6.3. In the morning, when Mr. Holmes leaves his house, he realizes that his grass is wet. He wonders whether it has rained during the night or whether he has forgotten to turn off his sprinkler. He looks at the grass of his neighbors, Dr. Watson and Mrs. Gibbon. Both lawns are dry, and he concludes that he must have forgotten to turn off his sprinkler.

(a) Identify the relevant variables a probabilistic network representing Mr. Holmes' reasoning problem. Also, identify the domains of the variables.
(b) Characterize the variables in terms of the taxonomy presented in Sect. 6.2.2 on page 152 and specify the causal links of your model using the prototypical causal structure shown in Fig. 6.3.
(c) If you were to construct the model using the idioms approach, which idiom(s) would you use?
(d) Verify that your model is consistent with the following dependence and independence statements:

   1. Information about the states of the lawns (i.e., wet or dry) is independent if we know that it has rained; otherwise, they are dependent.
   2. Information about the state of rain and information about the state of Holmes' sprinkler are dependent if the state of Holmes' lawn is known; otherwise, they are independent.

Exercise 6.4. Consider the inference problem in Example 2.4 on page 25, which is stated as follows:

Mr Holmes is working in his office when he receives a phone call from his neighbor Dr. Watson, who tells him that Holmes' burglar alarm has gone off. Convinced that a burglar has broken into his house, Holmes rushes to his car and heads for home. On his way, he listens to the radio, and in the news, it is reported that there has been a small earthquake in the area. Knowing that earthquakes have a tendency to turn burglar alarms on, he returns to his work.

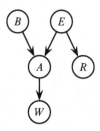

- W: Phone call from Watson
- A: Burglary alarm
- B: Burglary
- R: Radio news
- E: Earthquake

Fig. 6.26 Structure of a Bayesian network for the "burglary or earthquake" inference problem

The structure of a Bayesian network for this inference problem is shown in Fig. 6.26:

(a) Classify the variables in Fig. 6.26 according to the taxonomy in Sect. 6.2.2 on page 152.
(b) Verify that the structure in Fig. 6.26 is consistent with the prototypical causal structure shown in Fig. 6.3 on page 155.

Exercise 6.5. Consider the structure of the extended chest clinic model in Fig. 6.19 on page 172:

(a) Which kinds of idioms could have been applied in eliciting the structure?
(b) Give examples of substructures of the structure that could have been derived using the kinds of idioms suggested in part (a).

Exercise 6.6. Consider the task of providing your subjective probabilities of who is going to win the next World Cup in soccer:

(a) Provide your probability that Brazil wins.
(b) Consider the following gambles:

   1. If Brazil wins, you receive $10.
   2. If you draw a red ball from an urn with n red balls and $100 - n$ white balls, you receive $10.

   For which value of n are the two gambles equally attractive to you?
(c) Compare your original subjective probability that Brazil wins with $n/100$ from (b).

Chapter 7
Modeling Techniques

In this chapter we introduce a set of modeling methods and techniques for simplifying the specification of a probabilistic network.

The construction of a probabilistic network may be a labor-intensive task to perform. The construction involves a sequence of steps such as identifying variables, identifying states of variables, encoding dependence and independence relations as an acyclic, directed graph, and eliciting (conditional) probabilities and utilities as required by the structure of the acyclic, directed graph.

There are many reasons for considering the utilization of modeling techniques in the model development process. Modeling techniques may be applied in order, for instance, to simplify knowledge elicitation and model specification, capture certain properties of the problem domain that are not easily captured by an acyclic, directed graph, to reduce model complexity and improve efficiency of inference in the model, and so on.

Section 7.1 considers modeling techniques for adjusting the structure of a probabilistic network. This includes, in particular, modeling techniques that captures certain structural properties of the problem domain that help reduce the complexity of a model. Section 7.2 considers modeling techniques for the specification of conditional probability distributions. This includes modeling techniques for capturing uncertain information and for reducing the number of parameters to specify. Finally, Sect. 7.3 considers modeling techniques for influence diagram models. This includes modeling techniques of capturing properties of a problem domain that seemingly do not fulfill the underlying assumptions of influence diagrams.

7.1 Structure-Related Techniques

In this section we consider modeling techniques related to the structure of a probabilistic network. In particular, we consider parent divorcing, temporal transformation, the representation of structural and functional uncertainty, undirected dependence links, bidirectional relations, and the naive Bayes model.

U.B. Kjærulff and A.L. Madsen, *Bayesian Networks and Influence Diagrams: A Guide to Construction and Analysis*, ISS 22, DOI 10.1007/978-1-4614-5104-4_7, © Springer Science+Business Media New York 2013

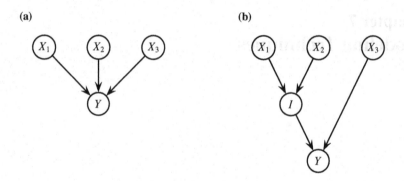

Fig. 7.1 (a) X_1, X_2, and X_3 are direct parents of Y. (b) X_3 is a direct parent of Y, while the combined influence of X_1 and X_2 is mediated through I

7.1.1 Parent Divorcing

The modeling techniques referred to as *parent divorcing* (Olesen, Kjærulff, Jensen, Jensen, Falck, Andreassen & Andersen 1989) is a commonly used modeling technique for reducing the complexity of a model by adjusting the structure of the graph of a probabilistic network. The technique of parent divorcing can be applied to reduce the complexity of specifying and representing the direct impact of a relatively large number of variables $X_1, \ldots, X_n$, referred to as the *cause variables*, on a single variable Y, referred to as the *effect variable*.

The basic idea of parent divorcing is to introduce layers of intermediate variables between the effect variable Y and its direct causes $X_1, \ldots, X_n$ such that each intermediate variable I captures the impact of its parents on the child variable. The parents of I may consists of a subset of intermediate variables and cause variables.

Figure 7.1a illustrates a model structure where the variable Y has three direct parent causes X_1, X_2, X_3. Parent divorcing applied to Y and its direct causes X_1, X_2, X_3 amounts to introducing a mediating variable I between Y and a subset of its parents X_1, X_2, X_3. Let the subset of parents be X_1 and X_2 such that Y after parent divorcing has parents X_3 and I while X_1 and X_2 are parents of I. The result of this process is as illustrated in Fig. 7.1b. Notice that X_1 and X_2 are *divorced* from the remaining parents of Y.

The following example illustrates how the use of parent divorcing may reduce the size of a conditional probability distribution significantly by exploiting structure within conditional probability distributions.

Example 7.1 (Parent Divorcing).   Consider Fig. 7.1 and assume Y is defined as the disjunction (denoted $\lor$) of its three parents X_1, X_2, and X_3. This implies that the conditional probability distribution $P(Y \mid X_1, X_2, X_3)$ is defined as shown in Table 7.1.

Table 7.1 The conditional probability distribution $P(Y \mid X_1, X_2, X_3)$

			Y	
X_1	X_2	X_3	false	true
false	false	false	1	0
false	false	true	0	1
false	true	false	0	1
false	true	true	0	1
true	false	false	0	1
true	false	true	0	1
true	true	false	0	1
true	true	true	0	1

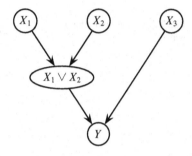

Fig. 7.2 Parent divorcing applied to the distribution for $Y = X_1 \vee X_3 \vee X_3$

Table 7.2 The conditional probability distribution $P(X_1 \vee X_2 \mid X_1, X_2)$

		$X_1 \vee X_2$	
X_1	X_2	false	true
false	false	1	0
false	true	0	1
true	false	0	1
true	true	0	1

By inspection of Table 7.1, it is clear that the conditional probability distribution $P(Y \mid X_1, X_2, X_3)$ has a lot of structure. This structure can be exploited to reduce the size of the largest conditional probability distribution using parent divorcing.

The structure in Fig. 7.1a defines Y as $Y = X_1 \vee X_2 \vee X_3$ disregarding the fact that disjunction ($\vee$) is a binary operator. On the other hand, the structure shown in Fig. 7.2 defines Y as $Y = (X_1 \vee X_2) \vee X_3$ by introducing a mediating variable Table 7.2 (the distribution $P(Y \mid X_1 \vee X_2, X_3)$ is equivalent).

Thus, instead of having one distribution of size 16 we have two tables of size 8. The reduction in size of the largest conditional probability table may seem insignificant. However, if there is a large number of parents, the reduction is significant. The reduction may make an otherwise intractable task tractable. □

The fundamental idea of parent divorcing is that through the utilization of mediating variables, it may be possible to divorce subsets of parents of the effect

Table 7.3 The conditional probability distribution $P(Y \mid X_1, \ldots, X_m)$

					Y		
X_1	X_2	X_3	$\cdots$	X_m	y_1	$\cdots$	y_n
x_1	x_2	x_3	$\cdots$	x_m	z_1	$\cdots$	z_n
$\vdots$					$\vdots$		
x_1	x_2'	x_3	$\cdots$	x_m	z_1'	$\cdots$	z_n'
$\vdots$					$\vdots$		
x_1'	x_2	x_3	$\cdots$	x_m	z_1'	$\cdots$	z_n'
$\vdots$					$\vdots$		
x_1'	x_2'	x_3	$\cdots$	x_m	z_1	$\cdots$	z_n

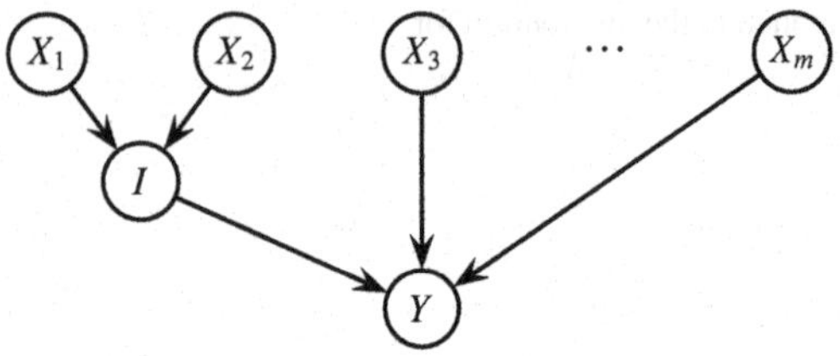

Fig. 7.3 Parent divorcing in general

variable in order to limit the size of parent sets. Parent divorcing is almost only used when the relation among parent variables can be expressed as a chain of associative binary operations such as $\vee, \wedge$, min, max, $+, -, \ldots$.

In general, the underlying assumption of parent divorcing is that the configurations of (X_1, X_2), that is, pairs of instantiations of X_1 and X_2, can be partitioned into sets $i_1, \ldots, i_m$ such that different configurations $(x_1, x_2), (x_1', x_2') \in i_i$ if and only if for all y:

$$P(y \mid x_1, x_2, x_3, \ldots, x_n) = P(y \mid x_1', x_2', x_3, \ldots, x_n).$$

Table 7.3 shows how the conditional distribution $P(Y \mid X_1, \ldots, X_m)$ may support the use of parent divorcing. For different configurations of X_1 and X_2, the child variable Y has the same distribution, for example, for configurations (x_1, x_2) and (x_1', x_2'), the distribution of Y is $z_1, \ldots, z_n$.

An intermediate variable I may be introduced in order to exploit of the structure of $P(Y \mid X_1, \ldots, X_m)$. Figure 7.3 illustrates how the intermediate variable I is introduced as a parent of Y and a child of X_1 and X_2.

The conditional probability distribution of the intermediate variable I is often a deterministic function in configurations of its parents. Table 7.4 shows the conditional probability distribution $P(I \mid X_1, X_2)$.

Since I replaces X_1 and X_2 as parents of Y, the conditional probability distribution of Y changes. Table 7.5 shows the conditional probability distribution $P(Y \mid I, X_3, \ldots, X_m)$.

Table 7.4 The conditional probability distribution $P(I \mid X_1, X_2)$

| | | I | |
X_1	X_2	i_1	i_2
x_1	x_2	1	0
x_1	x_2'	0	1
x_1'	x_2	0	1
x_1'	x_2'	1	0

Table 7.5 The conditional probability distribution $P(Y \mid I, X_3, \ldots, X_m)$

| | | | | | Y | | |
I	X_3	$\cdots$	X_m	y_1	$\cdots$	y_n
i_1	x_3	$\cdots$	x_m	z_1	$\cdots$	z_n
$\vdots$				$\vdots$		
i_2	x_3	$\cdots$	x_m	z_1'	$\cdots$	z_n'

The above property is captured by introducing a mediating variable I as parent of Y with X_1 and X_2 as parents. Parent divorcing is particularly useful in situations where the state space size of the intermediate variable is (significantly) smaller than the combined state space of its parents. Example 7.1 on page 192 shows one situation where parent divorcing improves the efficiency of a model. That is, parent divorcing is (representationally) efficient if $\|I\| < \|X_1\| \cdot \|X_2\|$, that is, if the number of subsets is less than the combined state space size of X_1 and X_2.

Parent divorcing may be considered as a relevant modeling technique when specifying $P(Y \mid X_1, \ldots, X_n)$ is a significant or even intractable knowledge acquisition task or when the size of n makes probabilistic inference intractable.

Notice that parent divorcing can always be applied to a variable and its parents. If the intermediate variable I in Fig. 7.1b has one state for each configuration of its parents, then the conditional probability distribution $P(Y \mid X_3, I)$ can be considered as equivalent to $P(Y \mid X_1, X_2, X_3)$. In this case, nothing has been gained from applying parent divorcing with respect to reducing the complexity of the model or improving efficiency of the model.

How to Implement This Technique

The parent divorcing modeling technique is implemented as follows:

1. Let $\mathcal{X}_W \subset \text{pa}(Y)$ be the subset of parents of Y to be divorced from $\text{pa}(Y) \setminus \mathcal{X}_W$.
2. Create an intermediate node I as a common child of $\mathcal{X}_W$ and a new parent of Y replacing $\mathcal{X}_W$ as parents of Y.
3. Let I have one state for each subset of $\mathcal{X}_W$ mapping to the same distribution on Y.
4. Define the distribution of I given $\mathcal{X}_W$ such that each subset of $\mathcal{X}_W$ mapping to the same distribution on Y maps to the same state of I.
5. Repeat the above steps for each subset $\mathcal{X}_W \subset \text{pa}(Y)$ to be divorced from $\text{pa}(Y) \setminus \mathcal{X}_W$.

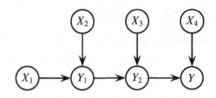

7.1.2 Temporal Transformation

In this section we focus on applying the temporal transformation to adjust the
network structure to capture structure within a conditional probability distribution
of an effect variable Y given a set of causes $X_1, \ldots, X_n$ expressing a *temporal (or
causal) order* on the impact of the cause variables on the effect variable Y. Instead
of combining causes pairwise as in parent divorcing, the influence of causes on the
effect variable is taken into account one cause at a time in their causal or temporal
order.

The method of temporal transformation is best illustrated by an example.
Figure 7.4 shows the result of applying the temporal order method on the conditional
probability distribution of a variable Y given cause variables X_1, X_2, X_3, and X_4.
Notice the temporal order on the causal impacts of the cause variables on the effect
variable. The intermediate variables Y_1 and Y_2 have the same state spaces as Y.

The variables in Fig. 7.4 may represent causes X_1, X_2, and X_3 of a disease Y and
a medicament X_4 for curing the disease. The causes X_1, X_2, and X_3 add to the level
of the disease Y independently, while the medicament cures the disease no matter
the level of the disease. In this example, it is important that X_4 is the last variable in
the temporal order of the causes. The example could be extended such that X_1, X_2,
and X_3 represent different risk factors of the disease that have a temporal order.

The temporal transformation method was introduced by Heckerman (1993)
and refined by Heckerman & Breese (1994). A temporal order of the causal
impacts of $X_1, \ldots, X_n$ on Y is not necessary for applying the method of temporal
transformation. In addition to representing a temporal order of causal influence, the
temporal transformation method can be used as an alternative to parent divorcing.
The parent divorcing method described in the previous section also captures internal
structure of a conditional probability distribution of an effect variable given a set of
cause variables. The parent divorcing method often constructs a (balanced) binary
tree combining causes pairwise recursively, while the temporal transformation
method constructs an unbalanced binary tree as illustrated in Fig. 7.4.

The *temporal transformation* was introduced by Heckerman (1993) as a method
for representing causal independence between a set of cause variables $X_1, \ldots, X_n$
with a common effect E. The model structure in Fig. 7.1a on page 192 does not
capture the property that cause variables $X_1, \ldots, X_n$ impact the effect variable E
independently. Temporal transformation can be used to implement independence of
causal influence as defined in Sect. 7.2.5.

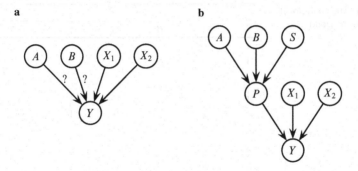

Fig. 7.5 (**a**) Should A or B be the parent of Y? (**b**) Modeling structure and functionality uncertainty

How to Implement This Technique

The temporal transformation modeling technique is implemented as follows.

1. Let $(X_1, \ldots, X_n)$ be an ordering of the parents pa(Y) of Y.
2. For $i = 2, \ldots, n - 1$ create an intermediate node Y_i with the same state space as Y as a child of X_i and a parent of Y_{i+1} where $Y_n = Y$.
3. Add X_1 as a parent of Y_2.
4. Define the distribution of Y_i for $i = 2, \ldots, n$ such that it captures the combined impact of its parents on Y.

7.1.3 Structural and Functional Uncertainty

When modeling certain domains as a probabilistic network, it may be difficult or even seem impossible to specify the set of independence and dependence assumptions using a DAG. It may seem impossible to specify a static DAG for a problem domain where dependence relations change or are uncertain in the sense that they are not known at the time of model development. Similarly, it may be that the functional dependence relation between a variable and (a subset of) its parents is uncertain.

Figure 7.5a shows an example where A and B may both be considered as parent of Y. However, due to the nature of the problem domain, only one of the two is parent of Y at any given time. This is an example of what we term *structure uncertainty*. Figure 7.5b shows how this behavior may be represented as a DAG where S is a selector variable specifying P as taking on the value of A or B. The nodes A, B, and P are assumed to have the same domain, that is,

$$\text{dom}(A) = \text{dom}(B) = \text{dom}(P) = (z_1, \ldots, z_n).$$

Table 7.6 The conditional probability distribution $P(P \mid A, B, S)$

			P			
S	A	B	z_1	z_2	$\cdots$	z_n
A	z_1	z_1	1	0	$\cdots$	0
A	z_1	z_2	1	0	$\cdots$	0
$\vdots$			$\vdots$			
A	z_2	z_1	0	1	$\cdots$	0
A	z_2	z_2	0	1	$\cdots$	0
$\vdots$			$\vdots$			
B	z_1	z_1	1	0	$\cdots$	0
B	z_1	z_2	0	1	$\cdots$	0
$\vdots$			$\vdots$			
B	z_2	z_1	1	0	$\cdots$	0
B	z_2	z_2	0	1	$\cdots$	0
$\vdots$			$\vdots$			

The prior distribution $P(S = A) = 1 - P(S = B)$ specifies the prior belief in A being the true parent of Y. Table 7.6 shows the conditional probability distribution $P(P \mid A, B, S)$. We can define $P(P \mid A, B, S)$ compactly through

$$P = \begin{cases} A & \text{if } S = A \\ B & \text{if } S = B \end{cases}$$

The following example illustrates how structure uncertainty between a variable Ann and two causes George and Henry may be represented.

Example 7.2 (Paternity). In addition to maintaining his orchard, Jack Fletcher breeds horses. Assume Jack—by mistake—placed a group of mares with two stallions (instead of a single stallion) for breeding. After some time, the foal Ann is born. It is clear that the sire of Ann is one the stallions. The question is which one.

The two stallions are Henry and George. Soon after birth, it is discovered that Ann is suffering from a disease caused by a certain genotype aa.

This implies that one of the stallions is a carrier of the gene making its offspring unsuitable for breeding. A carrier of the disease has genotype aA, while a pure horse has genotype AA. A stallion with the disease or carrying the disease should not be used in future breeding. For this reason, it is important to determine the paternity of Ann. The graph shown in Fig. 7.6 captures the properties of the problem. Each variable (except S) species the genotype of a horse where Sire denotes the *true* father of Ann.

The selector variable S specifies either Henry or George as the true father, and its domain is $\mathrm{dom}(S) = (\text{Henry}, \text{George})$. Thus, the conditional probability distribution $P(\text{Sire} \mid \text{George}, \text{Henry}, S)$ is defined as

$$\text{Sire} = \begin{cases} \text{Henry} & \text{if } S = \text{Henry} \\ \text{George} & \text{if } S = \text{George} \end{cases}$$

Fig. 7.6 Either George
or Henry is the true sire
of Ann

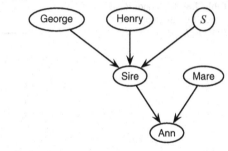

Table 7.7 The conditional
probability distribution
$P(\text{Sire}\,|\,\text{Henry}, \text{George}, S)$

S	Henry	George	Sire		
			aa	AA	aA
Henry	aa	aa	1	0	0
Henry	aa	AA	1	0	0
Henry	aa	aA	1	0	0
⋮			⋮		
George	aa	aa	1	0	0
George	aa	AA	0	1	0
George	aa	aA	0	0	1
⋮			⋮		

Fig. 7.7 Either
$Y = X_1 \vee X_2$
or $Y = X_1 \wedge X_2$

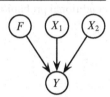

This construction can be generalized for more complex pedigrees with multiple generations and offspring of the stallion. Table 7.7 shows the conditional probability distribution $P(\text{Sire}\,|\,\text{Henry}, \text{George}, S)$ where aa, aA, and AA are the three different genotypes. □

The situation where a variable Y is a function of a subset of its parents such that the state of Y is either one or another (known) function of its parents is termed *functional uncertainty*. Functional uncertainty is similar to structure uncertainty. The following example illustrates how functional uncertainty between a variable Y and two causes X_1 an X_2 may be represented.

Example 7.3 (Functional Uncertainty). Consider two Boolean variables X_1 and X_2. Assume we know that there is a direct impact of configurations of X_1 and X_2 on the Boolean variable Y. Assume further that we know that either $Y = X_1 \vee X_2$ or $Y = X_1 \wedge X_2$ and that the first case is known to appear twice as frequently as the other.

This situation can be captured by a simplified version of the structure shown in Fig. 7.5b as illustrated in Fig. 7.7.

Table 7.8 The conditional probability distribution $P(Y \mid F, X_1, X_2)$

			Y	
F	X_1	X_2	false	true
$\vee$	false	false	1	0
$\vee$	false	true	0	1
$\vee$	true	false	0	1
$\vee$	true	true	0	1
$\wedge$	false	false	1	0
$\wedge$	false	true	1	0
$\wedge$	true	false	1	0
$\wedge$	true	true	0	1

Fig. 7.8 Functional uncertainty on the height of a person

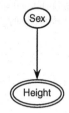

The state space of F is $\mathrm{dom}(F) = (\vee, \wedge)$ such that $P(F) = (2/3, 1/3)$. The conditional probability distribution of Y is defined as

$$Y = \begin{cases} \text{true} & \text{if } X_1 \vee X_2 \text{ and } F = \vee \\ \text{true} & \text{if } X_1 \wedge X_2 \text{ and } F = \wedge \\ \text{false} & \text{otherwise.} \end{cases}$$

This structure captures the uncertainty related to the impact of X_1 and X_2 on Y. Table 7.8 shows the resulting conditional probability distribution $P(Y \mid F, X_1, X_2)$. □

Example 7.4 (Functional Uncertainty: Guessing Game). In Example 4.10, we have implicitly used the functional uncertainty modeling technique. In the example, we assumed that the average height of a male person is greater than the average height of a female person. If the sex of a person is unknown to us when we want to reason about the height of the person, the situation is modeled using a simple variant of functional uncertainty as illustrated in Fig. 7.8.

The example may be extended by assuming there is a correlation between height and weight as illustrated in Fig. 7.9.

For each configuration of Sex, we define a linear function between weight and height. □

Fig. 7.9 Functional
uncertainty on the height of a
person

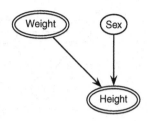

How to Implement This Technique

The functional uncertainty modeling technique is implemented as follows.

1. Let Y be a random variable with parents $pa(Y) = \{X_1, \ldots, X_n\}$ such that Y is a function of $pa(Y)$.
2. Assume the functional dependence relation between Y and $pa(Y)$ is uncertain such that the alternatives and their relative frequencies are known.
3. Create a discrete random variable F with one state for each possible functional dependence relation between Y and $pa(Y)$.
4. Define the prior probability distribution $P(F)$ such that it encodes the relative frequency of the possible functional dependence relations.

The structure uncertainty modeling technique is implemented similarly to the way functional uncertainty is implemented.

7.1.4 Undirected Dependence Relations

The DAG structure of a probabilistic network specifies a set of dependence and independence relations between variables. These dependence and independence relations are specified using directed links between pairs of variables only. When capturing a set of dependence relations between variables using a DAG, it is not unusual to encounter the problem of how (most efficiently) to represent a dependence relation which by nature is undirected.

Let X_1, X_2, and X_3 be discrete variables with the same set of states. Assume configurations where all variables are in the same state are illegal. This is a typical example of an *undirected dependence relation* over a set of variables. This type of undirected relation is referred to as a *constraint*.

A constraint over a subset of variables may be enforced by introducing an auxiliary variable referred to as the *constraint variable* with an appropriate number of states as a child of the variables to be constrained. Often the constraint variable is Boolean, but it may have more than two states. Configurations of the parent variables are mapped to states of the child, and the constraint is enforced using evidence on the constraint variable. For instance, assume that we want to enforce a prior

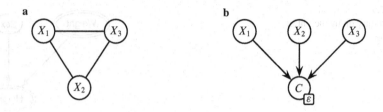

Fig. 7.10 (a) A functional relation $f(X_1, X_2, X_3)$ is to be enforced. (b) A constraint over X_1, X_2, and X_3 is enforced by instantiating C to on

joint probability potential $f(X_1, X_2, X_3)$ over variables X_1, X_2, and X_3. The joint probability can be enforced using a Boolean constraint node C with a conditional probability distribution defined as

$$P(C = \text{on} \mid X_1, X_2, X_3) = f(X_1, X_2, X_3), \tag{7.1}$$

$$P(C = \text{off} \mid X_1, X_2, X_3) = 1 - f(X_1, X_2, X_3). \tag{7.2}$$

The constraint is enforced by instantiating C to on.

Figure 7.10 illustrates how constraints over configurations of variables X_1, X_2, and X_3 are enforced by introducing an auxiliary variable C with two states. One state reflects legal configurations of variables X_1, X_2, and X_3, while the other state reflects illegal configurations of variables X_1, X_2, and X_3. In the example, all configurations where the three variables are not in the same state are legal, while the remaining configurations where all variables are in the same state are illegal. The constraint is enforced by instantiating the variable C to the state corresponding to legal configurations.

The following example illustrates the application of the modeling technique described above to an everyday problem.

Example 7.5 (Washing Socks (Jensen 1996)).   Two pairs of socks have been washed in the washing machine. The washing has been rather hard on the colors and patterns of the socks. One pair of socks is the pair of socks usually worn to play golf, while the other is the pair of socks usually worn during long airplane trips. The airplane socks help to improve blood circulation, while the golf socks have improved respiration. For this reason, it is important to pair the socks correctly.

The airplane socks are blue, while the golf socks are black. The patterns of two pairs of socks are also similar (at least after the washing).

A model for distinguishing the socks of different types has to capture the undirected relation over the four socks. The relation enforces the fact that there are exactly two airplane socks and two golf socks.

The model has four variables $S_1, \ldots, S_4$. Each S_i represents a sock and has domain $\text{dom}(S_i) = (\text{airplane}, \text{golf})$. The undirected relation $R(S_1, \ldots, S_4)$ is a constraint over configurations of the $S_1, \ldots, S_4$. Figure 7.11 illustrates the model

Fig. 7.11 The constraint over $S_1, \ldots, S_4$ is enforced by instantiating C to on

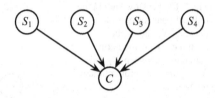

Table 7.9 The conditional probability distribution $P(C \mid S_1, \ldots, S_4)$

				C	
S_1	S_2	S_3	S_4	off	on
airplane	airplane	airplane	airplane	1	0
airplane	airplane	airplane	golf	1	0
airplane	airplane	golf	airplane	1	0
airplane	airplane	golf	golf	0	1
airplane	golf	airplane	airplane	1	0
airplane	golf	airplane	golf	0	1
airplane	golf	golf	airplane	0	1
airplane	golf	golf	golf	1	0
golf	airplane	airplane	airplane	1	0
golf	airplane	airplane	golf	0	1
golf	airplane	golf	airplane	0	1
golf	airplane	golf	golf	1	0
golf	golf	airplane	airplane	0	1
golf	golf	airplane	golf	1	0
golf	golf	golf	airplane	1	0
golf	golf	golf	golf	1	0

structure, while Table 7.9 shows the conditional probability distribution $P(C \mid S_1, \ldots, S_4)$. The conditional probability distribution $P(C \mid S_1, \ldots, S_4)$ may be defined as

$$P(\mathsf{C} = \mathsf{on} \mid s_1, s_2, s_3, s_4) = \begin{cases} 1 & \text{if } |\{s_i = \mathsf{airplane}\}| = 2 \\ 0 & \text{otherwise.} \end{cases}$$

The constraint is enforced by instantiating C to on. □

In the description above we have focused on binary constraint variables. In the general case the constraint variable may have more than two states. In this case multiple states of the constraint variable specifying legal configurations can be enforced using likelihood evidence assigning the value zero to all states specifying illegal configurations and one to all states specifying legal configurations.

Notice that enforcing a constraint may change the marginal distribution on parent variables. One approach to avoid this is described by Fenton, Neil & Lagnado (2011).

Fig. 7.12 (**a**) How should the bidirectional correlation between X_1 and X_2 be captured? (**b**) A mediating variable Y between X_1 and X_2 captures the bidirectional relation

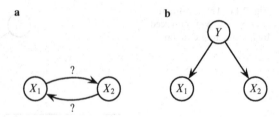

How to Implement This Technique

The undirected directions modeling technique is implemented as follows:

1. Let $\{X_1, \ldots, X_n\}$ be the set of variables over which the prior joint probability distribution $f(X_1, \ldots, X_n)$ is to be enforced.
2. Create a binary constraint node C with states off and on.
3. Add each $X \in \{X_1, \ldots, X_n\}$ as a parent of C.
4. Define the conditional probability distribution $P(C \mid X_1, \ldots, X_n)$ as specified in (7.1) and (7.2).
5. Instantiate C to state on enforcing the constraint.

7.1.5 Bidirectional Relations

Section 7.1.4 describes how an undirected dependence relation over a subset of variables can be enforced using a constraint variable. The introduction of a constraint variable is necessary in order to represent the undirected relation as a DAG. In this section we consider the similar problem of representing what seems to be a bidirectional relation between a pair of variables. That is, when a pair of variables is dependent, it is not always evident which direction the connecting link should have.

Figure 7.12a illustrates the situation where a pair of variables X_1 and X_2 should be connected by a link as there seems to be a direct dependence relation between X_1 and X_2, but it is not possible to identify the direction of the link. Should the link be directed from X_1 to X_2 or vice versa? An insufficient set of variables for capturing the dependence and independence properties of a problem domain as a DAG is a common cause of this type of difficulty in identifying the direction of a link. Figure 7.12b illustrates how a mediating variable Y may be used to capture the bidirectional relation. The mediating variable Y is introduced as a common cause of X_1 and X_2.

The following example illustrates how an insufficient set of variables for capturing the dependence properties of the problem domain can imply difficulties in determining the direction of links.

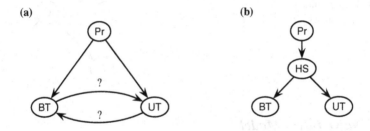

Fig. 7.13 (**a**) How should the bidirectional correlation between BT and UT be captured? (**b**) The bidirectional correlation between BT and UT is captured by the mediating variable HS

Example 7.6 (Insemination (Jensen 1996)).   Consider the task of monitoring the pregnancy state of a cow (also considered in Example 6.5 on page 153). Assume we have the options to perform a blood test and a urine test to determine the pregnancy state of the cow. Both the blood test and the urine test are indicators for the pregnancy state of the cow. Furthermore, we argue that there is a dependence relation between the results of the two tests (if either is positive (negative), we would expect the other test to be positive (negative) as well).

We know there is a correlation between blood test and urine test, but we cannot identify one test as a cause of the other test. This is indicated in Fig. 7.13a where Pr specifies the pregnancy state of the cow, while BT and UT specify the results of the blood and urine tests, respectively. We assume the blood test is not independent of the urine test given the pregnancy state of the cow.

Looking deeper into the properties of the problem domain, we identify some additional structure that alleviates the problem of a bidirectional relation between BT and UT. The two tests do not identify the pregnancy state of the cow directly. Instead the two tests identify the hormonal state of the cow. The resulting structure is shown in Fig. 7.13b where HS represents the hormonal state of the cow.

Notice that the structure shown in Fig. 7.13b correctly captures the conditional dependency of BT and UT given Pr.   □

How to Implement This Technique

The bidirectional relations modeling technique is implemented as follows.

1. Let X_1 and X_2 be a pair of variables which seems to have a bidirectional interaction.
2. Create a mediating variable Y such that it is the intermediate variable in a serial connection with X_1 and X_2.
3. The identification of the states of Y and the probability distribution of Y is domain dependent.

Fig. 7.14 The structure of
the naive Bayes model

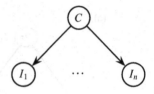

7.1.6 Naive Bayes Model

Restricted probabilistic graphical models are used or considered when low model complexity and high computational power are required. Low model complexity and high computational power are often required in classification-related problems. In a classification problem, the task is to assign a class label to an instance based on observations on properties of the instance.

The naive Bayes model is one of the simplest restricted probabilistic graphical models; see Friedman, Geiger & Goldszmidt (1997) who cite Duda & Hart (1973) and Langley, Iba & Thompson (1992). The naive Bayes model is a popular model due to its high representational and computational simplicity while maintaining an impressive performance on classification tasks.

Since the naive Bayes model is most commonly used for classification problems, we will describe the model from this point of view. We consider the task of classifying a set of instances into a predefined set of classes based on observations on properties of the instances. Let C be the class variable with one state for each possible class and let $\mathcal{I} = \{I_1, \ldots, I_n\}$ be the set of variables (also known as *attributes*, *indicators*, and *features*) where each variable represents a property that we can possibly observe and have decided to include in our model.

The structure of the naive Bayes model is the reason for the simplicity and efficiency of the model. The structure of the naive Bayes model is illustrated in Fig. 7.14 where the class variable C is the only parent of each attribute and no other structure is present in the graph. The naive Bayes model assumes conditional pairwise independence of the attributes given the class. This is a rather strong but often useful assumption.

The set of (conditional) probability distributions induced by the naive Bayes model consists of the prior distribution $P(C)$ on the class variable and the conditional probability distribution $P(I_i \mid C)$ on the attribute I_i given the class for all $i = 1, \ldots, n$. The naive Bayes model induces a joint probability distribution over the class and attributes as

$$P(\mathcal{X}) = P(C, I_1, \ldots, I_n) = P(C) \prod_{i=1}^{n} P(I_i \mid C).$$

Notice that this implies that the representational complexity of the model is linear in the number of attributes.

Fig. 7.15 A naive Bayes
model for classifying
mushrooms

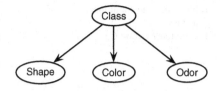

Probabilistic inference in a naive Bayes model consists of computing the conditional probability distribution $P(C\,|\,\varepsilon)$ where ε consists of observations on a subset of the attributes of the instance to be classified. For any set of observations $\varepsilon = \{i_1, \ldots, i_m\}$, we may calculate the likelihood of the class as

$$L(C\,|\,\varepsilon) = P(\varepsilon\,|\,C) = \prod_{i \in \varepsilon} P(i\,|\,C).$$

The posterior of the class is computed from the product of the prior and the likelihood by normalization $P(C\,|\,\varepsilon) = \alpha L(C\,|\,\varepsilon)P(C)$, where $\alpha = P(\varepsilon)^{-1} = (\sum_C L(C\,|\,\varepsilon)P(C))^{-1}$, or expressed via Bayes' rule as

$$P(C\,|\,\varepsilon) = \eta(P(\varepsilon\,|\,C)P(C)) = \frac{P(\varepsilon\,|\,C)P(C)}{P(\varepsilon)}.$$

Despite its simplicity and strong assumption of pairwise independence of the attributes given the class, the naive Bayes model has in practice been shown to have excellent performance on (many) classification tasks. This makes the naive Bayes model popular. The following example illustrates the most common application of the naive Bayes model.

Example 7.7 (Classification of Mushrooms). Consider the task of classifying a mushroom as either edible or poisonous based on observations on the shape, color, and odor of the mushroom. This is a classic classification problem. We make observations on the mushroom to identify it as either edible or poisonous.

Figure 7.15 shows a naive Bayes model for this classification task. The model has class variable Class and feature variables Color, Odor, and Shape. The class variable is the direct parent of each feature variable, and no other structure is present.

The class variable has states dom(Class) = (Edible, poisonous), while the feature variables have states dom(Odor) = (none, almond, spicy), dom(shape) = (flat, bell, convex), and dom(Color) = (brown, white, black).

The prior distribution on Class specifies the frequency of edible and poisonous mushrooms, while the conditional distribution of each feature variable specifies the distribution of the feature given the mushroom class. Table 7.10 shows the distribution $P(\text{Odor}|\text{Class})$. The distribution of each of the other feature variables is similar.

Table 7.10 The conditional		Odor		
probability distribution $P(\text{Odor}\,|\,C)$	C	none	almond	spicy
	edible	0.902	0.0979	0.0001
	poisonous	0.173	0.001	0.826

Each time a mushroom is picked up, the features of the mushroom are observed and entered into the model as evidence. After inference, the model returns the probability that the mushroom is edible.   □

There exist other classes of restricted probabilistic graphical models than the naive Bayes model. For instance, the tree-augmented naive Bayes model (Friedman et al. 1997) appears as a natural extension of the naive Bayes model, while the hierarchical naive Bayes model (Zhang 2004) is another extension of the naive Bayes model.

In Sect. 8.3.4, the task for learning structure restricted probabilistic graphical models such as the naive Bayes model and the tree-augmented naive Bayes model are described in detail.

How to Implement This Technique

The naive Bayes modeling technique is implemented as follows:

1. Let C be the class variable with one state for each possible class.
2. Let $\mathfrak{I} = \{I_1, \ldots, I_n\}$ be the set of feature variables.
3. Let C have no parents and let it be the only parent of each feature variable I_i. In this way, C becomes the intermediate variable in a serial connection with each pair of feature variables.
4. Define the prior probability distribution $P(C)$ such that it encodes the relative frequency of each class.
5. For each $I_i \in \mathfrak{I}$, define the conditional probability distribution $P(I_i\,|\,C)$ such that it encodes the relative frequency of each state of the feature given each state of the class variable.

If data are available, then it may be an advantage to estimate the prior and conditional probability distributions $P(C)$ and $P(I_1\,|\,C), \ldots, P(I_n\,|\,C)$ from data.

7.2 Probability Distribution-Related Techniques

In this section we consider modeling techniques related to the specification of probability distributions of a probabilistic network. In particular we consider measurement error, expert opinions, node absorption, setting a value by intervention, independence of causal influence, and mixture of Gaussian distributions.

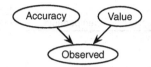

Fig. 7.16 The observed value of a phenomenon is a function of the accuracy of the measurement and the actual value of the measured phenomenon

Fig. 7.17 The measured temperature is a function of the quality of the thermometer and the actual temperature

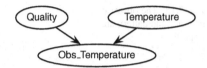

7.2.1 Measurement Uncertainty

Probabilistic networks are well-suited models for reflecting properties of problem domains with some kind of uncertainty. The sources of uncertainty may be many and diverse. In this section we consider a modeling technique for representing uncertainty related to measurements. Measurements and observations on the state of the problem domain such as, for instance, sensor readings and noisy observations are subject to uncertainty. In some situations it may be important to capture and represent the uncertainty explicitly in a probabilistic model.

Figure 7.16 illustrates a modeling technique that captures measurement uncertainty. The variable Value represents the actual value of the phenomenon being measured, the variables Observed and Accuracy represent the observed value of the phenomenon and the accuracy with which observations are made, respectively.

Example 7.8 (Explicit Representation of Uncertainty). Assume we would like to measure the temperature in a room. The *true* temperature is unknown, but we may use a thermometer to get an estimate of the temperature in the room. Assume we have two different thermometers to choose from: one thermometer of low quality and another thermometer of high quality. The high-quality thermometer offers more accurate estimates of the temperature.

Figure 7.17 shows the structure of a model with variables Obs_Temperature, Quality, and Temperature. Assume that the three variables have domains dom(Obs_Temperature) = (low, medium, high), dom(Quality) = (low, high), and dom(Temperature) = (low, medium, high). Table 7.11 shows the conditional probability distribution P(Obs_Temperature | Quality, Temperature).

Notice how the distribution over Obs_Temperature depends on the quality of the thermometer used to measure the temperature. This reflects the accuracy of each thermometer. □

Table 7.11 The conditional probability distribution $P(\text{Obs\_Temperature} \mid \text{Quality, Temperature})$

		Obs_Temperature		
Quality	Temperature	low	medium	high
low	low	0.6	0.3	0.1
low	medium	0.2	0.6	0.2
low	high	0.1	0.3	0.6
high	low	0.9	0.1	0
high	medium	0.05	0.9	0.05
high	high	0	0.1	0.9

Fig. 7.18 The observations on color and pattern are imperfect

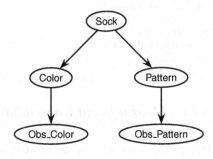

Table 7.12 The conditional probability distribution $P(\text{Obs\_Color} \mid \text{Color})$

	Obs_Color	
Color	blue	black
blue	0.9	0.1
black	0.05	0.95

An explicit representation of the accuracy with which observations are made is not always necessary.

Example 7.9 (Implicit Representation of Uncertainty). Example 7.5 on page 202 illustrates how to enforce the fact that there are two socks of each type (airplane and golf). To classify the four socks, we make observations on the color and pattern of each sock. Color and pattern are indicator variables for the type of sock.

The observations on color and pattern are imperfect due to the washing. Figure 7.18 shows the model structure for classifying a single sock based on (imperfect) observations on color and pattern.

The conditional probability distribution $P(\text{Obs\_Color} \mid \text{Color})$ is shown in Table 7.12. Notice that blue is observed as black in 10 % of the cases and black is observed as blue in 5 % of the cases.   □

The measure uncertainty modeling technique is closely related to the measurement idiom, see Sect. 6.3.2 on page 156.

How to Implement This Technique

The measurement uncertainty modeling technique is implemented as follows.

1. Let variable Value represent the actual value of the phenomenon being measured.
2. Create variables Observed and Accuracy representing the observed value of the phenomenon and the accuracy with which observations are made, respectively.
3. Let Value and Accuracy be the parents of Observed.
4. Let Observed have one state for each possible observation of Value.
5. Let Accuracy have one state for each possible level of accuracy of the observation on Value.
6. Define the prior probability distribution $P(\mathsf{Accuracy})$ such that it encodes the relative frequency of each possible level of accuracy.
7. Define the conditional probability distribution $P(\mathsf{Observation}\,|\,\mathsf{Accuracy},$ Value) such that it encodes the relative frequency of each possible observation given the level of accuracy and the actual value.

7.2.2   Expert Opinions

The specification of the parameters of a probabilistic network is often based on knowledge elicitation from problem domain experts. Typically, a knowledge engineer interviews one or more experts in order to assess the values of model parameters. In some cases, when the elicitation of parameters is based on assessments from a group of experts, it is advantageous that any differences in the assessed values are represented explicitly in the model. This is, for instance, useful when the group of experts are distributed physically and when the model is developed iteratively.

A conditioning (or auxiliary) variable can select among the opinions of different experts expressed in the probability assessments of a single variable. The conditioning variable is a parent of the variable of interest and has one state corresponding to each expert. The prior distribution of the auxiliary value will assign a weight to the experts represented in the auxiliary variable. Different auxiliary variables need not have the same set of states. The following example illustrates the modeling technique on a simple example.

Example 7.10 (Expert Opinions: Chest Clinic).   Consider the quantification of the chest clinic example (Example 4.2 on page 73). Assume the model is constructed by elicitation of knowledge from two experts Bill and John. Consider the elicitation of conditional probability distribution $P(\mathsf{Bronchitis}\,|\,\mathsf{Smoker})$, and assume that Bill and John have different opinions on this distribution.

To reflect the different opinions of the experts, we construct the model structure shown in Fig. 7.19 where dom(Experts) = (Bill, John) representing the two experts.

Table 7.13 shows the distribution $P(\mathsf{Bronchitis}\,|\,\mathsf{Smoker},\mathsf{Experts})$. The distribution encodes the different opinions of the experts on the conditional probability distribution, whereas the prior distribution $P(\mathsf{Experts})$ encodes the reliability of the experts.

Fig. 7.19 The variable
Experts has one state for each
expert

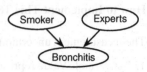

Table 7.13 The specification
of the conditional probability
distribution
P(Bronchitis | Smoker, Experts)

		Bronchitis	
Experts	Smoker	false	true
bill	false	0.7	0.3
bill	true	0.4	0.6
john	false	0.8	0.2
john	true	0.3	0.7

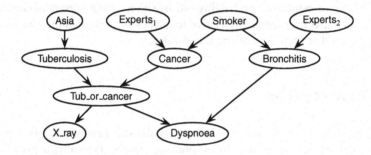

Fig. 7.20 A graph specifying the independence and dependence relations of the Asia example

The model captures the opinions of experts Bill and John using the Experts
variable to condition the conditional probability distribution they have different
opinions on.

One expert node is introduced for each conditional probability distribution
elicited from domain expert knowledge. Figure 7.20 illustrates how two groups
of experts have been consulted to elicit the conditional probability distributions
of Bronchitis and Cancer. By introducing multiple expert nodes, we assume the
elicitation of different conditional probability distributions to be independent. □

The model in Example 7.10 has an explicit representation of the opinions of the
two experts. In some situations, it is desirable not to have an explicit representation
of expert opinions in the model. This can be achieved by eliminating the variables
representing different experts from the model. This is described in Sect. 7.2.3.

How to Implement This Technique

The expert opinions modeling technique is implemented as follows:

1. Let $P(X | \text{pa}(X))$ be the conditional probability distribution assessed from a
   group of experts (one instance of $P(X | \text{pa}(X))$ is assessed from each expert).
2. Create a discrete random variable Experts with one state for each expert.

3. Let Experts be a parent of X.
4. Define the prior probability distribution $P(\text{Experts})$ such that it encodes the reliability of the experts.
5. Define the conditional probability distribution $P(X \,|\, \text{pa}(X), \text{Experts})$ such that for each state of Experts it encodes the assessment of $P(X \,|\, \text{pa}(X))$ given by the corresponding expert.

7.2.3 Node Absorption

Node absorption is the process of eliminating a variable from a model by arc reversals and barren variable eliminations. Recall that in Sect. 3.4.1.1 on page 56, we illustrated the application of Bayes' rule as an arc reversal operation, while in Sect. 5.1.1 on page 116, we considered repeated applications of arc reversal as an inference process.

The node absorption method may also be a useful tool in the model development process. Node absorption may be used to eliminate variables from a model which for one reason or another should not be included in the final model. If efficiency of probabilistic inference in a probabilistic network is of high priority, it may be worthwhile to eliminate variables that are neither observed nor the target of inference. In Sect. 5.1.1, we denoted a variable that is neither an evidence variable nor a target variable as a *nuisance variable*.

Example 7.11 (Node Absorption: Expert Opinions). Consider Example 7.10 on page 211 where $P(\text{Bronchitis} \,|\, \text{Smoker})$ has been elicited from the two experts Bill and John. From the example, we know that Bill and John disagree slightly on the strength of the dependence relation between Bronchitis and Smoker. This is captured by the graph of Fig. 7.19.

For different reasons (e.g., political), we would like to eliminate the intermediate variable Experts from the model while maintaining the underlying dependence relations between the remaining variables. This can be achieved using node absorption.

Since Experts has Bronchitis as its only child, a single arc reversal operation is sufficient to absorb Experts. Once the arc (Experts, Bronchitis) is reversed, Experts is barren and can therefore be removed from the graph without changing the dependence relations between the remaining variables in the graph.

If we assume that we have equal trust in the two experts, then Table 7.14 shows the conditional probability distribution $P(\text{Bronchitis} \,|\, \text{Smoker})$ after absorbing Experts from the distribution shown in Table 7.13.

The prior distribution $P(\text{Experts})$ can be interpreted as specifying our relative trust in the two experts. In the example, we have used a uniform distribution.   □

The order in which arcs are reversed may be constrained by the structure of the graph. That is, the sequence of arc reversals should be performed such that all intermediate graphs are acyclic. In addition, the order in which variables are

Table 7.14 The conditional probability distribution $P(\text{Bronchitis} \mid \text{Smoker})$ after absorbing Experts from the distribution shown in Table 7.13

Smoker	Bronchitis	
	false	true
false	0.75	0.25
true	0.35	0.65

absorbed and arcs are reversed may impact the size of the parent sets in the resulting graph.

How to Implement This Technique

The node absorption modeling technique is implemented as follows:

1. Let X be the variable to be eliminated by node absorption.
2. Let $\text{ch}(X)$ be the direct successors of X, that is, the children of X.
3. For each $Y \in \text{ch}(X)$ reverse the link (X, Y) according to the arc reversal operation. Traverse $\text{ch}(X)$ in topological order.
4. Eliminated X as a barren variable, that is, simply remove X and incoming links from the model.

Node absorption may be implemented as a single step operation in probabilistic network editor software.

7.2.4   Set Value by Intervention

An important distinction should be made between a passive observation of the state of a variable and an active action forcing a variable to be in a certain state. A passive observation of a variable impacts the beliefs of the ancestors of the variable, whereas an active action enforcing a certain state on a variable does not under the assumption of a causal ordering (see Sect. 4.2 on decision making under uncertainty). We refer to this type of active action as an *intervention*. When we make a passive observation on a variable, this produces a likelihood on the parents of the variable. This should not be the case when the value of a variable is set by intervention. The instantiation of a decision variable in an influence diagram is an example of this type of intervention.

In some situations it is undesirable to model active actions forcing a variable to be in a certain state as a decision in an influence diagram. Instead of modeling the situation using decision variables, a simple modeling technique may be used. The modeling technique is illustrated in Fig. 7.21.

Fig. 7.21 Modeling the
option of setting a value of B
by intervention

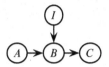

Table 7.15 (a) The
conditional probability
distribution $P(B \mid A)$. (b) The
conditional probability
distribution $P(B \mid I, A)$

a

A	B	
	false	true
false	0.9	0.1
true	0.2	0.8

b

I	A	B	
		false	true
no intervention	false	0.9	0.1
no intervention	true	0.2	0.8
false	false	1	0
false	true	1	0
true	false	0	1
true	true	0	1

In Fig. 7.21, we illustrate a situation where the value of the random variable B may be set by intervention. The causal properties of the example are such that the variable A has a causal impact on B which in turn has a causal impact on C. The variable I captures the property that the state of variable B may be set by intervention. Assuming Table 7.15(a) shows the conditional probability distribution $P(B \mid A)$, then Table 7.15(b) shows the distribution $P(B \mid A, I)$.

This construction of $P(B \mid I, A)$ implies that $C \perp\!\!\!\perp A \mid I = i$ where $i \neq$ no intervention, that is, setting I to a value different from no intervention makes A and C independent. Thus, if we enforce a certain state on B by selecting a state of I (different from no intervention), then observing C subsequently will not change the belief in A. In general, the conditional probability distribution $P(B \mid I, A)$ may be defined as

$$P(B = b \mid A, I = i) = \begin{cases} P(b \mid A) & \text{if } i = \text{no intervention} \\ 1 & \text{if } b = i \\ 0 & \text{otherwise} \end{cases} \tag{7.3}$$

where $\text{dom}(I) = \text{dom}(B) \cup \{\text{no intervention}\}$.

It is important to notice that when the state of B is observed, the observation is enforced by setting the state of B, whereas if the state of B is set by intervention, then I is instantiated to the corresponding state. When B is not observed, I is in the state no intervention.

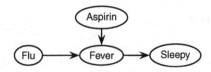

Fig. 7.22 Taking an aspirin forces the fever to a certain level. Subsequent observations on Sleepy should not change our belief in Flu

Example 7.12 (Set Value by Intervention).   Figure 7.22 depicts a simple model for reasoning about a common medical situation. The model captures the direct causal influence of flu on fever and the direct causal impact of fever on sleepiness. These events are represented by the variables Flu, Fever, and Sleepy, respectively.

In addition to the aforementioned variables, the model has the variable Aspirin. This variable represents the event that the patient takes an aspirin to reduce fever to a certain level. Once an aspirin has been taken, an observation on Sleepy will be uninformative with respect to Flu. This behavior may be enforced as described above.   □

How to Implement This Technique

The set value by intervention modeling technique is implemented as follows.

1. Let X be the random variable that may be set by intervention.
2. Create a random variable I with $\dom(I) = \dom(B) \cup \{$no intervention$\}$.
3. Let I be a parent of X.
4. Define the prior probability distribution $P(I)$ such that it encodes the relative frequency of setting each state of X and no intervention.
(5) Define the conditional probability distribution $P(X \,|\, \pa(X), I)$ according to (7.3).

7.2.5   Independence of Causal Influence

In this section we consider how a special kind of structure within a conditional probability distribution may be exploited to reduce the complexity of knowledge elicitation from exponential in the number of parents to linear in the number of parents. The property we consider is known as *independence of causal influence* (Heckerman 1993).

In an independence of causal influence model, the parent variables of a common child variable interact independently on the child. With a slight abuse of terms, the parents are sometimes said to be *causally independent*. All variables in an independence of causal influence model are assumed to be discrete random

Fig. 7.23 The causal
influence of C_i on E is
independent of the causal
influence of C_j on E (for
$i \neq j$)

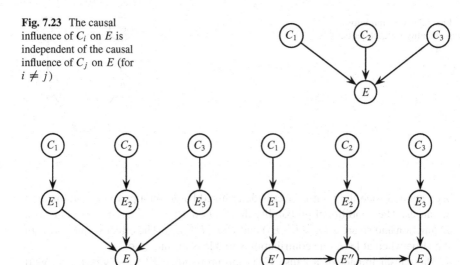

Fig. 7.24 Two model structures for capturing independence of causal influence

variables. The common child variable is denoted E, and it is referred to as the *effect variable*. The parent variables are denoted $C_1, \ldots, C_n$, and they are referred to as *cause variables* or *causes*, see Fig. 7.23.

The cause variables $C_1, \ldots, C_n$ may cause an abnormality modeled as the effect variable E to be present at a certain level. The states of the effect variable E specify different levels of the abnormality. The states of E are ordered in the sense that they represent different levels of abnormality and such that a *designated state* indicates that the abnormality is absent. Similarly, each of the causes have an absence state corresponding to *no impact* on the effect variable. The principle of an independence of causal influence model is that the causal impact of each cause is independent of the causal impact of all other causes.

In this section we consider the Boolean independence of causal influence model known as the *Noisy-OR model* (Pearl 1988). The Noisy-OR model is a commonly used example of a model for local conditional probability distributions that depends on fewer parameters than the total number of combinations of fa(E) (Laskey 1993).

In the Noisy-OR model, the effect variable E and the variables $C_1, \ldots, C_n$ are Boolean variables (i.e., binary discrete random variables with states false and true). The designated state is false. The causal impact of each cause C_i is independent of the causal impact of any other cause C_j for $i \neq j$. Figure 7.24 illustrates two different ways in which this independence may be modeled explicitly. Each variable E_i has the same state space as E, and it captures the contribution from cause C_i to the value of E. Each variable E_i is referred to as a *contribution variable* and $P(E_i | C_i)$ captures the causal impact of C_i on E. In the left part of the figure, the total impact on the effect variable is the disjunction of all causes, whereas in the right part of the figure, the temporal transformation modeling technique has

Fig. 7.25 One inhibitor
probability for each parent C_i
of the effect E

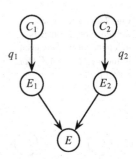

been applied such the total impact is determined based on a pairwise combination
of causes. The conditional probability distribution $P(E \mid E_1, \ldots, E_n)$ is defined as
disjunction and so are $P(E'' \mid E', E_2)$ and $P(E \mid E'', E_3)$. The effect variable E is in
state true when at least one contribution variable is in state true.

The causal impact of a cause C_i is the probability $P(E = \text{yes} \mid C_i = \text{yes})$
whereas $P(E = \text{yes} \mid C_i = \text{no}) = 0$. We denote $P(E = \text{yes} \mid C_i = \text{yes}) = 1 - q_i$
where q_i is referred to as the *inhibitor probability*, see Fig. 7.25.

In some domains there may be a need to have a *leak probability* $1 - q_0$
representing the probability of $E = \text{true}$ when all causes $C_1, \ldots, C_n$ are absent
where q_0 is known as the *default inhibitor*. A leak probability may be implemented
by introducing as a separate Boolean cause variable C_0 instantiated to state true.
The leak variable C_0 represents the set of causes not modeled explicitly in the
network. In this way, the leak variable can be used to enforce the closed-world
assumption. The leak probability is assigned as the probability that the effect will
occur in the absence of any of the causes $C_1, \ldots, C_n$ that are explicitly represented
in the network (Pradhan, Provan, Middleton & Henrion 1994).

Let us consider how the reduction from exponential to linear in the number
of parents is achieved. We may consider the conditional probability distribu-
tion $P(X_i \mid \text{pa}(X_i))$ as parameterized over a vector Θ_i of parameters θ_{ijk} with one
component for each possible value of X_i and combination of $\text{pa}(X_i)$ such that

$$P(x_{ijk} \mid \pi_{ij}, \Theta_i) = \frac{\theta_{ijk}}{\sum_k \theta_{ijk}},$$

where π_{ij} is the jth configuration of $\text{pa}(X_i)$. The above formula is an unrestricted
local conditional probability distribution. The distribution depends on $|\mathcal{X}_{\text{fa}(X_i)}|$
parameters.

In an independence of causal influence model, the conditional probability
distribution $P(X_i \mid \text{pa}(X_i))$ can be specified using a parameter vector Θ_i that grows
in size only linearly in the number of parents, $|\text{pa}(X_i)|$, rather than linearly in the
number of configurations of the parents, $|\mathcal{X}_{\text{pa}(X_i)}|$; that is, exponential in $|\text{pa}(X_i)|$.

For the Noisy-OR model, it is straightforward to determine the conditional
probability distribution $P(E \mid \text{pa}(E), \Theta_E)$ given a specific parameter vector Θ_E for
the model. In that case $P(E \mid C_1, \ldots, C_n)$ can be specified as

Fig. 7.26 Sore throat may be caused by both angina and cold

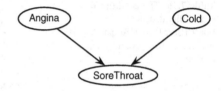

$$P(E = \text{true} \mid C_1 = c_1, \dots, C_n = c_n, \Theta_E) = 1 - \theta_0 \prod_{C_i = \text{true}} \theta_i,$$

where θ_0 is the default inhibitor, θ_i is the inhibitor for C_i, and $\Theta_E = \{\theta_0, \theta_1, \dots, \theta_n\}$. From this, it follows that

$$P(E = \text{false} \mid C_1 = c_1, \dots, C_n = c_n, \Theta_E) = \theta_0 \prod_{C_i = \text{true}} \theta_i.$$

The following example illustrates how independence of causal influence may be exploited to simplify the knowledge elicitation process.

Example 7.13 (Noisy-OR: Sore Throat). A physician wants to diagnose her patients with respect to diseases causing a sore throat. For simplicity of the example, we assume the physician is mainly interested in modeling the causal effect of cold and angina on sore throat. In addition to cold and angina, there are other potential causes of a sore throat. These other causes are not to be represented explicitly in the model though.

Thus, initially the model consists of three variables: SoreThroat, Angina, and Cold. All variables are Boolean with states false and true. Figure 7.26 shows the structure of the model.

The synergy between Angina and Cold with respect to their combined effect on SoreThroat is assumed to be minimal. Thus, we may use the Noisy-OR model to specify and represent the conditional probability distribution of SoreThroat given Angina and Cold where all other implicit causes of sore throat are captured by the background event.

First, assume the inhibitor probabilities are $q^{\text{Angina}} = 0.1$ and $q^{\text{Cold}} = 0.2$, while the default inhibitor is one (i.e., there are no other causes of sore throat). The combined impact of Angina and Cold on SoreThroat is computed as

$$P(\text{SoreThroat} = \text{false} \mid \text{Angina} = \text{false}, \text{Cold} = \text{false}, \Theta_E) = 1,$$

$$P(\text{SoreThroat} = \text{false} \mid \text{Angina} = \text{false}, \text{Cold} = \text{true}, \Theta_E) = 0.2,$$

$$P(\text{SoreThroat} = \text{false} \mid \text{Angina} = \text{true}, \text{Cold} = \text{false}, \Theta_E) = 0.1,$$

$$P(\text{SoreThroat} = \text{false} \mid \text{Angina} = \text{true}, \text{Cold} = \text{true}, \Theta_E) = 0.1 * 0.2.$$

Table 7.16 The conditional probability distribution $P(\text{SoreThroat} \mid \text{Angina}, \text{Cold})$ with a zero background event probability

		SoreThroat	
Angina	Cold	false	true
false	false	1	0
false	true	0.2	0.8
true	false	0.1	0.9
true	true	0.02	0.98

Table 7.17 The conditional probability distribution $P(\text{SoreThroat} \mid \text{Angina}, \text{Cold})$ with a nonzero background event probability

		SoreThroat	
Angina	Cold	false	true
false	false	0.95	0.05
false	true	0.19	0.81
true	false	0.095	0.905
true	true	0.019	0.981

Table 7.16 shows the distribution $P(\text{SoreThroat} \mid \text{Angina}, \text{Cold})$.

Next, assume the background inhibitor is 0.95 (i.e., the probability that sore throat is caused by the background event (other causes not represented in the model) is 0.05 such that $q_0 = 0.95$). The combined impact of Angina, Cold, and the background event on SoreThroat given is computed as

$$P(\text{SoreThroat} = \text{false} \mid \text{Angina} = \text{false}, \text{Cold} = \text{false}, \Theta_E) = 0.95,$$

$$P(\text{SoreThroat} = \text{false} \mid \text{Angina} = \text{false}, \text{Cold} = \text{true}, \Theta_E) = 0.95 * 0.2,$$

$$P(\text{SoreThroat} = \text{false} \mid \text{Angina} = \text{true}, \text{Cold} = \text{false}, \Theta_E) = 0.95 * 0.1,$$

$$P(\text{SoreThroat} = \text{false} \mid \text{Angina} = \text{true}, \text{Cold} = \text{true}, \Theta_E) = 0.95 * 0.1 * 0.2.$$

Table 7.17 shows the distribution $P(\text{SoreThroat} \mid \text{Angina}, \text{Cold})$.

By exploiting the independence of causal influence between Cold and Angina, the number of parameters to elicit has decreased from four to two. This may seem to be an insignificant reduction. However, if we consider the case where ten different causes of SoreThroat are to be represented explicitly in the model, then the number of parameters to elicit is reduced from 1,024 to 10.   □

The benefit of independence of causal influence becomes even more apparent when the effect has a large number of causes as the number of parameters grows linearly with the number of causes. The advantage of independence of causal influence is an exponential decrease in the number of parameters to elicit. The disadvantage of independence of causal influence is that any synergy between causes (with respect to their combined impact on the effect variable) is ignored.

Using independence of causal influence in conjunction with parent divorcing may reduce the complexity of inference exponentially.

Srinivas (1993) discusses a generalization of the Noisy-OR model to nonbinary variables. Pradhan et al. (1994) and Diez (1993) have considered in detail the Noisy-MAX model as a generalization of the Noisy-OR model to the case in which each variable is allowed to have a finite discrete state space. In the Noisy-MAX model, the max operator specifies the combination of the cause variables. In the Noisy-MAX model, the probability distribution of the effect variable E given its parent causes can be expressed as

$$P(E \mid C_1, \ldots, C_n) = \sum_{\max(E_1, \ldots, E_n)} \prod_{i=1}^{n} P(E_i \mid C_i).$$

One prerequisite for using the Noisy-MAX model is that the variable state spaces are ordered as, for example, (absent, mild, moderate, severe). In addition, each cause variable must have a *distinguished* (or absent) state designating an influence of the cause on the effect variable.

How to Implement This Technique

The independence of causal influence modeling technique is implemented as follows:

1. Let $\{C_1, \ldots, C_n\}$ be the set of causes of effect variable E.
2. Assume the impact of $C_1, \ldots, C_n$ on E can be modeled as a Noisy-OR model. Hence, $C_1, \ldots, C_n$ and E are Boolean variables.
3. Create one Boolean contribution variable E_i for each $C_i \in \{C_1, \ldots, C_n\}$.
4. Let each E_i be a child of C_i and a parent of E.
5. For each C_i, define the conditional probability distribution $P(E_i \mid C_i)$ such that $P(E_i = \text{true} \mid C_i = \text{true})$ is the probability of $E = \text{true}$ given $C_i = \text{true}$ and $C_j = \text{false}$ for $i \neq j$ and $P(E_i = \text{false} \mid C_i = \text{false}) = 1$.
6. Define the conditional probability distribution $P(E \mid E_1, \ldots, E_n)$ as disjunction (i.e., or).

Once the independence of causal influence modeling technique has been applied, it may be an advantage to use the parent divorcing modeling technique (see Sect. 7.1.1) to reduce the number of parents of the effect variable.

7.2.6 Mixture of Gaussian Distributions

When modeling problem domains with continuous entities, a decision on how to represent the continuous entities in a network has to be made. One option is to represent a continuous entity as a discrete variable with states representing intervals for the continuous entity. For instance, we may choose to represent temperature as a discrete variable with three states: low, medium, and high. In other cases, we may choose to approximate the distribution of a continuous entity using the conditional linear Gaussian distribution.

Fig. 7.27 Approximation of
a continuous distribution
using the MoGs modeling
technique

A third option is presented in this section. The third option is to approximate the continuous distribution of a variable using a mixture of Gaussian distributions (MoGs). This option is interesting as it is well known that mixtures of Gaussian distributions can approximate any probability distribution; see Shenoy (2006) who cites Titterington, Smith & Makov (1995).

An MoGs is a sum of Gaussian distributions where each component is weighted by a number between zero and one such that the sum of the weights is one, that is, the weights are probabilities. Assume X is a continuous variable with a probability distribution that can be approximated using the MoGs:

$$f(x) = \sum_{i=1}^{n} p_i \cdot \mathbb{N}(\alpha_i, \gamma_i), \qquad (7.4)$$

where $\alpha_i, \gamma_i \in \mathbb{R}$ and $0 \le p_i \le 1$ such that $\sum_i p_i = 1$ are the mean, variance, and weight of the i'th component in the mixture.

To approximate the probability distribution on X using (7.4), a selector variable S with n states is introduced as a parent of X. The variable X becomes a continuous variable with a conditional linear Gaussian distribution, see Sect. 4.1.2. Each state s_i of S corresponds to one component $p_i \cdot \mathbb{N}(\alpha_i, \gamma_i)$ in the mixture. The prior distribution on S is $P(S = s_i) = p_i$ while $X \mid s_i \sim \mathbb{N}(\alpha_i, \gamma_i)$. Figure 7.27 illustrates the use of the MoGs modeling technique on the distribution for X.

Using MoGs, the network becomes either a CLG Bayesian network or a CLQG influence diagram.

Example 7.14 (Mixture of Gaussian Distributions). A Gamma$(2, 2)$ distribution can, for instance, be approximated using a two-component mixture of MoGs such as

$$f(x) = 0.609 \cdot \mathbb{N}(4.57, 2.37) + 0.391 \cdot \mathbb{N}(1.93, 1.12). \qquad (7.5)$$

Figure 7.28 shows the result of approximating the Gamma$(2, 2)$ distribution with the above two-component mixture of Gaussian distributions.

The two-component approximation in (7.5) produces a reasonable fit to the Gamma$(2, 2)$ distribution. Whether or not the fit is of sufficient quality depends on the problem domain and application. $\square$

The MoGs modeling technique introduces a discrete random variable with one state for each component in the mixture. Approximating continuous distributions using the MoGs modeling technique is not necessarily simple and may produce networks where belief updating is computationally intensive.

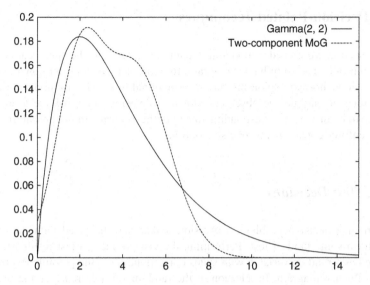

Fig. 7.28 A two-component mixture approximation of the Gamma(2, 2) distribution

How to Implement This Technique

The mixture of Gaussian distributions modeling technique is implemented as follows:

1. Let X be the variable of the probability distribution to approximate.
2. Assume the probability distribution of X can be approximated using the MoGs

$$f(x) = \sum_{i=1}^{n} p_i \cdot \mathbb{N}(\alpha_i, \gamma_i).$$

3. Create a discrete variable S with n states.
4. Let S be the parent of X in the network with $P(S = s_i) = p_i$.
5. For each state, s_i of S set $X \mid s_i \sim \mathbb{N}(\alpha_i, \gamma_i)$.

The process of identifying the number of components as well as the parameterization and weight of each component is not trivial.

In this section we have considered the case of approximating the prior distribution of a continuous variable with no parents. Shenoy (2006) presents a methodology for belief updating in hybrid Bayesian networks (i.e., Bayesian networks with both continuous and discrete variables and with no restrictions on the model structure) based on approximating distributions using MoGs. Shenoy (2006) gives examples on how to approximate different types of distributions using MoGs. This includes approximating the distribution of a discrete child of a continuous variable. Poland (1994) has presented an algorithm for identifying MoGs using the EM algorithm (see Sect. 8.5.1).

7.3   Decision-Related Techniques

In this section we consider modeling techniques related to the specification of a decision problem as an influence diagram. In particular, we consider how to model test decisions, how to exploit missing informational links, how to model variables which may or may not be observed prior to a decision, how to force a decision variable to be in a state corresponding to a hypothesis of maximum probability, and how to enforce constraints on decision options.

7.3.1   Test Decisions

As part of a decision problem, a decision maker may be faced with the option to perform some kind of test. Performing the test produces a test result which is modeled as a random variable with states corresponding to the possible test results in an influence diagram. In addition to the random variable representing the test result, the influence diagram has a decision variable with states representing whether or not the test is performed. If the test is performed, then the result of the test (usually) becomes available to the decision maker. If the test, on the other hand, is not performed, then no test result becomes available. The influence diagram may also have a utility function associated with the test specifying the cost of the test. Solving the influence diagram will produce a policy for when to perform the test.

The random variable representing the test result may be an informational parent of another decision variable in the influence diagram. If the test result variable is an informational parent of another decision in the influence diagram, then the variable must be observed prior to this decision. This, however, contradicts the fact that the test result is available only when the test is performed. In this section, we consider two examples that illustrate different approaches to alleviating this contradiction.

Example 7.15 (Oil Wildcatter (Raiffa 1968)). Example 4.5 on page 82 considers an oil wildcatter about to decide whether or not to drill for oil at a specific site. Prior to her decision on whether or not to drill for oil, the oil wildcatter has the option to take seismic soundings to better understand the geological structure of the site. The structure of the oil wildcatter model (Fig. 4.5 on page 82) is repeated in Fig. 7.29 for convenience.

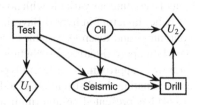

Fig. 7.29 The test result is only available after a test is performed

Table 7.18 The conditional probability distribution $P(\text{Seismic} | \text{Test}, \text{Oil})$

Test	Oil	Seismic		
		closed	open	diffuse
no	dry	0.1	0.3	0.6
no	wet	0.3	0.4	0.3
no	soaking	0.5	0.4	0.1
yes	dry	1/3	1/3	1/3
yes	wet	1/3	1/3	1/3
yes	soaking	1/3	1/3	1/3

Table 7.19 The conditional probability distribution $P(\text{Seismic} | \text{Test}, \text{Oil})$ where Seismic has a no result state

Test	Oil	Seismic			
		closed	open	diffuse	no result
no	dry	0.1	0.3	0.6	0
no	wet	0.3	0.4	0.3	0
no	soaking	0.5	0.4	0.1	0
yes	dry	0	0	0	1
yes	wet	0	0	0	1
yes	soaking	0	0	0	1

There are two informational links in the graph of Fig. 7.29. The link (Test, Drill) from Test to Drill and the link (Seismic, Drill) from Seismic to Drill are both informational links. The former link specifies whether or not the oil wildcatter decided to take seismic soundings prior to the drill decision. On the other hand, the latter link specifies that the value of Seismic is also known when making the drill decision. This cannot, however, be the case when the test is not performed.

We consider two alternative options to correct this problem. Both options consider the specification of the conditional probability distribution $P(\text{Seismic} | \text{Oil}, \text{Test})$.

One option is to specify $P(\text{Seismic} | \text{Oil}, \text{Test} = \text{no})$ as a uniform distribution. The corresponding distribution is shown in Table 7.18. If the oil wildcatter decides not to perform the test, then any observation on Seismic will not affect the belief in Oil (the likelihood potential over Oil induced by the observation on Seismic assigns equal likelihood to all states of Oil due to the uniform distribution).

The other option is to introduce an additional no result state in Seismic. The distribution $P(\text{Seismic} | \text{Oil}, \text{Test} = \text{no})$ is specified such that not performing the test instantiates Seismic to no result. The corresponding distribution is shown in Table 7.19. If the oil wildcatter decides not to perform the test, then Seismic is instantiated to no result.

The latter option is semantically more clear than the former option in the sense that it is easily understood that Seismic should be instantiated to no result when the test is not performed. On the other hand, the latter option increases the complexity of the model by introducing the additional no result state in the Seismic variable. □

Example 7.16 (Aspirin (Jensen 1996)). Example 7.12 on page 216 describes a simple model for reasoning about the effect of flu on fever and the effect of fever on sleepiness. Here we consider this example as a decision problem where the decision maker has to decide on whether or not to take an aspirin.

Fig. 7.30 Prior to deciding
on whether or not to take an
aspirin, we may measure the
temperature

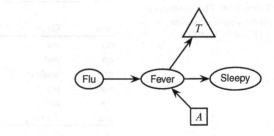

Fig. 7.31 The test for
temperature is modeled as a
decision value with a random
variable as a child specifying
the result of the test

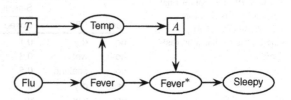

The level of fever may be reduced by taking an aspirin. This is represented by the
decision variable A. Notice that decision variable A is an intervening decision and
recall from the discussion on causality in Sect. 2.4 the important difference between
setting the state of a variable, that is, selecting an option for a decision variable,
and observing the state of a variable. Prior to taking an aspirin, there is the option
to measure temperature. This option is indicated using the triangular-shaped node
with label T in Fig. 7.30.

The test option indicated in Fig. 7.30 by the triangular-shaped node may be
represented using three nodes as indicated in Fig. 7.31. The three nodes represent
decision variable T and random variables Temp and Fever*. The decision variable T
represents whether or not the temperature is measured. The random variable Temp
specifies the temperature measured, and the random variable Fever* represents the
level of fever after taking an aspirin, while the random variable Fever represents the
level of fever prior to taking an aspirin.   □

How to Implement This Technique

The test decisions modeling technique is implemented as follows:

1. Let P be a discrete random variable representing the phenomenon that may be
   measured by a test.
2. Create a decision variable T with two states no test and test corresponding to not
   performing and performing the test, respectively.
3. Create a discrete random variable R representing the result of the test as a child
   of T and P.
4. Let R have one state for each possible test result and the state no result
   representing the event that the test is not performed, that is, $T = $ no test.

5. Define the conditional probability distribution $P(R \mid P, T)$ such that $P(R =$ no result$\mid P, T =$ no test$) = 1$ and $P(R \mid P, T =$ test$)$ encodes the probability of each possible test result given the actual value of phenomenon P.

Instead of using the state no result to specify no test result, a uniform distribution may be used. Furthermore, the modeling technique may be used in combination with the measurement uncertainty modeling technique described in Sect. 7.2.1.

7.3.2 Missing Informational Links

Informational links of an influence diagram define the points at which information is assumed to become available to the decision maker. An informational link (X, D) from a random variable X to a decision variable D specifies that the value of X is known to the decision maker when the decision corresponding to decision variable D is made. The informational links of an influence diagram induce a partial order over the random variables relative to the decision variables. The partial order over random variables is important for the solution of an influence diagram. In essence, the partial order over random variables induces a constraint on the order in which variables may be eliminated when solving the decision model; see Sect. 5.2 for details on solving decision models. Thus, correct specification of informational links is imperative.

When the influence diagram has only a single decision, then informational links can be ignored if the influence diagram is solved for each set of evidence. That is, the influence diagram is solved prior to making the decision each time the influence diagram is used. This implies that the optimal strategy is only implicitly available to the decision maker as the optimal decision is determined for each evidence scenario prior to making the decision. This can be particularly useful if the optimal policy for the decision has a large state space.

Example 7.17 (Missing Informational Links). In Example 7.6 on page 205, we considered the task of monitoring the pregnancy state of a cow. Assume that in addition to the blood and urine tests, we have the option to make a scanning of the cow. A scanning of the cow will produce a more accurate estimation of the pregnancy of the cow. The option to scan the cow introduces the variable Sc with states false and true as a child of Pr.

The pregnancy state of the cow is estimated 6 weeks after the initial insemination of the cow. Based on the observations and the probability distribution of the pregnancy state of the cow, we need to make a decision on whether or not to repeat the insemination of the cow or to wait for another 6 weeks before estimating the pregnancy state of the cow. This decision introduces the decision variable D with states Wait and repeat.

The cost of repeating the insemination is 65 no matter the pregnancy state of the cow. If the cow is pregnant, and we wait, it will cost us nothing, but if the cow is

Table 7.20 The utility
function $U(\text{Pr}, \text{D})$

Pr	D	
false	wait	−95
false	repeat	−65
true	wait	0
true	repeat	−65

(a)   (b)

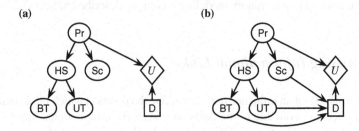

Fig. 7.32 (a) Informational links are unnecessary in influence diagrams with a single decision.
(b) Informational links only clutter up the graph

not pregnant, and we wait, it will cost us another 30 units plus the eventual repeated insemination (that makes a total of 95 for waiting). A blood test has the cost 1 and a urine test has the cost 2. This defines a utility function over variables Pr and D, see Table 7.20.

Figure 7.32a shows the resulting structure of the model. Notice that there are no informational links in the structure.

Since the structure in Fig. 7.32a does not contain any informational links, it does not properly reflect the test options available prior to deciding on whether or not to repeat the insemination.

To capture the three test options, we may introduce an additional no test state in each of the test result variables (BT, UT, and Sc). This would produce the structure shown in Fig. 7.32b.

Alternatively, we may use the fact that the decision problem contains a single decision variable. This allows us to leave out informational links and instantiate the random variables observed prior to the decision as the observations are made. This leaves us with Fig. 7.32a instead of the more cluttered Fig. 7.32b.   □

When informational links are included in the influence diagram, the solution will identify a decision policy specifying an optimal decision option for each configuration of the parents of the decision. Thus, the influence diagram is solved once and *off-line* in the sense that the influence diagram is solved before the decision maker has to make a decision and before any observations are made. On the other hand, it is necessary to resolve the influence diagram each time the decision maker has to make the decision when informational links are not included in the network. The solution process identifies an optimal decision option for the specific set of observations made prior to the decision. The influence diagram is solved *on-line* in

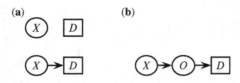

Fig. 7.33 (a) In some situations X is observed prior to D, while in others it is not. (b) By introducing an additional variable, we capture the situation where an observation on X may or may not be available

the sense that the influence diagram is solved when the decision maker has to make the decision and after observations have been made. Solving the influence diagram on-line is often a significantly simpler task than solving it off-line and may be the only option when the influence diagram is complex.

How to Implement This Technique

The missing informational links modeling technique is implemented as follows.

1. Let D be the decision under consideration.
2. Assume observations ε have been made prior to making decision D where $\mathrm{pa}(D) \subseteq \mathcal{X}(\varepsilon)$.
3. Insert ε as evidence and solve the influence diagram.
4. The expected utility associated with each state d of D is $\mathrm{EU}(d \,|\, \varepsilon)$, that is, the expected utility of decision option d given observations ε.

The above steps should be repeated each time observations are made prior to deciding on D.

7.3.3 Missing Observations

The structure of an influence diagram induces a partial order on the random variables of the model relative to the decision variables of the model. The partial order is induced by the informational links of the graph of the influence diagram. An informational link (X, D) from a node representing a random variable X to a node representing a decision variable D specifies that the value of X is observed when decision D is to be made. That is, the value of X is *always* observed prior to decision D.

Figure 7.33a illustrates a typical dilemma a knowledge engineer may be faced with when representing a decision problem as an influence diagram. In some situations, the random variable X is observed prior to the decision represented as D, and in other situations, it is not observed prior to making decision D. In this

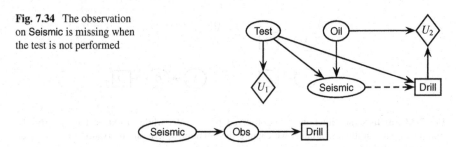

Fig. 7.34 The observation on Seismic is missing when the test is not performed

Fig. 7.35 This graph captures the situation where the result of seismic soundings may not be available

section we describe a modeling technique for solving the dilemma where a random variable X *may or may not* be observed prior to a decision D. This is a typical and frequently occurring situation when considering decision problems with sensor readings or other similar types of observations, which may, for some reason, be missing or lost.

Figure 7.33b illustrates the solution to the dilemma. An auxiliary variable O is introduced as a child of X and a parent of D. The random variable O has the state space of X extended with one additional state, for example, named none. Each state $o \in \text{dom}(O)$ corresponding to state $x \in \text{dom}(X)$ represents an observation of X to the state x, while the additional state none represents the event that X is not observed. The conditional probability distribution $P(O \,|\, X)$ is constructed such that $P(O = o \,|\, X = x) = p$ and $P(O = \text{none} \,|\, X = x) = 1 - p$ where p specifies the probability that the observation on X is made when X is in state x. The following example illustrates the use of this modeling technique.

Example 7.18 (Missing Observations: Oil Wildcatter). In Example 4.5 on page 82, the oil wildcatter has the option to take seismic soundings prior to the drill decision. In this example we will assume that the oil wildcatter is not in full control of the test option. This implies that the test event should be modeled as a random variable. Figure 7.34 shows the resulting structure.

The dashed link from Seismic to Drill indicates that Seismic is only observed when the test is performed. This property can be captured by the approach described above. Figure 7.35 shows the structure which captures the situation where the result of seismic soundings may not be available.

The conditional probability distribution of Obs is shown in Table 7.21. The variable Obs has one state for each state of Seismic and one additional state none representing the event that no result is available. The table specifies the probability that the result of seismic soundings is available to be 0.9.

The variable Obs is always observed. Either it instantiates Seismic to the state representing the seismic soundings result or it carries no information on the test result. $\square$

Table 7.21 The conditional probability distribution $P(\text{Obs}|\text{Seismic})$

Seismic	Obs			
	closed	open	diffuse	none
closed	0.9	0	0	0.1
open	0	0.9	0	0.1
diffuse	0	0	0.9	0.1

Fig. 7.36 Decision D selects a hypothesis of maximum probability

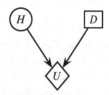

How to Implement This Technique

The missing observations modeling technique is implemented as follows:

1. Let X be the discrete random variable that may or may not be observed prior to decision D.
2. Create a discrete random variable O with state space $\text{dom}(O) = \text{dom}(X) \cup \{\text{none}\}$ representing the observation on X when it is observed and none when it is not.
3. Let X be the parent of O and let O be a parent of D.
4. Define the prior probability distribution $P(O|X)$ such that $P(O = o|X = x) = p$ and $P(O = \text{none}|X = x) = 1 - p$ where p specifies the probability that the observation on X is made when X is in state x.
5. Instantiate O to the state of X when X is observed and instantiate O to the state none when X is not observed.

7.3.4 Hypothesis of Highest Probability

An influence diagram is useful for solving problems of decision making under uncertainty. The variables of an influence diagram consist of a mixture of random variables and decision variables. The random variables are used for representing uncertainty, while the decision variables represent entities under the full control of the decision maker. The state of a random variable may be observable or hidden, while the state of a decision variable is under the full control of the decision maker.

As indicated above there is a fundamental difference between random variables and decision variables. Situations exist, however, where it is useful to have the decision maker select a decision option corresponding to the state of a random variable. In a medical diagnosis situation, for instance, it may be necessary to have the model suggest the most likely diagnosis as the disease with the maximum probability where the presence or absence of diseases are modeled as random variables. Figure 7.36 illustrates a simple modeling technique for representing this

Fig. 7.37 Decision D selects
a hypothesis of maximum
probability

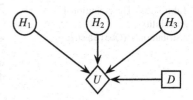

situation. Let D be the discrete decision variable under the full control of the
decision maker and let H be the hypothesis variable such that D and H have the
same (or equivalent) state spaces, that is, $\mathrm{dom}(D) = \mathrm{dom}(H)$. The goal is to assign
the maximum expected utility to the decision option d of D corresponding to the
hypothesis h of H with maximum probability. This is achieved by adding a utility
function U with domain $\mathrm{dom}(U) = \{D, H\}$, assigning utilities to configurations
of H and D as

$$U(h, d) = \begin{cases} 1 & \text{if } h = d \\ 0 & \text{otherwise.} \end{cases}$$

That is, all configurations where the decision variable D and the hypothesis
variable H are in the *same* state are assigned the value one, while all remaining
configurations are assigned the value zero. In effect, each state of D has expected
utility corresponding to (a linear transformation of) the probability of the hypothesis.
Since influence diagrams are solved by selecting the decision option with maximum
expected utility, the optimal decision policy for D will select a hypothesis with
maximum probability.

In the general case, each hypothesis h may be a configuration over a set of
variables such that the utility function has more than one hypothesis variable as
parent. Figure 7.37 illustrates this situation.

Example 7.19 (Hypothesis of Highest Probability). In the chest clinic example
(Example 4.2 on page 73) a physician is diagnosing her patients with respect to
lung cancer, tuberculosis, and bronchitis based on observations of symptoms and
possible causes of the diseases.

Assume the physician would like to select the single diagnosis with highest probability. Figure 7.38 shows the structure of a model where the decision variable D
selects the disease hypothesis with highest probability. The decision variable D has
states Bronchitis, cancer, and tuberculosis.

The utility function $U(T, L, B, D)$ encodes the behavior of the model, and it is
specified as

$$U(T, L, B, D) = \begin{cases} 1 & \text{if } B = \text{yes}, L = \text{no}, T = \text{no and } D = \text{bronchitis} \\ 1 & \text{if } B = \text{no}, L = \text{yes}, T = \text{no and } D = \text{cancer} \\ 1 & \text{if } B = \text{no}, L = \text{no}, T = \text{yes and } D = \text{tuberculosis} \\ 0 & \text{otherwise.} \end{cases}$$

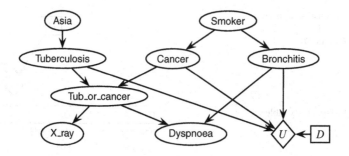

Fig. 7.38 Selecting a disease hypothesis with highest probability

This model will enforce the expected utility function over D to assign the hypothesis with the highest probability with the maximum expected utility. □

How to Implement This Technique

The hypothesis of highest probability modeling technique is implemented as follows:

1. Let H be the random variable for which the hypothesis (i.e., state) of highest probability is to be selected.
2. Create a decision variable D with the same state space as H, that is, such that $\mathrm{dom}(D) = \mathrm{dom}(H)$.
3. Create a utility function U with D and H as its parents.
4. Define the utility function $U(H, D)$ such that

$$U(h, d) = \begin{cases} 1 & \text{if } h = d \\ 0 & \text{otherwise,} \end{cases}$$

where h and d are states of H and D, respectively.

7.3.5 Constraints on Decisions

One of the underlying assumptions of representing and solving a decision making problem with uncertainty using influence diagrams is that the decision maker is in full control of her decision options. It is, however, common that a decision making problem is subject to certain constraints on the decision (and random) variables. We consider the situation where certain configurations of decision variables are illegal in the sense that such configurations should never be optimal.

Fig. 7.39 A constraint on configurations of decisions D_1 and D_2

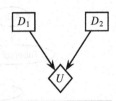

Table 7.22 The utility table $U(D_1, D_2)$

D_1	D_2	
don't sell	don't sell	-100
don't sell	sell	-100
sell	don't sell	-100
sell	sell	0

The basic idea of the approach considered here is to assign an infinitely large negative value to configurations of decision options that are illegal. Since influence diagrams are solved by maximizing the expected utility, decision options with infinitely large negative expected utilities will not be chosen.

It is, however, not possible to specify that a configuration of variables has infinitely large negative expected utility. Instead of using an infinitely large negative value, we may use zero (or a very large negative value). This implies that it may be necessary to make a linear utility transformation to avoid zero expected utilities for any configuration which is not illegal. This transformation of the utility function will preserve the optimal policy.

Example 7.20 (Constraints on Decisions). Assume that two decisions D_1 and D_2 specify two different points in time where the decision maker can choose to sell an old car that needs repair. If both decisions are to keep the car (i.e., don't sell), then a repair cost is induced. If the car is sold at decision D_1, then it is not an option to sell the car at decision D_2.

This implies that options available for the decision maker at decision D_2 are constrained by the decision made at decision D_1. This property can be encoded as a constraint over D_1 and D_2 as illustrated in Fig. 7.39 with utility function $U(D_1, D_2)$ as specified in Table 7.22.

In order to avoid problems related to decision options having zero expected utility due to illegal events, a linear transformation of the utility function can be made. In the example, we may add a constant greater than the numerical value of the cost of repairing the car to all utilities. This will force all utilities to be positive and zero expected utility to be assigned to illegal configurations only.   □

In the example we assume that D_1 and D_2 are temporally ordered in the enclosing model, that is, decision D_1 is made prior to decision D_2. This assumption has no impact on the model.

How to Implement This Technique

The constraints on decisions modeling technique is implemented as follows.

1. Let $\{D_1, \ldots, D_n\}$ be the set of decisions to be constrained.
2. Create a utility node U.
3. Add each $D \in \{D_1, \ldots, D_n\}$ as a parent of U.
4. Define the utility function $U(D_1, \ldots, D_n)$ such that all illegal configurations of $\{D_1, \ldots, D_n\}$ are assigned a large negative value.

If a linear transformation of the utility function is required, this should be performed subsequently.

7.4 Summary

In this chapter we have introduced modeling methods and techniques for adjusting the structure of a probabilistic network, for the specification of conditional probability distributions, and for influence diagram models.

The construction of a probabilistic network may be a labor-intensive task to perform. A sequence of steps related to knowledge acquisition and representation is performed in the process of constructing a probabilistic network. The steps include identifying variables, identifying states of variables, identifying and encoding dependence and independence relations among variables as an acyclic, directed graph, and eliciting the quantification of the model as required by the structure.

In Chap. 8 we discuss methods for data-driven modeling.

Exercises

Exercise 7.1. Assume that the causal influences of Angina, Cold, and Flu on SoreThroat can be assumed to be independent. Furthermore, assume that there is a *background* event that can cause the throat to be sore.

The probability of a sore throat being caused by other causes is 0.05, whereas the inhibitor probabilities for Angina, Cold, and Flu are 0.3, 0.4, and 0.25, respectively. The prior probabilities for Angina, Cold, and Flu are 0.4, 0.1, and 0.25, respectively.

(a) Construct a Bayesian network model representing the causal impact on Sore Throat.
(b) Compute the prior probability of SoreThroat.
(c) Apply the parent divorcing modeling technique to simplify the model.

Table 7.23 The utility function U(Surgery, Appendicitis)

	¬surgery	surgery
¬appendicitis	5	−5
appendicitis	−10	10

Exercise 7.2. Consider the Asia network in Fig. 7.20 on page 212.

(a) Perform a sequence of node absorption operations to remove the variables Experts$_1$ and Experts$_2$.
(b) Assume bronchitis can be cured by taking a certain type of medicine. Extend the network accordingly.

Exercise 7.3. Consider the naive Bayes network for classifying mushrooms in Fig. 7.15 on page 207. Assume no odor is perfectly observed, whereas almond is mistakenly observed as spicy in 10 % of the cases, while spicy is mistakenly observed as almond in 5 % of the cases. Extend the network accordingly.

Exercise 7.4. Consider the Asia network in Fig. 4.2 on page 74, see Example 4.2 on page 73.

(a) Perform a node absorption operation to remove the variable Tub_or_cancer.
(b) Apply the parent divorcing technique on the resulting network.

Exercise 7.5. Assume appendicitis may cause fever, pain, or both. If a patient has appendicitis, then the patient will have an increased white blood cells count. When a patient potentially has appendicitis, the physician may choose to carry out surgery right away or wait for a blood test result. Fever and pain are observed.

The prevalence of appendicitis is 0.15. The true positive rates are 0.98, 0.95, and 0.99 for fever, pain, and white_cells_count, respectively. The true negative rates are 0.5, 0.4, and 0.95 for fever, pain, and white_cells_count, respectively. The utilities of operating are shown in Table 7.23.

(a) Build a model for the diagnosis problem.
(b) Compute the expected utility of the scenario where the physician does not wait for the blood test result.
(c) Compute the expected utility of the scenario where the physician waits for the blood test result.
(d) Prior to deciding on whether or not to carry out surgery, the physician has the option to carry out a test for the white blood cell count.
    Extend the model to include a representation of the test decision.

Chapter 8
Data-Driven Modeling

In this chapter we introduce data-driven modeling as the task of inducing a Bayesian network by fusion of (observed) data and domain expert knowledge.

The data-driven modeling is illustrated in Fig. 8.1. The assumption is that some underlying process has generated a database of observed cases as well as domain expert experience and knowledge. The task of data-driven modeling is to fuse these information sources in order to induce a representative model of the underlying process. If the model is a good approximation of the underlying process, then it can be used to answer questions about properties of the underlying process.

In this book we consider the use of Bayesian networks to model the underlying process. The process of inducing a Bayesian network from a database of cases and expert knowledge consists of two main steps. The first step is to induce the structure of the model, that is, the DAG, while the second step is to estimate the parameters of the model as defined by the structure. In this chapter we consider only discrete Bayesian networks. Thus, the task of data-driven modeling is to construct a Bayesian network $\mathcal{N} = (\mathfrak{X}, \mathcal{G}, \mathcal{P})$ from the available information sources. In general, the problem of inducing the structure of a Bayesian network is NP-complete (Chickering 1996). Thus, heuristic methods are appropriate.

Section 8.1 gives some background on data-driven modeling and presents a set of assumptions underlying the presented approach to data-driven modeling. Sections 8.2 and 8.3 consider two different approaches to structure learning of Bayesian networks. In Sect. 8.2, a number of different constraint-based algorithms for structure learning are considered. We consider the PC, PC*, and NPC algorithms. In Sect. 8.3, a number of different search and score-based algorithms for structure learning are considered. This includes algorithms for learning structure restricted models from data. Section 8.4 provides a complete example on structure learning using (some of) the algorithms described in this chapter. In Sect. 8.5, we consider the expectation–maximization algorithm for parameter estimation. In addition to the two main steps of data-driven modeling, there is the step of sequential parameter learning. Structure learning and parameter estimation are performed during the model construction phase, whereas sequential parameter learning is

U.B. Kjærulff and A.L. Madsen, *Bayesian Networks and Influence Diagrams: A Guide to Construction and Analysis*, ISS 22, DOI 10.1007/978-1-4614-5104-4_8,
© Springer Science+Business Media New York 2013

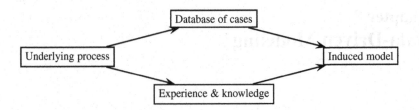

Fig. 8.1 We assume the underlying process generates a database of cases as well as experience and knowledge that can be fused for learning a model of the process

performed during model usage. In Sect. 8.6, we consider sequential parameter learning, which is the task of adjusting parameters of the model as the model is used, that is, as new cases occur.

Parts of this chapter have appeared in Madsen et al. (2005).

8.1   The Task and Basic Assumptions

Data-driven modeling is the task of identifying a Bayesian network model from a source of data. We assume the underlying process follows a probability distribution P_0 (referred to as *the underlying probability distribution of the process*). That is, we assume the data source can be adequately represented by sampling from P_0. The goal of data-driven modeling is to identify a model representation of P_0.

To simplify the task, the probability distribution P_0 is assumed to be a *DAG-faithful* probability distribution with underlying DAG $\mathcal{G}_0$. That is, we assume the distribution P_0 can be represented as a Bayesian network (if P_0 is not DAG-faithful, a Bayesian network may still be an excellent approximation).

The *faithfulness assumption* (also known as the *stability* assumption) says that the distribution P induced by $\mathcal{N} = (\mathcal{X}, \mathcal{G}, \mathcal{P})$ satisfies no independence relations beyond those implied by the structure of $\mathcal{G}$ (Spirtes, Glymour & Scheines 2000, Pearl 2000). A Bayesian network is faithful if and only if for every d-connection there is a corresponding conditional dependence, that is,

$$X \not\perp_G Y \mid \Rightarrow X \not\perp_p Y \mid Z.$$

We assume the underlying probability distribution P_0 to be DAG-faithful with DAG $\mathcal{G}_0$.

The database of cases generated by the underlying and unknown process (i.e., the data source for learning) is denoted $\mathcal{D} = \{c^1, \ldots, c^N\}$, where N is the number of cases in the database. We assume $\mathcal{D}$ consists of independent and identically distributed data cases drawn at random from the probability distribution P_0, that is, we assume cases are drawn at random and independently from the same probability distribution P_0.

Table 8.1 A database of cases

	X_1	X_2	...	X_n
c^1	blue	yes	...	low
c^2	green	no	...	low
c^3	red	N/A	...	high
...	...	...	...	...
c^N	red	no	...	high

Each case $c^i = \{x_1^i, \ldots, x_n^i\}$ in $\mathcal{D}$ specifies an assignment of a value x_j^i to each variable $X_j \in \mathcal{X}$. Some values in $\mathcal{D}$ may be missing, but missing values are assumed to be missing at random (MAR) or missing completely at random (MCAR), that is, the missing data mechanism is uninformative and can be ignored (Cowell et al. 1999). A variable never observed is called a *hidden* or a *latent* variable.

Example 8.1 (Data Cases). Table 8.1 shows a database of N cases $\mathcal{D}=\{c^1, \ldots, c^N\}$ over n variables $\mathcal{X}=\{X_1, \ldots, X_n\}$.

In case c^2, for instance, variable X_2 is observed to have value no, that is, $x_2^2 = $ no, while its value is missing in case c^3 (missing values are indicated using N/A). $\square$

We consider learning a Bayesian network as the task of identifying a DAG structure $\mathcal{G}$ and a set of conditional probability distributions $\mathcal{P}$ with parameters Θ on the basis of $\mathcal{D} = \{c^1, \ldots, c^N\}$ and possibly some domain expert background knowledge.

Applying Occam's Razor (the law of parsimony), see Sect. 6.6, to the problem of learning the structure of a Bayesian network from a database of cases suggests that the simplest model of a set of competing models is preferable. Why should we adhere to the Occam's Razor principle, that is, adhere to one specific selection bias? One argument is that we want models that generalize correctly with respect to subsequent data and it is unlikely that we by coincidence will find a simple model which fits the data as well as a very complex model.

Learning the structure of a sparse graph is computationally less involved than learning the structure of a dense graph where the number of edges is used as a measure of the density of the graph. Inducing a graph from a sample of cases that requires the induced graph to be dense is computationally more expensive than inducing a graph from a sample of cases that requires the induced graph to be sparse. In addition, domains that require the induced graph to be dense may be difficult to represent as a Bayesian network as inducing the graph is computationally expensive, representing a dense graph requires a lot of storage, and inference in dense graphs may be intractable.

The size of the space of possible DAGs grows super-exponentially with the number of vertices in the graph. Robinson (1977) gives the following recursive formula for calculating the number $f(n)$ of DAGs on n vertices:

$$f(n) = \sum_{i=1}^{n} (-1)^{i+1} \binom{n}{i} 2^{i(n-i)} f(n-i). \tag{8.1}$$

Table 8.2 The number of	Number of vertices (n)	Number of possible DAGs $(f(n))$
possible DAGs for values of n	1	1
from one to ten as calculated	2	3
using $f(n)$	3	25
	4	543
	5	29,281
	6	3,781,503
	7	1,138,779,265
	8	783,702,329,343
	9	1,213,442,454,842,881
	10	4,175,098,976,430,598,143

Table 8.2 shows the number of possible DAGs for values of n from one to ten (OEIS 2010). For example, $f(10) \approx 4.2 \cdot 10^{18}$.

8.1.1   Basic Assumptions

Under the following conditions, the structure learning algorithm considered will discover a DAG structure equivalent to the DAG structure of P_0 (Spirtes et al. 2000):

- The independence relationships have a perfect representation as a DAG. This is the *DAG faithfulness assumption*.
- The database consists of a set of independent and identically distributed cases.
- The database of cases is infinitely large.
- No hidden (latent) variables are involved.
- In the case of structure learning, data is complete.
- In the case of constraint-based learning algorithms, the statistical tests have no error.

Even though the basic assumptions may not in practice be satisfied for a specific database of cases or it may not be known if a database of cases satisfies the basic assumptions, the structure learning algorithms considered in this chapter can be applied to construct a knowledge representation of the data. A common approach to handle missing values in structure learning, for instance, is to ignore the missing values.

8.1.2   Equivalent Models

Two DAGs representing the same set of conditional dependence and independence relations (CDIRs) are equivalent in the sense that they can capture the same set of probability distributions. That is, two models M_1 and M_2 are statistically equivalent

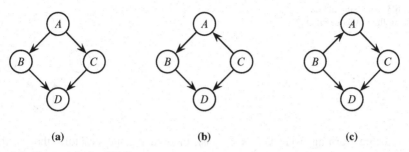

Fig. 8.2 Three equivalent DAGs

if and only if they contain the same set of variables and joint samples over them provide no statistical grounds for preferring one over the other.

The equivalence class of a DAG $\mathcal{G}$ is the set of DAGs with the same set of d-separation relations as $\mathcal{G}$. A *PDAG*—an *acyclic, partially directed graph*, that is, an acyclic graph with some edges undirected (also known as a *pattern* or *essential graph*)—can be used to represent the equivalence class of a set of DAG structures, that is, a maximal set of DAGs with the same set of d-separation relations (Pearl 2000).

Any two models M_1 and M_2 over the same set of variables, whose graphs $\mathcal{G}_1$ and $\mathcal{G}_2$, respectively, have the same *skeleton* $\mathcal{G}_S$ (i.e., undirected graph obtained by replacing directed edges with undirected edges) and the same v-structures are *equivalent*. That is, two DAGs $\mathcal{G}_1$ and $\mathcal{G}_2$ are equivalent if they have the same skeleton and the same set of uncovered *colliders* (i.e., $X \rightarrow Y \leftarrow Z$-structures where X and Z are not connected by a link, which are also known as *v-structures*) (Pearl 2000).

Example 8.2 (Equivalent Models). The models $A \rightarrow B \rightarrow C$ and $A \leftarrow B \leftarrow C$ and $A \leftarrow B \rightarrow C$ are equivalent, as they share the skeleton $A - B - C$ and have no v-structures.

Hence, based on observational data alone, we cannot distinguish $A \rightarrow B \rightarrow C$ and $A \leftarrow B \leftarrow C$ and $A \leftarrow B \rightarrow C$. These models can, however, be distinguished from $A \rightarrow B \leftarrow C$. □

An *equivalence class* is a maximal set of DAGs with the same set of independence properties.

Example 8.3 (Equivalence Class). The three DAGs in Fig. 8.2 all represent the same set of conditional independence and dependence relations.

Figure 8.3 shows the equivalence class of the three equivalent DAGs of Fig. 8.2. □

If structure is identified from data, then two DAGs $\mathcal{G}_i$ and $\mathcal{G}_j$ from the same equivalence class cannot be distinguished. Based on observational data alone, we can at most hope to identify a PDAG representing the equivalence class of the generating distribution P_0.

Fig. 8.3 The equivalence
class of the DAGs in Fig. 8.2

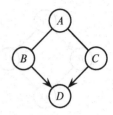

Structure learning from data is the task of inducing the structure, that is, the graph, of a Bayesian network from data. There exist different classes of algorithms for learning the structure of a Bayesian network such as constraint-based algorithms and search and score-based algorithms as well as combinations of the two. In the next section, we consider structure learning algorithms based on the constraint-based approach (Wermuth & Lauritzen 1983, Verma & Pearl 1992, Spirtes et al. 2000).

8.2   Constraint-Based Structure Learning

In the constraint-based approach, the DAG $\mathcal{G}$ of a Bayesian network $\mathcal{N} = (\mathcal{X}, \mathcal{G}, \mathcal{P})$ is considered as an encoding of a set of CDIRs $\mathcal{M}_{\mathcal{G}}$, which can be read off $\mathcal{G}$ using the d-separation criterion (Lauritzen et al. 1990b, Geiger, Verma & Pearl 1990). Structure learning is then the task of identifying a DAG structure that (best) encodes a set of CDIRs. The set of CDIRs may, for instance, be derived from the data source by statistical tests. Based on observational data $\mathcal{D}$ alone, we can at most hope to identify an equivalence class of graphs encoding the CDIRs of the generating distribution P_0.

A constraint-based structure learning algorithm proceeds by determining the validity of independence relations of the form $I(X, Y \mid S_{XY})$ (i.e., X is independent of Y given subset S_{XY} where $X, Y \in \mathcal{X}$ and $S_{XY} \subseteq \mathcal{X}$). The structure learning algorithm will work with any information source able to provide such information. We will consider the case where the validity of independence relations is determined by statistical hypothesis tests of independence based on a database of cases.

8.2.1   Statistical Hypothesis Tests

A set of CDIRs may be generated by statistical tests on the database of cases. In each test, the hypothesis tested is that of independence between a pair of variables.

Let X and Y be a pair of variables for which we would like to determine dependence by statistical hypothesis testing. First, we test for marginal independence, and subsequently we test for conditional independence given subsets of other variables. In the case of marginal independence testing between X and Y, the hypothesis H_0 (and alternative hypothesis H_1) to be tested is

$$H_0 : P(X, Y) = P(X)P(Y), \quad \text{that is,} \quad X \perp\!\!\!\perp_P Y$$
$$H_1 : P(X, Y) \neq P(X)P(Y).$$

Hence, the null hypothesis H_0 is $X \perp\!\!\!\perp_P Y$, while the alternative hypothesis H_1 is $X \not\perp\!\!\!\perp_P Y$.

In order to test the hypothesis, we may use the likelihood G^2 test statistic. Under the null hypothesis H_0, the likelihood G^2 test statistic has an asymptotic χ^2 distribution with the appropriate *degrees of freedom* denoted df. The likelihood G^2 test statistic is computed as

$$G^2 = 2 \sum_{x,y} N_{xy} \log\left(\frac{N_{xy}}{\mathbb{E}_{xy}}\right),$$

where $\mathbb{E}_{xy} = \frac{N_x N_y}{N}$ and N_{xy} specifies the number of cases in $\mathcal{D}$ where $X = x$ and $Y = y$.

In the case of conditional independence testing between X and Y given a subset S_{XY}, the hypothesis H_0 (and alternative hypothesis H_1)to be tested is

$$H_0 : P(X, Y \mid S_{XY}) = P(X \mid S_{XY})P(Y \mid S_{XY}), \quad \text{that is,} \quad X \perp\!\!\!\perp_P Y \mid S_{XY}$$
$$H_1 : P(X, Y \mid S_{XY}) \neq P(X \mid S_{XY})P(Y \mid S_{XY}).$$

The null hypothesis H_0 is $X \perp\!\!\!\perp_P Y \mid S_{XY}$, while the alternative hypothesis H_1 is $X \not\perp\!\!\!\perp_P Y \mid S_{XY}$. In the case of conditional independence testing, the likelihood G^2 test statistic is computed as

$$G^2 = 2 \sum_{x,y,z} N_{xyz} \log\left(\frac{N_{xyz}}{\mathbb{E}_{xyz}}\right),$$

where $\mathbb{E}_{xyz} = \frac{N_{xz} N_{yz}}{N_z}$ and z is a configuration of S_{XY}.

If the test statistic G^2 is sufficiently small, that is, $G^2 < c$, then the null hypothesis H_0 is not rejected. Since the value of c is unknown, the probability distribution of G^2 under H_0 and a significance level α are used. The significance level α is the probability of rejecting a true hypothesis and is typically set to 0.05 (or 0.01 or 0.001). Not rejecting a hypothesis does not imply that data support independence. A hypothesis is not rejected when there is no evidence in the data against the hypothesis.

Under the null hypothesis H_0 (i.e., (conditional) independence of X and Y), the likelihood G^2 test statistic has, as mentioned above, an asymptotic χ^2 distribution

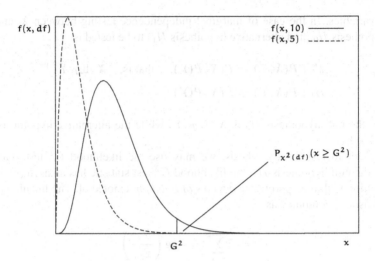

Fig. 8.4 The χ^2 density function for different degrees of freedom

with an appropriate number of degrees of freedom denoted df. The value of df is
defined as

$$\text{df} = (\|X\| - 1)(\|Y\| - 1) \prod_{Z \in S_{XY}} \|Z\|,$$

where $\|X\|$, $\|Y\|$, and $\|Z\|$ are the number of distinct values of X, Y, and Z,
respectively, in $\mathcal{D}$.

If the tail probability of the χ^2 distribution at G^2 is less than α, then H_0 is
rejected. Otherwise, it is not rejected. Thus, the hypothesis H_0 is rejected in favor
of the alternative hypothesis H_1 when $P_{\chi^2(\text{df})}(x \geq G^2) < \alpha$, see Fig. 8.4 for an
illustration.

In the figure $f(x, 5)$ and $f(x, 10)$ are χ^2 density functions with five and ten
degrees of freedom, respectively. The solid line specifies the density function with
ten degrees of freedom, while G^2 specifies the value of the likelihood G^2 test
statistic. The tail of the distribution $P_{\chi^2(df)}(x \geq G^2)$ is the area indicated in the
figure and is often referred to as the p-value. If the tail is less than the significance
level, then the independence hypothesis is rejected. It is clear from the figure that it
is important to use the correct value of df when considering the tail probability of
the distribution.

Example 8.4 (Statistical Test).   Consider the statistical test for (marginal) independence between a pair of variables X and Y with states n and y given the contingency
table shown in Table 8.3.

The hypothesis to be tested is $H_0 : X \perp\!\!\!\perp_P Y$ under the distribution induced by
the contingency table. Computing the test statistic G^2 proceeds as

Table 8.3 Contingency table for testing marginal independence of X and Y

	Y		
X	n	y	
n	12	1	13
y	84	3	87
	96	4	100

$$G^2 = 2 \sum_{x,y} N_{xy} \log \left(\frac{N_{xy}}{\mathbb{E}_{xy}} \right)$$

$$= 2 \left(12 \log \left(\frac{12}{\frac{13*96}{100}} \right) + 1 \log \left(\frac{1}{\frac{13*4}{100}} \right) + 84 \log \left(\frac{84}{\frac{87*96}{100}} \right) + \right.$$

$$\left. 3 \log \left(\frac{3}{\frac{87*4}{100}} \right) \right)$$

$$= 0.2194.$$

Since $G^2 \sim \chi^2(1)$ under H_0, we obtain a p-value of 0.64 (i.e., $P_{H_0}(X \geq G^2) = 0.64$). At a significance level $\alpha = 0.01$, we cannot reject the hypothesis H_0. Hence, X and Y are assumed to be independent. □

The value of df is computed as the sum of $(\|X\| - 1)(\|Y\| - 1)$ over all configurations of S_{XY} correcting for marginal zero counts (i.e., $N_x = 0$ or $N_y = 0$). The value of $\|X\|$ (or $\|Y\|$) is decreased by one for each marginal count equal to zero. This means that the degrees of freedom are reduced when a variable is never observed to be in a specific state.

It is common to perform tests $X \perp\!\!\!\perp Y \mid S_{XY}$ for $|S_{XY}| = 0, 1, 2, 3$ as the tests become unreliable (for finite data sets) when the size of S_{XY} exceeds three as the number of counts N_{xyz} become too small.

If we *reject* H_0 when it is *true*, we incur a *Type I error*. On the other hand, if we *do not reject* H_0 when it is *false*, we incur a *Type II error*. In the constraint-based approach to structure learning, the relative frequency of Type I and Type II errors can (to some extent) be controlled by varying the significance level used in the statistical tests for conditional independence. The lower the significance level, the lower the probability of incurring a Type I error.

8.2.2 Structure Constraints

Prior to the testing phase, background knowledge of domain experts in the form of constraints on the structure of the DAG can be specified. It is possible to specify the presence and absence of edges, the orientation of edges, and a combination.

If the background knowledge is assumed to be consistent with the underlying DAG $\mathcal{G}_0$ of the generating distribution P_0, then it is not necessary to test the validity of the background knowledge. Hence, specifying background knowledge may reduce the number of statistical tests. Unfortunately, this may, in practice, produce unwanted behavior of the edge-orientation algorithm (as described later). This means that background knowledge should often be used with caution.

Domain expert knowledge can be used to guide and speed up the structure learning process. If the database of cases contains a large number variables, then the space of possible DAG structures is huge (as indicated in Table 8.2). In this case, the use of domain expert knowledge to define a set of constraints on the network structure may increase the time efficiency significantly. Domain expert knowledge may also be used in combination with structure restricted algorithms and the edge-orientation algorithm.

8.2.3 PC Algorithm

The PC algorithm (Spirtes & Glymour 1991, Spirtes et al. 2000) (which is similar to the IC algorithm (Verma & Pearl 1992, Pearl 2000)) is a constraint-based algorithm for learning the structure of a Bayesian network. The main steps of the PC algorithm are:

(1) Test for (conditional) independence between each pair of variables represented in $\mathcal{D}$ to derive $\mathcal{M}_\mathcal{D}$, the set of CDIRs
(2) Identify the skeleton of the graph induced by $\mathcal{M}_\mathcal{D}$
(3) Identify colliders
(4) Identify derived directions

The PC algorithm produces a PDAG representing an equivalence class. Each step of the PC algorithm is described in the following sections where the task is to identify a graph $\mathcal{G}$ representing the independence model of the underlying process generating the database of cases.

Step 1: Test for (Conditional) Independence

We try to determine the validity of the conditional independence statement $X \perp\!\!\!\perp Y \mid S_{XY}$ by a statistical hypothesis test as explained in Sect. 8.2.1.

The independence hypothesis is tested for conditioning sets S_{XY} of cardinality $0, 1, 2, 3$ in that order. If the hypothesis $X \perp\!\!\!\perp Y \mid S_{XY}$ cannot be rejected based on some preselected significance level α, then the search for an independence relation between X and Y is terminated.

Example 8.5 (Independence Tests). Assume $\mathcal{D}$ is a database of cases generated from the Burglary or Earthquake network in Fig. 5.5 on page 119. If the sample is

Fig. 8.5 Skeleton representing the CDIRs $\mathcal{M}_\mathcal{D}$ of (8.2) and (8.3)

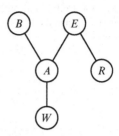

sufficiently large, the conditional independence tests will generate the set $\mathcal{M}_\mathcal{D} = \mathcal{M}_{\not\perp\!\!\!\perp} \cup \mathcal{M}_{\perp\!\!\!\perp}$ of CDIRs where

$$\mathcal{M}_{\perp\!\!\!\perp} = \{B \perp\!\!\!\perp E, B \perp\!\!\!\perp R, B \perp\!\!\!\perp W \mid A, A \perp\!\!\!\perp R \mid E, E \perp\!\!\!\perp W \mid A,$$
$$R \perp\!\!\!\perp W \mid A\} \tag{8.2}$$

$$\mathcal{M}_{\not\perp\!\!\!\perp} = \{B \not\perp\!\!\!\perp A, B \not\perp\!\!\!\perp A \mid \{E\}, B \not\perp\!\!\!\perp A \mid \{R\}, B \not\perp\!\!\!\perp A \mid \{W\}, B \not\perp\!\!\!\perp A \mid \{E, R\},$$
$$B \not\perp\!\!\!\perp A \mid \{E, W\}, B \not\perp\!\!\!\perp A \mid \{R, W\}, B \not\perp\!\!\!\perp A \mid \{E, R, W\}, A \not\perp\!\!\!\perp E, \ldots,$$
$$A \not\perp\!\!\!\perp W, \ldots, E \not\perp\!\!\!\perp R, \ldots\}. \tag{8.3}$$

We will continue this example in the following subsections describing the steps of the PC algorithm. $\qquad\square$

Step 2: Identify the Skeleton

The *skeleton* of an acyclic, directed, or partially directed graph $\mathcal{G}$ is the undirected graph $\mathcal{G}'$ obtained from $\mathcal{G}$ by removing the direction on all directed edges. The skeleton of the graph induced from $\mathcal{M}_\mathcal{D}$ is constructed from the conditional dependence and independence statements of $\mathcal{M}_\mathcal{D}$ generated by the statistical test in Step (1) of the PC algorithm.

For each pair of variables X and Y where no independence statements $X \perp\!\!\!\perp Y \mid S_{XY}$ exist, the undirected edge $\{X, Y\}$ is created in the skeleton.

Example 8.6 (Example 8.5, cont.). From $\mathcal{M}_\mathcal{D}$, the skeleton of $\mathcal{G}$ is generated. The skeleton of $\mathcal{G}$ generated from $\mathcal{M}_\mathcal{D}$ is shown in Fig. 8.5.

Comparing the skeleton of Fig. 8.5 with the skeleton of the graph of Fig. 5.5 on page 119, we see a perfect match.

In addition, it is obvious that the graph of Fig. 8.5 is a more intuitive and compact representation of the dependence and independence model than that of (8.2) and (8.3). $\qquad\square$

Fig. 8.6 Colliders identified
from $\mathcal{M}_{\mathcal{D}}$ and the skeleton of
Fig. 8.5

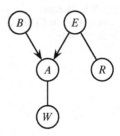

Fig. 8.7 Rule $\mathcal{R}_1$ for identifying derived directions

Step 3: Identify Colliders

Once the skeleton has been identified, colliders in the skeleton are identified. Based
on the skeleton, we search for subsets of variables $\{X, Y, Z\}$ such that X and Y are
neighbors, Z and Y are neighbors, while X and Z are not neighbors. For each such
subset, a collider $X \to Y \leftarrow Z$ is created when $Y \notin S_{XZ}$ for any S_{XZ} satisfying
$X \perp\!\!\!\perp Z \,|\, S_{XZ}$ in $\mathcal{M}_{\mathcal{D}}$. This is referred to as rule $\mathcal{R}_0$.

Example 8.7 (Example 8.6, cont.). From the skeleton $\mathcal{G}_S = (V_S, E_S)$ (see Fig. 8.5)
and $\mathcal{M}_{\mathcal{D}}$ (see (8.2) and (8.3)), a single collider $B \to A \leftarrow E$ is identified, see
Fig. 8.6. This collider is identified as $\{B, A\}, \{E, A\} \in E_S, \{B, E\} \notin E_S$, and $A \notin$
S_{BE} for any $B \perp\!\!\!\perp E \,|\, S_{BE}$.

Notice that the collider $B \to A \leftarrow W$ is not identified as $A \in S_{BW}$
for $B \perp\!\!\!\perp W \,|\, S_{BW}$ where $S_{BW} = \{A\}$. A similar argument holds for the potential
collider $E \to A \leftarrow W$.   □

Step 4: Identify Derived Directions

After identifying the skeleton and the colliders of $\mathcal{G}$, derived directions are identified.
The direction of an edge is said to be *derived* when it is a logical consequence of
(the lack of) previous actions (i.e., since the edge was not directed in a previous step
and it should have been in order to have a certain direction, then the edge must be
directed in the opposite direction).

Starting with any PDAG including all valid colliders, a maximally directed
PDAG can be obtained following four necessary and sufficient rules (Verma &
Pearl 1992, Meek 1995). That is, by repeated application of these four rules, all
edges common to the equivalence class of $\mathcal{G}$ are identified. The four rules $\mathcal{R}_1$ to $\mathcal{R}_4$
are illustrated in Figs. 8.7–8.10.

Fig. 8.8 Rule $\mathcal{R}_2$ for
identifying derived directions

Fig. 8.9 Rule $\mathcal{R}_3$ for
identifying derived directions

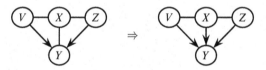

Fig. 8.10 Rule $\mathcal{R}_4$ for
identifying derived directions.
(The *dashed line* between X
and V indicates that X and V
are adjacent, i.e., connected
by an edge.)

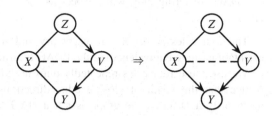

Rules $\mathcal{R}_1$ to $\mathcal{R}_4$ direct edges under the assumption that $\mathcal{G}$ is a valid DAG, that is, they ensure that no directed cycle is created in the graph and no additional colliders are created.

Rule $\mathcal{R}_1$ as illustrated in Fig. 8.7 follows from the fact that the collider $X \rightarrow Y \leftarrow Z$ was not identified as a valid collider. Since the edge between Y and Z is not part of the aforementioned collider, it must be directed from Y to Z.

Rule $\mathcal{R}_2$ as illustrated in Fig. 8.8 follows from the fact that directing the edge between X and Z from Z to X will induce a directed cycle in the graph. Thus, the edge must be directed from X to Z.

Rule $\mathcal{R}_3$ as illustrated in Fig. 8.9 follows from the fact that directing the edge between X and Y from Y to X will inevitable produce an additional collider $V \rightarrow X \leftarrow Z$ or a directed cycle. Hence, the edge must be directed from X to Y.

Rule $\mathcal{R}_4$ as illustrated in Fig. 8.10 follows from the fact that directing the edge between X and Y from Y to X will inevitable produce an additional collider $Y \rightarrow X \leftarrow Z$ or a directed cycle. Hence, the edge must be directed from X to Y. The dashed lines used to illustrate the fourth rule indicate that X and V are connected by an edge (either directed or not).

The fourth rule is not necessary if the orientation of the initial PDAG is limited to containing colliders only. The initial PDAG may contain non-colliders when expert knowledge on edge directions is included in the graph.

Example 8.8 (Example 8.7, cont.). As neither the collider $B \rightarrow A \leftarrow W$ nor the collider $E \rightarrow A \leftarrow W$ were identified as a collider of $\mathcal{G}$, the edge between A and W must be directed from A to W. This is an application of rule $\mathcal{R}_1$.

Fig. 8.11 Derived directions
identified from the skeleton
and the colliders identified in
Example 8.7

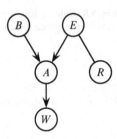

Figure 8.11 shows the equivalence class of $\mathcal{M}_\mathcal{D}$. The equivalence class contains two DAGs differing only with respect to the orientation of the edge between E and R.   □

The four rules $\mathcal{R}_1$ to $\mathcal{R}_4$ are necessary and sufficient for achieving maximal orientation (up to equivalence) of the PDAG returned by the PC algorithm.

We use these four rules repeatedly until no edge can be given an orientation. Notice that the result of closing edge directions under rules $\mathcal{R}_1$ to $\mathcal{R}_4$ is not necessarily a DAG. If the graph is not a DAG, then expert knowledge may be appropriate in order to direct an edge. Once an edge has been directed by use of expert knowledge, derived directions should be identified. This process may be repeated until a DAG structure is obtained. Experience shows that most edges are directed using $\mathcal{R}_1$ and that $\mathcal{R}_3$ is only rarely used.

Since the goal of structure learning is to induce a DAG structure over the variables in the data, a decision has to be made on how to handle directed cycles and additional colliders induced by rules $\mathcal{R}_1$ to $\mathcal{R}_4$. In a practical implementation of the algorithm as part of a tool, we suggest to give the user a warning and to enforce the constraint that the induced graph must be acyclic with the possible implication that edges may be reversed after the application of a rule in order to enforce acyclicity.

Example 8.9 (Example 8.8, cont.).   There are two possible completions of the PDAG shown in Fig. 8.11 into a DAG. Either the edge between E and R is directed from E to R or vice versa. The two DAGs induce the same set of CDIRs.

Since, based on observational data alone, we cannot determine the direction of the edge between E and R, the direction can either be selected at random or we can exploit expert knowledge, if available. From our knowledge of the problem domain and the underlying process, we may argue that if there is an edge between E and R, then it should be directed from E to R. An earthquake may cause a report on the radio reporting the earthquake. A report on the radio definitely cannot cause an earthquake. Figure 8.12 shows the result.

Once the edge between E and R has been given a direction, the resulting graph is a DAG. This completes the structure learning process. The next step in the learning process is to determine or estimate the parameters of the model in order to obtain a fully specified Bayesian network.   □

Fig. 8.12 The result of
structure learning

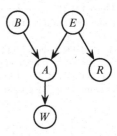

8.2.4 PC* Algorithm

To speed up the structure learning process, various heuristic improvements of the
straightforward incremental testing scheme have been developed (Spirtes et al.
2000).

One of the main improvements is to identify the conditioning set S_{XY} using
an undirected graph specifying pairs of variables that have been identified as
(conditional) independent, given previous test results. Thus, an undirected graph
describing the current set of neighbors of each variable is maintained.

This neighbor graph may be updated each time an independence statement
is identified (or after the completion of the sequence of tests performed for a
fixed cardinality of the conditioning set), that is, each independence test not
rejecting the hypothesis. Hence, the conditional independence of X and Y is
only tested conditional on subsets of the neighbors of X and Y in the undirected
graph. This can significantly reduce the number of independence tests performed.
With the improvement described above, the algorithm is referred to as the PC*
algorithm (Spirtes et al. 2000).

Similarly, the order in which we try out the possible conditioning sets of a fixed
cardinality may be selected according to how likely they are to cause independence
for the edge under consideration. For instance, the heuristic rule that the variables of
the conditioning set should be strongly correlated with both endpoints of the edge
being tested may be used.

Due to the nature of the testing scheme, the conditioning set S_{XY} for an identified
independence relation $X \perp\!\!\!\perp Y \mid S_{XY}$ is minimal in the sense that no proper subset
of S_{XY} makes X and Y independent. This is an important property that is exploited
by the NPC algorithm.

8.2.5 NPC Algorithm

The NPC algorithm (Steck & Tresp 1999) is an extension of the PC algorithm. The
additional feature of the NPC algorithm over the PC algorithm is the introduction
of the notion of a *necessary path condition* (Steck & Tresp 1999) for the absence of
an edge.

Fig. 8.13 The necessary path condition says that in order for $X \perp\!\!\!\perp Y \mid \{Z_1, \ldots, Z_n\}$ to be valid, there should be for each $Z_i, i = 1, \ldots, n$ exist a path between X and Z_i not crossing Y and vice versa

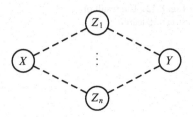

Necessary Path Condition

Informally, the necessary path condition for the absence of an edge says that in order for two variables X and Y to be independent (in a DAG faithful data set) conditional on a minimal set S_{XY}, there must exist a path between X and every $Z \in S_{XY}$ (not crossing Y) and between Y and every $Z \in S_{XY}$ (not crossing X), see Fig. 8.13. Otherwise, the inclusion of each Z in S_{XY} is unexplained. Thus, in order for an independence relation to be valid, a number of edges (or paths) are required to be present in the graph. This is the *necessary path condition*.

The necessary path condition introduces the concept of an *ambiguous edge*. An edge $\{X, Y\}$ is ambiguous if the absence of $\{X, Y\}$ depends on the presence of an edge $\{X', Y'\}$ and vice versa. In that case, $\{X, Y\}$ and $\{X', Y'\}$ are said to be interdependent. An ambiguous edge indicates inconsistency in the set of independence relations derived by the statistical tests. A maximal set of interdependent ambiguous edges is denoted an *ambiguous region*. The necessary path condition is probably better explained by the following example.

Example 8.10 (Necessary Path Condition).   Assume we are given the set of independence relations $\mathcal{M}_{\perp\!\!\!\perp}$ over {Tub_or_cancer, Tuberculosis, Cancer, X_ray} from the Asia example (Example 4.2 on page 73) where

$$\mathcal{M}_{\perp\!\!\!\perp} = \Big\{ \text{X\_ray} \perp\!\!\!\perp \text{Tub\_or\_cancer} \mid \{\text{Cancer, Tuberculosis}\},$$

$$\text{X\_ray} \perp\!\!\!\perp \text{Tuberculosis} \mid \text{Tub\_or\_cancer},$$

$$\text{X\_ray} \perp\!\!\!\perp \text{Cancer} \mid \text{Tub\_or\_cancer} \Big\}. \tag{8.4}$$

The set $\mathcal{M}_{\perp\!\!\!\perp}$ specifies the independence relations induced by the quantification of the model (i.e., by the conditional probability distributions on the variables of the model given their parents). The CDIR

$$\text{X\_ray} \perp\!\!\!\perp \text{Tub\_or\_cancer} \mid \{\text{Cancer, Tuberculosis}\}$$

Fig. 8.14 The condition
graph over $\mathcal{M}_{\perp\!\!\!\perp}$

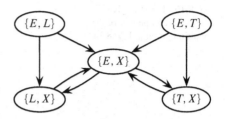

follows from the fact that Tub_or_cancer is a deterministic function of the variables
Tuberculosis and Cancer. That is, whenever the states of Tuberculosis and Cancer
are given, the state of Tub_or_cancer is known and hence independent of the state
of X_ray. Assume further that the collider

$$\text{Tuberculosis} \rightarrow \text{Tub\_or\_cancer} \leftarrow \text{Cancer}$$

is known to be present, see, for example, Fig. 8.16.

The set $\mathcal{M}_{\perp\!\!\!\perp}$ consists of three independence statements, which are inconsistent
(in the rest of this section we use E as short for Tub_or_cancer, L as short for Cancer,
and T as short for Tuberculosis). The absence of the edge between E and X depends
on the presence of the edges $\{E, L\}$, $\{E, T\}$, $\{X, L\}$, and $\{X, T\}$ according to the
necessary path condition. Contrary to this, the absence of the edge between X and T
depends on the presence of the edges $\{E, X\}$ and $\{E, T\}$. Similarly, the absence of
the edge between X and L depends on the presence of the edges $\{E, X\}$ and $\{E, L\}$.
□

The interdependencies between edges induced by the set of conditional inde-
pendence statements $\mathcal{M}_{\perp\!\!\!\perp}$ may be displayed as a directed graph $\mathcal{G}_{\mathcal{M}_{\perp\!\!\!\perp}} = (V, E)$,
where each vertex $v \in V$ corresponds to an edge $\{X, Y\}$ in $\mathcal{G}$ and each directed
edge $(u, v) \in E$ specifies that the absence of v in $\mathcal{G}$ depends on the presence of u
in $\mathcal{G}$. The graph $\mathcal{G}_{\mathcal{M}_{\perp\!\!\!\perp}}$ is referred to as the *condition graph*.

Example 8.11 (Condition Graph). Figure 8.14 shows the condition graph over
$\mathcal{M}_{\perp\!\!\!\perp}$ in (8.4). Notice that vertices $\{L, X\}$ and $\{E, X\}$ as well as $\{T, X\}$ and $\{E, X\}$
are connected by two directed edges.

The condition graph shows how the absence of each edge depends on the
presence of other edges. □

When a vertex v in $\mathcal{G}_{\mathcal{M}_{\perp\!\!\!\perp}}$ does not have a parent, it means that the absence of
the edge represented by v does not depend on the presence of any other edges.
Hence, the independence statement related to the absence of the edge represented
by v satisfies the necessary path condition.

The set of ambiguous regions can be identified as the strongly connected
components of $\mathcal{G}_{\mathcal{M}_{\perp\!\!\!\perp}}$, where a *strongly connected component* is a maximal subgraph
in which every vertex is connected to every other vertex by a directed path.

Fig. 8.15 The strongly
connected component of the
condition graph in Fig. 8.14

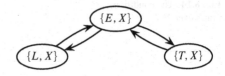

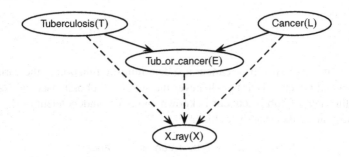

Fig. 8.16 Ambiguous edges in the skeleton due to the deterministic relation
between Cancer, Tuberculosis, and Tub_or_cancer

Fig. 8.17 The two possible
resolutions of the ambiguous
region

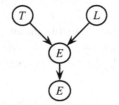 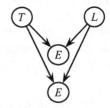

Example 8.12 (Strongly Connected Component). The strongly connected component of the condition graph of Fig. 8.14 is shown in Fig. 8.15. The component
consists of vertices $\{L, X\}$, $\{E, X\}$, and $\{T, X\}$. This set of vertices represents an
ambiguous region over the corresponding edges.

The graph of Fig. 8.16 illustrates an alternative graphical representation of the
ambiguous region consisting of three edges. The graph does not, however, illustrate
how the absence of an edge depends on the presence of another set of edges.

The two possible resolutions of the ambiguous region are shown in Fig. 8.17. The
ambiguous region may be resolved by including either edge $\{E, X\}$ or edges $\{L, X\}$
and $\{T, X\}$ in the graph. The minimal resolution is $\{E, X\}$.   □

An ambiguous region is resolved by including a minimal number of ambiguous
edges in order to satisfy a maximal number of independence relations. In a
graphical representation, ambiguous regions should, for instance, have different
colors as they consist of independent sets of ambiguous edges. A resolution of an
ambiguous region is a minimal set of edges which will remove all ambiguous edges.

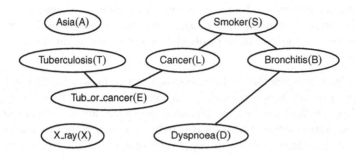

Fig. 8.18 Skeleton representing the CDIRs $\mathcal{M}_\mathcal{D}$ generated from $\mathcal{D}$ by the PC algorithm

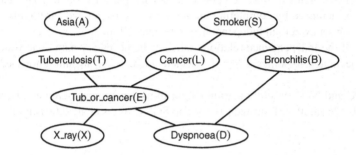

Fig. 8.19 Skeleton representing the CDIRs $\mathcal{M}_\mathcal{D}$ generated from $\mathcal{D}$ by the NPC algorithm after selecting minimal resolutions

A resolution is to include some of the ambiguous edges in the graph in order to be able to make as many independence relations as possible fulfill the necessary path condition.

Example 8.13 (Skeleton (PC Algorithm)). Figure 8.18 shows the skeleton of the graph identified by the PC algorithm based on a (sufficiently large) sample $\mathcal{D}$ generated from the Asia network (Fig. 4.2 on page 74).

Comparing the skeleton of Fig. 8.18 with the skeleton of the acyclic, directed graph of Fig. 4.2, we notice that three edges seem to be missing in Fig. 8.18. These are the edges {Asia, Tuberculosis}, {X_ray, Tub_or_cancer}, and {Dyspnoea, Tub_or_cancer}. The edge {Asia, Tuberculosis} is too weak not to be rejected by the hypothesis test, whereas the edges {X_ray, Tub_or_cancer} and {Dyspnoea, Tub_or_cancer} are absent due to the (deterministic) relation between Tub_or_cancer, Tuberculosis, and Cancer as explained in Example 8.10.

Figure 8.19 shows the skeleton of the graph identified by the NPC algorithm based on the same data set $\mathcal{D}$. The edge {Asia, Tuberculosis} is missing for the reasons explained above, whereas the (deterministic) relation between Tuber_or_ cancer, Tuberculosis, and Cancer has induced two ambiguous regions; one ambiguous region with ambiguous edges {Cancer, X_ray}, {Tuber_or_cancer, X_ray}, and {Tuberculosis, X_ray} and another ambiguous region with ambiguous edges {Cancer, Dyspnoea}, {Tuber_or_cancer, Dyspnoea}, and {Tuberculosis, Dyspnoea}.

The ambiguous regions can be resolved by selecting the minimal resolution in each region.   □

In the above presentation we have assumed that at most a single conditional independence statement is generated for each pair of variables. If multiple conditional independence statements are generated, it is necessary to introduce a more complicated graphical notion where it is possible to represent the fact that an edge may depend on different subsets of edges (one subset of edges corresponding to each independence statement).

In order to increase reliability and stability of the NPC algorithm, multiple independence statements may be generated for each pair of variables. This can, for instance, be achieved by completing the iteration step for a fixed cardinality of the conditioning set even if an independence statement is found.

If one of the independence relations satisfies the necessary path condition, then the independence hypothesis is not rejected. Otherwise, an ambiguous edge is created.

The PC and NPC structure learning algorithms can be considered as extensions of the WL (Wermuth & Lauritzen 1983) and SGS (Spirtes et al. 2000) algorithms.

8.3 Search and Score-Based Structure Learning

Search and score-based algorithms are a different approach to learning structure. It is an optimization-based search approach requiring a scoring function and a search strategy. In this section, we consider structure learning algorithms based on the search and score-based approach (Cooper & Herskovits 1992, Heckerman 1998, Chickering 2002).

In the search and score-based approach, each DAG $\mathcal{G} = (V, E)$ of a Bayesian network $\mathcal{N} = (\mathcal{X}, \mathcal{G}, \mathcal{P})$ has a score that is computed from the data $\mathcal{D}$ using a score function $s(\mathcal{G} : \mathcal{D})$. Structure learning is then the task of identifying a DAG structure that has an optimal score. The score is a selection criterion that measures *goodness* of the model as a representation of the data $\mathcal{D}$. The task is solved by searching through the space of possible structures computing a score for each candidate and selecting the structure with the highest score of all structures visited in the search.

8.3.1 Space of Structures

In search and score-based learning, the structure search space can, for instance, be the space of DAGs or the space of equivalence classes. Here, we consider only the space of DAGs as the structure search space. This means that we search through the set of all possible DAGs over the observed variables X in the complete data D, that is, we assume no hidden variables.

One method to identifying a DAG structure with an optimal quality score is to use brute-force search where we systematically generate each candidate structure and compute the score of the structure, that is, an exhaustive search in the space of structures. An optimal scoring DAG is selected as a structure with the highest score computed. This approach is, however, infeasible in practice as the number of possible DAGs grows super-exponentially with the number of vertices in the graph as illustrated in Table 8.2. Even for DAGs with as few as ten vertices, an exhaustive search would be infeasible as the number of DAGs is approximately $4.2 \cdot 10^{18}$.

As a result of the super-exponential growth in the number of possible DAGs, we need to use a different method for navigating the space of possible DAGs structures, that is, a transition function between neighboring network structures.

8.3.2   Search Procedures

There are a number of different search procedures that can be used to navigate in the space of structures. An exhaustive search is one (infeasible) option. Another option is to use a local search strategy based on *greedy hill climbing* where we in each step select as the next candidate the neighboring structure of the current candidate with the highest score. Given a candidate DAG structure $G = (V, E)$, the search procedure makes incremental changes to G aimed at improving the quality of the structure as determined by the quality score. Each modification to G changes the score of the structure.

The incremental changes considered are to add a directed edge, to reverse a directed edge, or to delete a directed edge under the constraint that the structure remains a DAG. This means that the operator is not allowed to make the structure cyclic, that is, after performing the operation, the structure should remain a DAG.

Figure 8.20 illustrates the three operations of incremental change considered where (a) is the current candidate G, (b) is the candidate obtained by adding the edge (X_1, X_4) to G, (c) is the candidate obtained by removing the edge (X_1, X_2) from G and (d) is the candidate obtained by reversing the edge (X_1, X_2) in G.

The search procedure for navigating the space of DAG structures can be described as a hill climbing algorithm:

(1) Start with an initial structure G.
(2) Repeat.

   (a) Generate a set G of candidate structures from G using the search operators for modifying a structure.
   (b) Compute the score of each candidate structure in G.

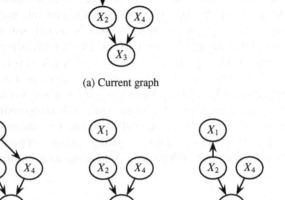

(a) Current graph

(b) Add edge   (c) Remove edge   (d) Reverse edge

Fig. 8.20   Three operations of incremental change

(c) Select $\mathcal{G}^* \in G$ with highest score.

(d) Use $\mathcal{G}^*$ as the candidate $\mathcal{G}$ in the next iteration.

Until the score cannot be improved.

(3) Return $\mathcal{G}$.

The search procedure moves between neighboring candidate structures using the score function to select the next structure. As described above, the search moves deterministically in the search space by greedily selecting the highest scoring neighbor to the current candidate where the neighborhood is defined with respect to the incremental change operators, that is, adding, removing, or reversing an edge. The greedy search procedure can move to neighbors stochastically, and it can be restarted several times with different initial structures in an attempt to improve the quality score and avoid local maxima of the score function.

The search procedure may be invoked with different initial structures, for example, a DAG with no edges, a DAG constructed from domain knowledge, a random DAG, or a Chow–Liu tree (see Sect. 8.3.4). A variety of different heuristics can be applied to the search space in addition to the greedy hill climbing heuristic. This includes randomized walks with multiple restarts.

8.3.3   Score Functions

The objective in the search and score-based approach to structure learning is to identify the network structure with the highest quality score as a representation of

X	Y
Table 8.4 A set of values (x_i, y_i) for variables X and Y	
1	38
2	52
3	140
4	150

data $\mathcal{D}$. Let $s(\mathcal{G} : \mathcal{D})$ be the score function where $\mathcal{G}$ is the DAG and $\mathcal{D}$ is data. The score function $s(\mathcal{G} : \mathcal{D})$ is a measure of the quality or goodness of $\mathcal{G}$ as a representation of $\mathcal{D}$. The task is to find a structure $\mathcal{G}$ with the optimal score.

There are a number of possible candidates to choose from when selecting the score function to be used by the search procedure, see, for example, Darwiche (2009). One option would be to use the likelihood function. In this case, the score function $s(\mathcal{G} : \mathcal{D})$ is based on the best guess of the associated conditional probability distributions given the structure $\mathcal{G}$ and the data $\mathcal{D}$. That is, to compute the score, we need to estimate the parameters of the conditional probability distributions defined by $\mathcal{G}$. This is done by maximum likelihood estimation as described below.

The log-likelihood function $l(\mathcal{G} : \mathcal{D})$ of data $\mathcal{D} = \{c^1, \ldots, c^N\}$ given structure $\mathcal{G}$ is computed as

$$l(\mathcal{G} : \mathcal{D}) = \sum_{l=1}^{N} \log P(c^l) = \sum_{l=1}^{N} \sum_{i=1}^{|V|} \log P(X_i = x_i^l \mid \mathrm{pa}(X_i) = x_{\mathrm{pa}(X_i)}^l, c^l),$$

where $(x_i^l, x_{\mathrm{pa}(X_i)}^l)$ are the values of $(X_i, \mathrm{pa}(X_i))$ in the lth case c^l of $\mathcal{D}$. The log-likelihood $l(\mathcal{G} : \mathcal{D})$ of the data $\mathcal{D}$ given structure $\mathcal{G}$ is the sum of the contributions from each variable for each case. Notice that the log-likelihood function $l(\mathcal{G} : \mathcal{D})$ grows linearly with N.

The log-likelihood function $l(\mathcal{G} : \mathcal{D})$ is, however, not an appropriate score function to use as it often leads to *overfitting* the model to the data $\mathcal{D}$. Overfitting refers to the use of a representation that is too complex, that is, has too many parameters compared to the available data. The result of data overfitting will be a complete DAG as adding a parent always increases the log-likelihood score.

To cope with overfitting, we use a score function that takes the complexity of the representation into account. The more complex the structure, the higher the gain in model quality as measured by the log-likelihood $l(\mathcal{G} : \mathcal{D})$ is required. This means a trade-off between model quality and complexity.

Example 8.14 (Overfitting). A classical example for illustrating the problem of overfitting a model to a data set is that of finding a polynomial that fits perfectly a set of value pairs $(x_1, y_1), (x_2, y_2), \ldots, (x_n, y_n)$ (Darwiche 2009).

Assume we are given the set of value pairs shown in Table 8.4. The data suggests a positive linear relationship between X and Y where values of X close to zero produce values of Y close to zero. Figure 8.21 shows a plot of the data (x, y), a linear function $f_l(x)$, and a polynomial $f_p(x)$ of degree three fitted to the data.

The plot suggests that the relationship between X and Y is linear, that is, $y = \alpha \cdot x + \beta$. The fit between the linear function and the data is not perfect in the

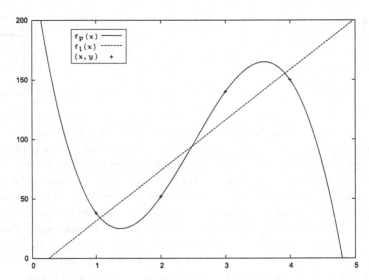

Fig. 8.21 Four data points and a linear function and a three-degree polynomial fitted to the data to illustrate overfitting

sense that the line does not pass through each data point. We can, on the other hand, achieve a perfect fit by using a polynomial of degree three. A polynomial of degree three will guarantee a perfect fit to the data, but it will also overfit the data as illustrated in Fig. 8.21. The problem is that the model does not generalize the data very well. For instance, for values of x close to zero, $f_p(x)$ goes to infinity.

In this example, we could use Occam's Razor (the law of parsimony) to select the simple model (a linear function) based on the argument that it is unlikely that we by coincidence have found a simple model that fits the data (almost) as good as a more complex model.   □

The Jeffreys–Schwarz criterion, also called the Bayesian Information Criterion (BIC), is an example of a score function that penalizes the log-likelihood score with a term defined by the structure of the network. The BIC score is computed as $l(\mathcal{G} : \mathcal{D}) - \frac{\kappa}{2} \log N$, where κ is the number of free parameters in the network (also known as degrees of freedom) and N is the number of cases in the database. The first term measures how well the parameterized model predicts the data, while the second term penalizes complexity of the model. Notice that the second term grows logarithmically with N, while the first term grows linearly with N. This means that the complexity penalty decreases as the size of the database increases. The log-likelihood term will dominate the penalty score when the size grows sufficiently large.

The Akaike's Information Criterion (AIC) is another example of a function score that penalizes the log-likelihood score with a term defined by the structure of the network. The AIC score is computed as $l(\mathcal{G} : \mathcal{D}) - \kappa$. Notice that the BIC score penalizes model complexity more than the AIC score. In the following paragraphs, we will use BIC as the score function.

We may say that a function is decomposable if it decomposes into the combination of a set of simpler functions, for example, a function may decompose into the sum over a set of simpler functions. A decomposable score function leads to a significant gain in the computational efficiency over a non-decomposable score function as it enables us to compute the score $s(\mathcal{G} : \mathcal{D})$ of $\mathcal{G}$ given $\mathcal{D}$ as a function of local scores $s(X | \mathrm{pa}(X) : \mathcal{D})$, that is,

$$s(\mathcal{G} : \mathcal{D}) = \sum_{X \in \mathcal{X}} s(X | \mathrm{pa}(X_i) : \mathcal{D}).$$

When the scoring function decomposes according to the structure of the DAG, we can maximize the contribution to the score of each variable independently under the assumption of parameter independence, that is, the parameters for each variable are assumed to be independent of the parameters of other variables.

The BIC score is an example of a decomposable score. This simplifies the calculation of the score of a neighbor of the current candidate DAG for which the score is known. If the change operation (i.e., add or reverse) for an edge (X, Y) is valid, then it is necessary only to consider the families of X and Y. For instance, for addition of edge (X, Y), we can compute the difference as

$$\Delta(X, Y) = s(Y | \mathrm{pa}(Y) \cup \{X\} : \mathcal{D}) - s(Y | \mathrm{pa}(Y) : \mathcal{D}).$$

This reduces the number of calculations significantly. The BIC score $BIC(\mathcal{G} | \mathcal{D})$ of structure $\mathcal{G}$ given complete data $\mathcal{D}$ is computed as

$$BIC(\mathcal{G} | \mathcal{D}) = l(\mathcal{G} : \mathcal{D}) - \frac{\kappa}{2} \log N$$

$$= \sum_{l=1}^{N} \sum_{i=1}^{|V|} \log P(X_i = x_i^l | \mathrm{pa}(X_i) = x_{\mathrm{pa}(X_i)}^l, c^l) - \frac{\kappa}{2} \log N$$

$$= \sum_i \sum_k \sum_j N_{ijk} \log \frac{N_{ijk}}{N_{ik}} - \frac{\kappa}{2} \log N$$

where N_{ijk} is the number of cases in $\mathcal{D}$ where $X_i = x_k$ and $\mathrm{pa}(X_i)$ is in configuration j and N_{ik}/N_{ijk} is the maximum likelihood estimate for $P(X_i = x_k | \mathrm{pa}(X_i) = j, \mathcal{D})$. The number of free parameters κ is defined as

$$\kappa = \sum_{X \in V} \kappa(X_i) = \sum_{X \in V} (\|X\| - 1) \prod_{Y \in \mathrm{pa}(X)} \|Y\|.$$

The BIC decomposes into one term for each variable. The local *family* BIC score is

$$BIC(\mathrm{fa}(X_i) | \mathcal{D}) = \sum_k \sum_j N_{ijk} \log \frac{N_{ijk}}{N_{ik}} - \frac{\kappa(X_i)}{2} \log N.$$

Table 8.5 Joint probability distributions over X, Y, and Z

X	Y	Z	$P(x, y, z)$
x_0	y_0	z_0	0.02
x_0	y_0	z_1	0.005
x_0	y_1	z_0	0.0075
x_0	y_1	z_1	0.0675
x_1	y_0	z_0	0.576
x_1	y_0	z_1	0.144
x_1	y_1	z_0	0.018
x_1	y_1	z_1	0.162

Table 8.6 Data $\mathcal{D}$ over X, Y, and Z

c^l	X	Y	Z	$P_{\mathcal{G}_1}(c^l)$	$P_{\mathcal{G}_2}(c^l)$	$P_{\mathcal{G}_3}(c^l)$
c_1	x_0	y_0	z_0	0.02	0.0155	0.02
c_2	x_0	y_1	z_0	0.075	0.0466	0.075
c_3	x_1	y_0	z_1	0.144	0.2725	0.144
c_4	x_1	y_1	z_1	0.162	0.0681	0.162
c_5	x_1	y_0	z_0	0.576	0.4478	0.576
c_6	x_0	y_1	z_1	0.0675	0.0284	0.0675

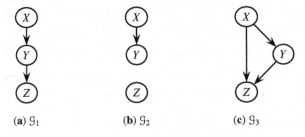

(a) $\mathcal{G}_1$ (b) $\mathcal{G}_2$ (c) $\mathcal{G}_3$

Fig. 8.22 Three different network structures for encoding the joint probability distribution in Table 8.5

If data is incomplete, then the calculation of the log-likelihood is considerably more complicated, and it is necessary to perform parameter estimation, for instance, using the expectation–maximization (EM) algorithm (Lauritzen 1995). Parameter estimation using the EM algorithm is described in Sect. 8.5.

Example 8.15 (Log-Likelihood vs. BIC). Let X, Y, and Z be three binary variables with state spaces $\{x_0, x_1\}$, $\{y_0, y_1\}$, and $\{z_0, z_1\}$, respectively. Assume we know the underlying probability distribution over X, Y, and Z and that we can use this probability distribution to identify a parameterization of possible DAG structure representations of a sample over X, Y, and Z. Table 8.5 shows the joint probability distribution over X, Y, and Z. The aim then is to identify the DAG structure that best represents a small random sample $\mathcal{D}$ of six cases over X, Y, and Z shown in Table 8.6.

Figure 8.22 shows three different DAG structures $\mathcal{G}_1, \mathcal{G}_2$, and $\mathcal{G}_3$ over X, Y, and Z. We want to evaluate how well each of these structures (when quantified

Table 8.7 The conditional probability distributions $P(Y\,|\,X)$ and $P(Z\,|\,Y)$

$P(Y\,|\,X)$	$X = x_0$	$X = x_1$		$P(Z\,|\,Y)$	$Y = y_0$	$Y = y_1$
$Y = y_0$	0.25	0.8		$Z = z_0$	0.8	0.1
$Y = y_1$	0.75	0.2		$Z = z_1$	0.2	0.9

Table 8.8 The conditional probability distribution $P(Z\,|\,X, Y)$

		Z	
X	Y	z_0	z_1
x_0	y_0	0.8	0.2
x_0	y_1	0.1	0.9
x_1	y_0	0.8	0.2
x_1	y_1	0.1	0.9

using the joint probability distribution in Table 8.5) represents the sample $\mathcal{D}$. From the distribution in Table 8.5, the content of the conditional probability distributions $P(X)$, $P(Y\,|\,X)$, $P(Z\,|\,Y)$, $P(Z)$, and $P(Z\,|\,X, Y)$ associated with $\mathcal{G}_1$, $\mathcal{G}_2$, and $\mathcal{G}_3$ can be computed to be $P(X) = (0.1, 0.9)$, $P(Z) = (0.6215, 0.3785)$ and as shown in Tables 8.7 and 8.8.

We want to evaluate the quality of $\mathcal{G}_1$, $\mathcal{G}_2$ and $\mathcal{G}_3$ as representations of $\mathcal{D}$ using log-likelihood and BIC as the quality scores. For candidate $\mathcal{G}_1$, we compute the log-likelihood score $s_l(\mathcal{G}_1 : \mathcal{D})$ and BIC score $s(\mathcal{G}_1 : \mathcal{D})$ as

$$s_l(\mathcal{G}_1 : \mathcal{D}) = \sum_{i=1}^{6} \log P_{\mathcal{G}_1}(c^i)$$

$$= \log(0.02) + \log(0.075) + \log(0.144) + \log(0.162)$$

$$+ \log(0.576) + \log(0.0675) = -13.51,$$

$$s(\mathcal{G}_1 : \mathcal{D}) = s_l(\mathcal{G}_1 : \mathcal{D}) - \frac{5}{2}\log(6) = -13.51 - 4.48 = -17.99.$$

For candidate $\mathcal{G}_2$, we compute scores

$$s_l(\mathcal{G}_2 : \mathcal{D}) = -15.59,$$

$$s(\mathcal{G}_2 : \mathcal{D}) = s_l(\mathcal{G}_2 : \mathcal{D}) - \frac{4}{2}\log(6) = -15.59 - 3.58 = -19.17.$$

For candidate $\mathcal{G}_3$, we compute scores

$$s_l(\mathcal{G}_3 : \mathcal{D}) = s_l(\mathcal{G}_1 : \mathcal{D}) = -13.51,$$

$$s(\mathcal{G}_3 : \mathcal{D}) = s_l(\mathcal{G}_1 : \mathcal{D}) - \frac{7}{2}\log(6) = -13.51 - 6.27 = -19.78.$$

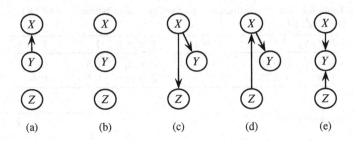

Fig. 8.23 Five different neighbors of the DAG in Fig. 8.22b

$\mathcal{G}_i$	$BIC(\mathcal{G}_i : \mathcal{D})$
$\mathcal{G}_1$ (Fig. 8.22a)	−17.99
$\mathcal{G}_2$ (Fig. 8.23a)	−19.17
$\mathcal{G}_3$ (Fig. 8.23b)	−19.23
$\mathcal{G}_4$ (Fig. 8.23c)	−21.19
$\mathcal{G}_5$ (Fig. 8.23d)	−21.19
$\mathcal{G}_6$ (Fig. 8.23e)	−20.05

Table 8.9 BIC scores for neighbors of $\mathcal{G}$ in Fig. 8.22b

Thus, $s(\mathcal{G}_1 : \mathcal{D}) > s(\mathcal{G}_2 : \mathcal{D}) > s(\mathcal{G}_3 : \mathcal{D})$. Therefore, the best structure (of the ones considered) to represent $\mathcal{D}$ based on the BIC score is $\mathcal{G}_1$. Based on the log-likelihood, score structures $\mathcal{G}_1$ and $\mathcal{G}_3$ are equally good, but $\mathcal{G}_3$ has a higher complexity penalty than $\mathcal{G}_1$ caused by the (unnecessary) edge (X, Z). The edge (X, Z) is unnecessary as X is independent of Z given Y which is clear from Table 8.8.   □

Example 8.16 (Hill Climbing). Assume the graph in Fig. 8.22b is the current candidate structure $\mathcal{G}$ in the hill climbing algorithm described above. The BIC score was computed as $s(\mathcal{G} : \mathcal{D}) = -19.16$ in Example 8.15.

In the second step of the hill climbing algorithm, the set G of candidate structures is generated using the incremental change operators of adding, reversing, and deleting a single edge in the current candidate. Figure 8.22a shows the candidate obtained by adding the edge (Y, Z) to $\mathcal{G}$, while Fig. 8.23 shows the remaining neighbors of $\mathcal{G}$ obtained by adding, deleting, or reversing a single edge.

The greedy hill climbing algorithm selects as the next candidate structure the DAG with the highest score. The scores of neighboring DAGs are shown in Table 8.9.

Recall from Example 8.15 that $BIC(\mathcal{G} : \mathcal{D}) = -19.17$. This means that the only neighbor with a strictly higher score is $\mathcal{G}_1$. Notice that $\mathcal{G}_4$ and $\mathcal{G}_5$ have the same score as they are equivalent models. The remaining neighbors have lower BIC scores than $\mathcal{G}$. Hence, $\mathcal{G}_1$ is selected as the candidate in the next iteration of the hill climbing algorithm.   □

8.3.4 Learning Structure Restricted Models

The complexity of a Bayesian network learned from data may become infeasible due to too large conditional probability distributions caused by high density of the learned graph or by high complexity of the junction tree representation used for belief update. Some score functions such as BIC aim at reducing the complexity of the model by including a model complexity penalty. This may be insufficient in some cases, for example, when building probabilistic graphical models for classification involving a large number of variables or when data are sparse. As an alternative, it may be useful to consider learning probabilistic graphical models with restricted structure. In this section, we consider three simple approximations of a joint probability distribution, namely, naive Bayes models, Chow–Liu trees, and tree-augmented naive Bayes models.

Naive Bayes Model

The naive Bayes model introduced in Sect. 7.1.6 is one of the simplest restricted probabilistic graphical models there is. As mentioned in the aforementioned section, the naive Bayes model is popular due to its high representational and computational simplicity while maintaining an impressive performance on classification tasks. The structure of the naive Bayes model is fixed with one variable R defined as root (or target variable) and the remaining variables $X \in \mathfrak{X} \setminus \{R\}$ defined as children of R. The children of R are referred to as feature variables. In this way the naive Bayes model encodes conditional independence between each pair of feature variables $X, Y \in \mathfrak{X} \setminus \{R\}$ given R and therefore ignores any correlation between X and Y when R is known.

Learning a naive Bayes model $\mathcal{N} = (\mathfrak{X}, \mathcal{G}, \mathcal{P})$ from data $\mathcal{D}$ is simple. Let $R \in \mathfrak{X}$ be the selected root variable. Construct the DAG $\mathcal{G}$ of $\mathcal{N}$ such that there is one edge from R to each feature variable $X \in \mathfrak{X}$ and define the conditional probability distributions of $\mathcal{P}$ as the frequencies in $\mathcal{D}$ in the case of complete data or estimate the distributions using the EM algorithm (see Sect. 8.5.1) in case of incomplete data.

Example 8.17 (Naive Bayes Model). Assume we need to construct a naive Bayes model for classifying patients as suffering or not suffering from lung cancer, represented by the variable L with states no and yes. To perform the classification, we have four indicator variables (or attributes) where A represents whether or not the patient has recently been to Asia, D represents whether or not the patient has dyspnoea, S represents whether or not the patient smokes, and X represents whether or not the patient has a positive X-ray. The structure of the corresponding naive Bayes model is shown in Fig. 8.24.

To complete the model specification, we need to quantify the links in the graph, that is, the conditional probability distribution of each attribute given the

Fig. 8.24 A naive Bayes
model for classifying patients

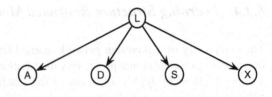

Table 8.10 Contingency
table for L and A

	A		
L	yes	no	
yes	3	510	513
no	81	9,406	9,487
	84	9,916	10,000

Table 8.11 The conditional
probability
distribution $P(A\mid L)$

	A	
L	yes	no
yes	3/513	510/513
no	81/9,487	9,406/9,487

class variable L. We assume a complete database of cases is available. Let $\mathcal{D} = \{c^1, \ldots, c^N\}$ be the complete database with $N = 10,000$. Each case is assumed to contain a value for the class variable L and each attribute.

The data frequencies for L are $N_{L=\text{yes}} = 513$ and $N_{L=\text{no}} = 9,487$ with $N = N_{L=\text{yes}} + N_{L=\text{no}}$. This produces the prior probability distribution

$$P(L) = \left(\frac{513}{10,000}, \frac{9,487}{10,000} \right) = (0.0513, 0.9487).$$

Let Table 8.10 be the contingency table computed from $\mathcal{D}$ for the pair A and L. From Table 8.10, we can compute the conditional probability distribution $P(A\mid L)$ as specified in Table 8.11.

The remaining conditional probability distributions of the model can be computed in the same way using contingency tables computed from $\mathcal{D}$. □

Example 7.7 on page 207 describes another example of a naive Bayes model.

Chow–Liu Trees

A Chow–Liu tree is the *best possible* tree-shaped Bayesian network approximation $\mathcal{N} = (\mathcal{X}, \mathcal{G}, \mathcal{P})$ to a probability distribution $P(\mathcal{X})$ such that all edges are directed away from the root of the tree. The quality of the approximation is measured using the Kullback–Leibler distance between the *true* distribution $P(\mathcal{X})$ and the distribution defined by the Chow–Liu tree represented as $\mathcal{N}$. When we learn from data, the *true* distribution is defined by the frequencies of the observations.

Table 8.12 Contingency
table for D and S

		D	
S	yes	no	
yes	2,744	2,295	5,039
no	1,539	3,422	4,961
	4,283	5,717	10,000

Chow & Liu (1968) showed that the optimum tree can be found as the maximum-weight spanning tree over all variables, where the weight of each edge is defined as the mutual information $I(X, Y)$ between the variables X and Y connected by the edge. A Chow–Liu tree can be constructed as follows:

(1) Calculate the mutual information $I(X, Y)$ (a measure of marginal dependence)

$$I(X,Y) = \sum_X P(X) \sum_Y P(Y \mid X) \log \frac{P(Y, X)}{P(Y)P(X)}$$

for each pair X and Y in $\mathcal{X}$.
(2) Build a maximal-weight spanning tree $\mathcal{G}$ for the complete I-weighted graph.
(3) Direct edges by selecting a root $R \in \mathcal{X}$ and direct edges away from R.

In the second step, the complete I-weighted graph is a complete graph where the edge between any pair X and Y is annotated with the mutual information $I(X, Y)$.

The result of the procedure above is a maximum-weight spanning tree $\mathcal{G}$ with root R. To complete the process, conditional probability distributions defined by the structure of $\mathcal{G}$ should be estimated, in our case, from data $\mathcal{D}$.

The maximal-weight spanning tree can be constructed efficiently using Prim's greedy algorithm. Prim's algorithm constructs the spanning tree by greedily adding vertices to the tree. It starts with a tree over one vertex and at each step greedily selects the next edge to include as an edge with maximal weight such that it connects a node in the current tree to a node not in the current tree. The initial vertex can be chosen arbitrarily, and the algorithm terminates when all vertices are connected in a tree structure. A node is selected as root at random, and edges are directed away from the root.

When constructing a Chow–Liu tree from data, the probability distributions $\mathcal{P}$ involved in the calculation of the mutual information $I(X, Y)$ are computed as frequencies in the data $\mathcal{D}$. In order to complete the specification of $\mathcal{N} = (\mathcal{X}, \mathcal{G}, \mathcal{P})$, the conditional probability distributions $\mathcal{P}$ are estimated from $\mathcal{D}$.

Example 8.18 (Chow–Liu Tree). Assume we are interested in determining the maximal-weight spanning tree over the attributes listed in Example 8.17 in order to better understand the dependence relations between the four attributes. That is, we want to construct a Chow–Liu tree over the four variables A, D, S, and X.

The first step of the algorithm is to compute the mutual information $I(X, Y)$ for each pair of variables. Let us consider the computation of $I(D, S)$. The contingency table for D and S are shown in Table 8.12.

Table 8.13 Mutual information $I(X, Y)$ for each pair

	A	D	S
D	0.0001		
S	0.0	0.0283	
X	0.0	0.3628	0.0089

Fig. 8.25 A maximal-weight spanning tree

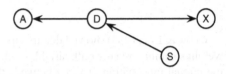

From Table 8.12, the mutual information $I(D, S)$ is computed as

$$I(D, S) = \sum_{D,S} P(D, S) \log \frac{P(D, S)}{P(D)P(S)}$$

$$= \frac{3{,}422}{10{,}000} * \log \frac{3{,}422/10{,}000}{5{,}717/10{,}000 \cdot 4{,}961/10{,}000}$$

$$+ \frac{2{,}295}{10{,}000} * \log \frac{2{,}295/10{,}000}{5{,}717/10{,}000 \cdot 5{,}039/10{,}000}$$

$$+ \frac{1{,}539}{10{,}000} * \log \frac{1{,}539/10{,}000}{4{,}283/10{,}000 \cdot 4{,}961/10{,}000}$$

$$+ \frac{2{,}744}{10{,}000} * \log \frac{2{,}744/10{,}000}{4{,}283/10{,}000 \cdot 5{,}039/10{,}000}$$

$$= 0.0283.$$

The mutual information $I(X, Y)$ for each pair of variables is shown in Table 8.13. The algorithm needs to select three edges to construct a maximal-weight spanning tree. The first two edges selected are $\{X, D\}$ and $\{S, D\}$ as they are the two highest scoring edges. The pair with the third highest mutual information score is $\{S, X\}$. However, selecting this edge would cause a cycle in the graph. Instead, the edge $\{A, D\}$ is selected. A maximal-weight spanning tree has now been constructed.

To complete the model construction, we need to select a root vertex and direct edges away from it. Assume we select (at random) S as root. The resulting maximal-weight spanning tree is shown in Fig. 8.25.

The conditional probability distributions of the model can be determined as frequency counts as illustrated in Example 8.17.   □

Notice that the mutual information $I(X, Y)$ is related to the likelihood test statistic G^2 used for constraint-based structure learning considered in Sect. 8.2. The relation is $I(X, Y) = G_{X,Y}^2 / (2N)$ where N is the number of cases in the database.

Tree-Augmented Naive Bayes

The simplicity and strong independence assumptions of the naive Bayes model may not be desirable in all cases. In this section, we consider the tree-augmented naive Bayes model as an alternative to the naive Bayes model that supports conditional dependence relations between feature variables.

The tree-augmented naive Bayes model is a naive Bayes model augmented with a spanning tree over the feature variables. The algorithm for constructing a tree-augmented naive Bayes considered here is based on the Chow–Liu algorithm. We may refer to the learned probabilistic graphical model as a conditional Chow–Liu tree. A conditional Chow–Liu tree can be constructed as follows:

(1) Let $R \in \mathcal{X}$ be the root vertex.
(2) Calculate the conditional mutual information $I(X, Y \mid R)$ (a measure of conditional marginal dependence)

$$I(X, Y \mid R) = \sum_R P(R) \sum_X P(X \mid R) \sum_Y P(Y \mid X, R) \log \frac{P(Y, X \mid R)}{P(Y \mid R) P(X \mid R)}$$

for each pair $X, Y \in \mathcal{X} \setminus \{R\}$.
(3) Build a maximal-weight spanning tree $\mathcal{T}$ for the complete I-weighted graph over $\mathcal{X} \setminus \{R\}$.
(4) Direct edges in $\mathcal{T}$ by selecting a root in $\mathcal{X} \setminus \{R\}$ and direct edges away from it.
(5) Construct the graph corresponding to a Naive Bayes model over $\mathcal{X}$ with R as root and add the edges of $\mathcal{T}$ to it.

In the third step, the complete I-weighted graph is a complete graph where the edge between any pair X and Y is annotated with the mutual information $I(X, Y \mid R)$.

The tree-augmented naive Bayes model is useful for solving classification tasks where a specific variable of the model is the target of the reasoning. The target variable is used to construct a conditional Chow–Liu tree (i.e., a Chow–Liu tree overall variables except the selected target) with the target as root. Notice that all feature variables except one have two parents in the graph, that is, the root and one other feature variable.

When constructing a conditional Chow–Liu tree from data, the probability distributions $\mathcal{P}$ involved in the calculation of the conditional mutual information $I(X, Y \mid R)$ are computed as frequencies in the data $\mathcal{D}$. In order to complete the specification of $\mathcal{N} = (\mathcal{X}, \mathcal{G}, \mathcal{P})$, the conditional probability distributions $\mathcal{P}$ are estimated from $\mathcal{D}$.

Example 8.19 (Tree-Augmented Naive Bayes).   Continuing Example 8.17, assume we want to construct a tree-augmented naive Bayes model for classifying patients as suffering or not suffering from lung cancer (L) based on observations on smoking (S), recent visit to Asia (A), dyspnoea (D), and a X-ray result (X). That is, we want to construct a tree-augmented naive Bayes model over the set of five variables $\mathcal{X} = \{A, D, L, S, X\}$ with L as the target vertex of the tree.

Table 8.14 Data frequencies
for D, L, and S

L	D	S	
yes	yes	yes	393
yes	yes	no	24
yes	no	yes	85
yes	no	no	11
no	yes	yes	2,351
no	yes	no	1,515
no	no	yes	2,210
no	no	no	3,411
			10,000

The first step of the algorithm is to compute the conditional mutual information $I(X, Y \mid L)$ for each pair of variables conditional on target L. Let us consider the computation of $I(D, S \mid L)$. The data frequencies for L, D, and S are shown in Table 8.14.

From Table 8.14, the conditional mutual information $I(D, S \mid L)$ is computed as

$$I(D, S \mid L) = \sum_{L,D,S} P(L, D, S) \log \frac{P(D, S \mid L)}{P(D \mid L) P(S \mid L)}$$

$$= \frac{3,411}{10,000} * \log \frac{3,411/9,487}{5,621/9,487 \cdot 4,926/9,487}$$

$$+ \frac{2,210}{10,000} * \log \frac{2,210/9,487}{5,621/9,487 \cdot 4,561/9,487}$$

$$+ \frac{1,515}{10,000} * \log \frac{1,515/9,487}{3,866/9,487 \cdot 4,926/9,487}$$

$$+ \frac{2,351}{10,000} * \log \frac{2,351/9,487}{3,866/9,487 \cdot 4,561/9,487}$$

$$+ \frac{11}{10,000} * \log \frac{11/513}{96/513 \cdot 35/513}$$

$$+ \frac{85}{10,000} * \log \frac{85/513}{96/513 \cdot 478/513}$$

$$+ \frac{24}{10,000} * \log \frac{24/513}{417/513 \cdot 35/513}$$

$$+ \frac{393}{10,000} * \log \frac{393/513}{417/513 \cdot 478/513}$$

$$= 0.0259.$$

Notice that the conditional mutual information $I(D, S \mid L)$ is slightly lower than the mutual information $I(D, S)$ computed in Example 8.18.

Table 8.15 Conditional mutual information $I(X, Y \mid L)$ for each pair

	A	D	S
D	0.0002		
S	0.0004	0.0259	
X	0.0002	0.449	0.0015

Fig. 8.26 A tree-augmented naive Bayes model

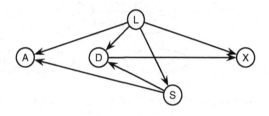

The conditional mutual information $I(X, Y \mid L)$ for each pair of variables is shown in Table 8.15. The algorithm needs to select three edges to construct a maximal-weight spanning tree over $\mathcal{X} \setminus \{L\}$. The first two edges selected are $\{X, D\}$ and $\{S, D\}$ as they are the two highest scoring edges. The pair with the third highest conditional mutual information score is $\{S, X\}$. However, selecting this edge would cause a cycle in the graph. Instead, the edge $\{A, S\}$ is selected. A maximal-weight spanning tree T has now been constructed over $\mathcal{X} \setminus \{L\}$.

To complete the model construction, we need to select a root vertex and direct edges away from it in the maximal-weight spanning tree T. Assume we select (at random) S as root. The resulting tree-augmented naive Bayes model is shown in Fig. 8.26.

The conditional probability distributions of the model can be determined as frequency counts as illustrated in Example 8.17.

Notice the difference in structure over $\mathcal{X} \setminus \{L\}$ between Figs. 8.26 and 8.25. By conditioning on L, the pairwise mutual information scores change, and the two structures are not identical. ☐

8.4 Worked Example on Structure Learning

This section provides a worked example on learning the structure of a (discrete) Bayesian network from a database of cases. The example illustrates the use of different algorithms and discusses the difference between the algorithms.

For the example, a (sufficiently large) database of cases $\mathcal{D}$ is generated from the Asia network (Fig. 4.2 on page 74) by sampling. We assume the sample $\mathcal{D} = \{c^1, \ldots, c^N\}$ is a complete database with $N = 10,000$, that is, there are no missing values in the data. Notice that the sample $\mathcal{D}$ satisfies the basic assumptions of Sect. 8.1.1 related to the data.

If $\mathcal{D}$ would have contained missing values, then the learning algorithms would have ignored the (parts of the) cases where values would have been missing. Since the database is complete, the parameters of each learned Bayesian network can

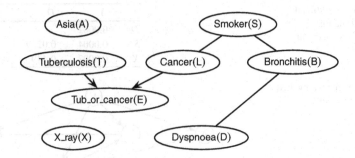

Fig. 8.27 The PDAG produced by the PC algorithm

be determined using maximum likelihood estimation. Had $\mathcal{D}$ been incomplete, the parameter estimation could have been performed using the EM algorithm (see Sect. 8.5.1).

In the example, we assume the goal is to construct an efficient Bayesian network knowledge representation of $\mathcal{D}$. For each learned Bayesian network $\mathcal{N} = (\mathcal{X}, \mathcal{G}, \mathcal{P})$, the quality of the model as a representation of $\mathcal{D}$ is assessed using different score functions on the form $s(\mathcal{G} : \mathcal{D})$.

For ease of presentation, we use abbreviations to represent the variables. Each variable in the data is represented by the first letter, for example, variable Tuberculosis is represented by letter T, except variables Tub_or_cancer and Cancer that are represented by letters E and L, respectively.

In the example, we assume no prior knowledge on the structure, that is, no constraints are placed on the structure learning algorithms, and no prior knowledge on the parameter values.

8.4.1   PC Algorithm

When using the constraint-based PC algorithm for learning the structure of a Bayesian network, see Sect. 8.2.3, we need to decide on a significance level α to use in the statistical tests. A significance level $\alpha = 0.05$ is not uncommon and will be used in this example.

The PC algorithm performs a sequence of statistical tests for pairwise (conditional) independence. Based on the test results, a skeleton is identified, see Fig. 8.27 (and ignore the direction of edges).

The skeleton produced by the statistical tests has a subset set of variables $\{T, L, E\}$ such that T and E are neighbors, L and E are neighbors, while T and L are not neighbors. For this subset, the collider $T \rightarrow E \leftarrow L$ is created as $E \notin S_{LT} = \emptyset$ for any S_{LT}, where S_{LT} is a conditioning set for conditional independence between L and T, satisfying $T \perp\!\!\!\perp L \,|\, S_{LT}$ as T and L are assumed marginally independent based on the test results, see Sect. 8.2.3 for a description of rule $\mathcal{R}_0$ for identifying colliders. The resulting PDAG is shown in Fig. 8.27.

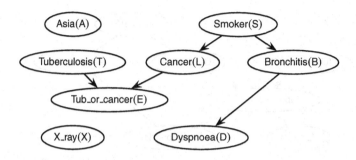

Fig. 8.28 The DAG $\mathcal{G}_{PC}$ produced using the PC algorithm

The direction of undirected edges $\{\{S, L\}, \{S, B\}, \{B, D\}\}$ can either be set at random under the constraint that no new v-structures are created in the process or set based on domain expert knowledge.

Based on experience and knowledge of the problem domain, we may argue that the undirected edge $\{S, B\}$ should be directed as (S, B) and the undirected edge $\{S, L\}$ should be directed as (S, L).

Once the undirected edge $\{S, B\}$ is directed as (S, B), the undirected edge $\{D, B\}$ is directed as (B, D) using rule $\mathcal{R}_1$, see Sect. 8.2.3 on page 248 for a description of the four rules identifying derived directions.

The resulting DAG $\mathcal{G}_{PC}$ is shown in Fig. 8.28. Since data is complete, the parameters of the conditional probability distributions defined by the DAG $\mathcal{G}_{PC}$ can be determined using maximum likelihood estimation. Using the score functions defined in Sect. 8.3.3, the model quality is computed as $s_{ll}(\mathcal{G}_{PC} : \mathcal{D}) = -24,034.3$, $s_{AIC}(\mathcal{G}_{PC} : \mathcal{D}) = -24,098.8$, and $s_{BIC}(\mathcal{G}_{PC} : \mathcal{D}) = -24,048.3$.

The PC algorithm can be characterized as aggressive as it does not perform further tests for a pair of variables when a single conditional independence hypothesis has not been rejected. The algorithm focuses on not rejecting the independence assumption as soon as possible, that is, the aim is to find an independence hypothesis that is not rejected by the statistical tests as soon as possible in order to reduce the connectivity of the graph. Since the results are used to guide the tests, the resulting skeleton may be sensitive to errors in the tests for pairwise independence. This means that the PC algorithm is usually fast in producing sparsely connected DAGs, that is, DAGs with few edges.

8.4.2   NPC Algorithm

When using the constraint-based NPC algorithm for learning the structure of a Bayesian network, see Sect. 8.2.5, we need to decide on a significance level α to use in the statistical tests. A significance level $\alpha = 0.05$ is not uncommon and will be used in this example.

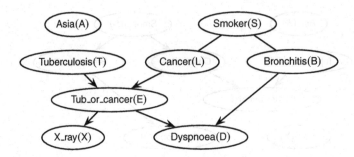

Fig. 8.29 The PDAG produced by the NPC algorithm after resolving ambiguous regions

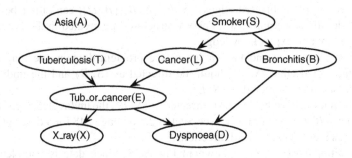

Fig. 8.30 The DAG $\mathcal{G}_{NPC}$ produced using the NPC algorithm

As the PC algorithm, the NPC algorithm performs a sequence of statistical tests for pairwise (conditional) independence. The necessary path condition ensures that (some) inconsistencies in the test results are presented to the user as ambiguous regions.

Based on the test results, a skeleton is identified, see Fig. 8.29 (and ignore the direction of edges), where ambiguous regions have been resolved as explained in Example 8.13.

Once the ambiguous regions have been resolved by including edges $\{E, D\}$ and $\{E, X\}$, the collider with edges (E, D) and (B, D) is created and rule $\mathcal{R}_1$ is applied to direct $\{E, X\}$ as (E, X), respectively.

The undirected edges $\{\{S, L\}, \{S, B\}\}$ are directed at random under the constraint that no new colliders are created in the process or by using domain expert knowledge.

The resulting DAG $\mathcal{G}_{NPC}$ is shown in Fig. 8.30. Since data is complete, the parameters of the conditional probability distributions defined by $\mathcal{G}_{NPC}$ can be determined using maximum likelihood estimation. Using the score functions defined in Sect. 8.3.3, the model quality is computed as $s_{ll}(\mathcal{G}_{NPC} : \mathcal{D}) = -22{,}309.4$, $s_{AIC}(\mathcal{G}_{NPC} : \mathcal{D}) = -22{,}326.4$, and $s_{BIC}(\mathcal{G}_{NPC} : \mathcal{D}) = -22{,}387.7$.

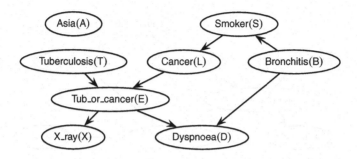

Fig. 8.31 The DAG $\mathcal{G}_{GSS}$ produced using the greedy search and score-based algorithm

The NPC algorithm is less efficient than the PC algorithm as a larger number of conditional independence tests are performed. On the other hand, the NPC algorithm is more robust. The necessary path condition ensures that some inconsistencies in the test results can be reported to the user. The user has the option to make decisions on the edges to include in the skeleton in order to resolve ambiguous regions.

8.4.3 Search and Score-Based Algorithm

When using the greedy search and score algorithm (GSS) for learning the structure of a Bayesian network, see Sect. 8.3, we need to decide on a score function to use in the search procedure for the optimal structure and an upper limit on the number of parents in the resulting DAG. In the example, we use the BIC score and no upper limit on the number of parents.

The search procedure is started from an initial structure. In the example, we start the search from the empty graph, but other graphs can be used, for instance, a Chow–Liu tree may be a possible candidate for an initial structure.

The GSS algorithm performs a sequence of local greedy hill climbing operations to improve the quality score. At each step, the next neighboring candidate with the highest score is selected. The incremental changes considered are to add a directed edge, to reverse a directed edge, or to delete a directed edge under the constraint that the structure remains a DAG.

The resulting DAG $\mathcal{G}_{GSS}$ produced by the GSS algorithm is shown in Fig. 8.31. Notice that $\mathcal{G}_{GSS}$ is equivalent to the *true* DAG $\mathcal{G}_0$ except for the edge $\{A, T\}$.

Since data are complete, the parameters of the conditional probability distributions defined by the DAG $\mathcal{G}_{GSS}$ can be determined using maximum likelihood estimation. Using the score functions defined in Sect. 8.3.3, the model quality is computed as $s_{ll}(\mathcal{G}_{GSS} : \mathcal{D}) = -22,309.4$, $s_{AIC}(\mathcal{G}_{GSS} : \mathcal{D}) = -22,326.4$, and $s_{BIC}(\mathcal{G}_{GSS} : \mathcal{D}) = -22,387.7$.

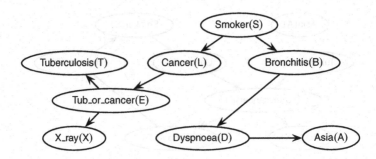

Fig. 8.32 The DAG $\mathcal{G}_{CL}$ produced using the Chow–Liu algorithm

In practice, the GSS algorithm is often slower than the PC and NPC algorithms and more sensitive to missing values in the data. It is possible to launch the GSS algorithm with an initial structure to guide the search procedure. The initial structure could be a Chow–Liu tree.

8.4.4 Chow–Liu Tree

The Chow–Liu algorithm is in principle a search and score-based algorithm for learning the structure of a Bayesian network as described in Sect. 8.3.4. It determines the *best possible* tree-shaped Bayesian network approximation $\mathcal{N} = (\mathcal{X}, \mathcal{G}, \mathcal{P})$ to a probability distribution $P(\mathcal{X})$ such that all edges are directed away from the root of the tree where in the case of learning from data $\mathcal{D}$, the probability distribution $P(\mathcal{X})$ is defined by the frequencies of the observations in $\mathcal{D}$.

When using the Chow–Liu algorithm to learn the structure of a Bayesian network, we need to select a root variable. In the example, we will use variable S as root. This means that edges are directed away from S.

The Chow–Liu algorithm determines the skeleton of the DAG as a maximal-weight spanning tree $\mathcal{G}$ for the complete I-weighted graph, that is, each edge between any pair X and Y is annotated with the mutual information $I(X, Y)$ computed using frequency distributions of $\mathcal{D}$. Once the skeleton of the DAG has been identified, edges are directed away from the selected root S. The resulting DAG $\mathcal{G}_{CL}$ is shown in Fig. 8.32. Notice that the skeleton is missing the edge $\{E, D\}$ as this would create a cycle in the graph and that the edge $\{T, A\}$ is replaced by the edge $\{A, D\}$.

Since data is complete, the parameters of the conditional probability distributions defined by the DAG $\mathcal{G}_{CL}$ can be determined using maximum likelihood estimation. Using the score functions defined in Sect. 8.3.3, the model quality is computed as $s_{ll}(\mathcal{G}_{CL} : \mathcal{D}) = -22{,}830.9$, $s_{AIC}(\mathcal{G}_{CL} : \mathcal{D}) = -22{,}845.9$, and $s_{BIC}(\mathcal{G}_{CL} : \mathcal{D}) = -22900$.

The Chow–Liu algorithm is based on computing the mutual information score $I(X, Y)$ between each pair of variables X and Y. This operation can be performed

Table 8.16 Different model quality scores for the five DAGs considered in the example

Graph ($\mathcal{G}$)	$s_{ll}(\mathcal{G}, \mathcal{D})$	$\kappa(\mathcal{G})$	$s_{AIC}(\mathcal{G}, \mathcal{D})$	$s_{BIC}(\mathcal{G}, \mathcal{D})$
$\mathcal{G}_{PC}$	−24,034.3	14	−24,048.3	−24,098.8
$\mathcal{G}_{CL}$	−22,830.9	15	−22,845.9	−22,900
$\mathcal{G}_0$	−22,318.5	18	−22,338.5	−22,401.4
$\mathcal{G}_{NPC}$	−22,309.4	17	−22,326.4	−22,387.7
$\mathcal{G}_{GSS}$	−22,309.4	17	−22,326.4	−22,387.7

fast compared to the operations performed by other structure learning algorithms. Using the Chow–Liu algorithm, it is efficient to create a maximum-weight spanning tree of the variables represented in the data $\mathcal{D}$. A maximum-weight spanning tree could, for instance, be used as the initial starting point for the GSS structure learning algorithm.

8.4.5 Comparison

The above subsections have illustrated how four different structure learning algorithms can be applied to induce a Bayesian network from data $\mathcal{D}$. It is clear from the descriptions above that there are differences in the DAGs produced by the algorithms. The PC and Chow–Liu algorithms produce the most sparse graphs, while the NPC and GSS algorithms produce DAGs that are equivalent to the underlying DAG $\mathcal{G}_0$ of the Bayesian network from which $\mathcal{D}$ is sampled except for the missing edge $\{A, T\}$.

Table 8.16 compares the log-likelihood (ll), degrees of freedom (κ), and AIC and BIC scores produced by the DAG candidates produced by the four different structure learning algorithms and $\mathcal{G}_0$ the underlying DAG structure used to generate the data. The rows are organized in increasing order from the top.

Notice that $\mathcal{G}_{SS}$ and $\mathcal{G}_{NPC}$ are equivalent. Therefore, the scores for the two DAGs are the identical. In practice, the NPC and the search and score-based algorithms will sometimes produce different results. This is due to the data not satisfying the basic assumptions described in Sect. 8.1.1.

The structure produced by the NPC and GSS algorithms has only slightly larger scores than the underlying DAG $\mathcal{G}_0$ (the difference originates from sampling noise due to the weak dependence relation between A and T specified on the edge (A, T) in $\mathcal{G}_0$ and not in $\mathcal{G}_{NPC}$ and $\mathcal{G}_{NPC}$). In practice, the underlying DAG $\mathcal{G}_0$ will not be known. In fact, the goal of structure learning is to uncover a DAG equivalent to DAG $\mathcal{G}_0$.

Notice that the quality scores are relatively robust with respect to different DAG structures, and since $\frac{\log N}{2} > 1$, the BIC score is greater than the AIC score for each candidate DAG structure.

In the example, $\mathcal{D}$ is assumed to have no missing values. If values are missing in the data, these data values are usually ignored in the structure learning algorithms described here, but other approaches to handling missing values in structure learning do exist (e.g., completion of data or using the structure EM algorithm (Friedman 1998)).

Cowell (2001) describes the conditions under which conditional independence and scoring methods lead to identical selection of Bayesian network models. He argues that under the conditions of a given vertex ordering and complete data, the distinction between conditional independence tests and searching the model space using a score metric is largely a myth. In practice, however, the algorithms may produce different results, and using both to find the best structure is advised.

8.5   Batch Parameter Learning

Parameter estimation in a Bayesian network $\mathcal{N} = (\mathcal{X}, \mathcal{G}, \mathcal{P})$ is the task of estimating the values of parameters Θ corresponding to DAG structure $\mathcal{G}$ and distributions $\mathcal{P}$ from a database of cases $\mathcal{D} = \{c^1, \dots, c^N\}$.

Let $\mathcal{N} = (\mathcal{X}, \mathcal{G}, \mathcal{P})$ be a Bayesian network with parameters $\Theta = \{\Theta_i\}$ where $\Theta_i = \{\Theta_{ij}\}$ and $\Theta_{ij} = \{\theta_{ijk}\}$ such that $\theta_{ijk} = P(X_i = k \mid \mathrm{pa}(X_i) = j)$ for each i, j, k. Batch parameter learning from data is to estimate the value of θ_{ijk} from $\mathcal{D}$.

When each case $c^i \in \mathcal{D}$ is *complete*, maximum likelihood parameter estimation is simple (a case c^i is complete when c^i assigns a value to each variable $X_i \in \mathcal{X}$). The basic idea of parameter learning is illustrated in the following example.

Example 8.20 (Maximum Likelihood Estimate).   Table 8.3 on page 245 shows the contingency table for testing marginal independence of X and Y in Example 8.4. From this example, we may determine a maximum likelihood estimate $\hat{P}(X = y \mid Y = y)$ of the conditional probability of $X = y$ given $Y = y$ as follows:

$$\hat{P}(X = y \mid Y = y) = \frac{\hat{P}(X = y, Y = y)}{\hat{P}(Y = y)}$$

$$= \frac{\frac{n(X=y, Y=y)}{N}}{\frac{n(Y=y)}{N}}$$

$$= \frac{n(X = y, Y = y)}{n(Y = y)}.$$

From Table 8.3, we have $n(X = y) = 87$ and $n(X = y, Y = y) = 3$. Estimating the conditional probability parameter $\hat{\theta}_{Y=y \mid X=y} = \hat{P}(Y = y \mid X = y)$ proceeds as

$$\hat{P}(Y = y \mid X = y) = \frac{n(Y = y, X = y)}{n(X = y)} = \frac{3}{87} = 0.034.$$

The remaining parameters of $P(Y \mid X)$ are estimated in a similar way.   □

8.5.1 Expectation–Maximization Algorithm

Parameter estimation in the case of missing values may be performed using the expectation–maximization (EM) algorithm (Lauritzen 1995). The EM algorithm is well suited for calculating maximum likelihood (ML) and maximum a posterior (MAP), that is, an estimate incorporating prior knowledge on the parameters, estimates in the case of missing data. The EM algorithm proceeds by iterating two steps: the expectation step (E-step) and the maximization step (M-step).

Let $\mathcal{N} = (\mathcal{X}, \mathcal{G}, \mathcal{P})$ be a Bayesian network for which we would like to estimate the parameters Θ of $\mathcal{P}$ from a database of cases $\mathcal{D}$. The estimation of the parameters Θ from $\mathcal{D}$ proceeds, as mentioned above, by iterating the E-step and the M-step. Given an initial assignment to the parameters Θ, the E-step is to compute the expected data frequencies under Θ, while the subsequent M-step is to maximize the log-likelihood of the parameters under the expected data frequencies. These two steps are alternated iteratively until a stopping criterion is satisfied.

In the case of missing data, the log-likelihood function of the parameters is a linear function in the sufficient marginals (Lauritzen 1995). The log-likelihood function $l(\Theta)$ of the parameters Θ given the data $\mathcal{D} = \{c^1, \ldots, c^N\}$ and DAG $\mathcal{G}$ is

$$l(\Theta) = \sum_{l=1}^{N} \log P(c^l \,|\, \Theta)$$

$$= \sum_{l=1}^{N} \sum_{i=1}^{|V|} \log P(X_i = x_i^l \,|\, \mathrm{pa}(X_i) = x_{\mathrm{pa}(X_i)}^l, \Theta_i, c^l)$$

$$= \sum_{i=1}^{|V|} l(\Theta_i),$$

where $l(\Theta_i) = \sum_{l=1}^{N} \log P(X_i = x_i^l \,|\, \mathrm{pa}(X_i) = x_{\mathrm{pa}(X_i)}^l, \Theta_i, c^l)$ assuming the parameters Θ_i to be independent and $(x_i^l, x_{\mathrm{pa}(X_i)}^l)$ are the values of $(X_i, \mathrm{pa}(X_i))$ in the lth (possibly incomplete) case c^l of $\mathcal{D}$.

For Bayesian networks, the E-step of the EM algorithm is to compute expected counts (expected data frequencies for a complete database), where expectation is taken with respect to the joint distribution over V under the current parameter values Θ and observed data $\mathcal{D}$:

$$\mathbb{E}_{\Theta}(N_{ijk}) = \sum_{l=1}^{N} P(X_i = k, \mathrm{pa}(X_i) = j \,|\, c^l, \Theta_i, G),$$

where N_{ijk} is the count for $(X_i, \mathrm{pa}(X_i)) = (k, j)$ and c^l is the lth case of $\mathcal{D}$. Next, the M-step computes new estimates θ_{ijk}^* of θ_{ijk} interpreting the expected data frequencies as actual data frequencies from a complete database of cases:

$$\theta^*_{ijk} = \frac{\mathbb{E}_\Theta(N_{ijk})}{\sum_{k=1}^{||X_i||} \mathbb{E}_\Theta(N_{ijk})}.$$

The E-step and M-step are iterated until convergence of $l(\Theta)$ (or until a limit on the number of iterations is reached).

We say convergence is achieved when the difference between the log-likelihoods of two consecutive iterations is less than or equal to a log-likelihood threshold, δ, times the numerical value of the log-likelihood, that is,

$$l_i(\Theta) - l_{i+1}(\Theta) \leq \delta |l_{i+1}(\Theta)|$$

where $l_i(\Theta)$ is the log-likelihood of Θ after the ith iteration and $l_{i+1}(\Theta)$ is the log-likelihood of Θ after the $(i + 1)$st iteration.

Alternatively, an upper limit on the number of iterations can be specified in order to ensure that the procedure terminates.

Example 8.21 (Toss of a Coin).   Consider the task of predicting the toss of a coin. Having no additional knowledge of the coin, we would assume it to be fair. Hence, we assume heads and tails to be equally likely as the result of tossing the coin. Let X be a discrete random variable with two states tails and heads representing the outcome of a toss of the coin.

If we have observed the result of ten tosses of the coin, we can use this data to predict the result of a subsequent toss of the coin. Assume we make the following sequence of observations on previous tosses of the coin:

tails, tails, heads, tails, heads, tails, , tails, tails, heads,

where the observation on the outcome of the seventh throw is missing, that is, we know the coin was tossed, but for some (random and uninformative) reason, we do not have access to the result. From this sequence of observations, we want to estimate the distribution of X.

Since we have no extra knowledge about the coin, we assume a uniform prior distribution on X. Hence, the initial parameter assignment is set to $P(X) = (0.5, 0.5)$, that is, $\Theta = \{\theta_{\mathsf{tails}} = 0.5, \theta_{\mathsf{heads}} = 0.5\}$.

The estimated distribution after running the EM algorithm with the data and parameter setting described above is $P(X) = (0.74992, 0.25008)$ with an experience count $\alpha = 10$. This distribution is the result of five iterations of the EM algorithm with $\delta = 0.0001$ and

$$l_1(\Theta) = -5.54518$$

$$l_2(\Theta) = -4.548$$

$$l_3(\Theta) = -4.50078$$

$$l_4(\Theta) = -4.49877$$

$$l_5(\Theta) = -4.49868.$$

The parameter estimation is completed after five iterations of the EM algorithm as described above.

Taking the observations on the coin into account, we predict tails to be approximately three times as likely as heads. □

In the EM algorithm, the log-likelihood function $l(\Theta)$ of the model given data is used as a quality measure to compare different parameterizations of the same network structure. When the increase in quality of the parameterization between subsequent iteration steps is below a threshold, the EM algorithm terminates. The log-likelihood function is well suited for this purpose. However, the log-likelihood quality measure does not take network complexity into account. Thus, the log-likelihood measure is not well suited for model selection due to overfitting from using a too-complex network structure. The log-likelihood measure will take its maximum value for a complete graph.

Instead of using log-likelihood for model selection, the AIC or BIC score may be used. The scores are described in Sect. 8.3.3.

8.5.2 Penalized EM Algorithm

When both data and domain expert knowledge are available, both of these two sources of knowledge should be taken into consideration by the parameter estimation algorithm. This can be achieved using the *penalized* EM algorithm to achieve a MAP estimate.

Domain expert knowledge on the parameters of a conditional probability distribution is specified in the form of a Dirichlet probability distribution and an *experience table*. For each variable X_i, the distribution $P(X_i | \text{pa}(X_i)) = \{P(X_i = k | \text{pa}(X_i) = j)\}$ and the *experience counts* $\alpha_{i1}, \ldots, \alpha_{im}$, where $m = \|\text{pa}(X_i)\|$, associated with X_i are used to specify the prior expert knowledge associated with $P(X_i | \text{pa}(X_i))$. The size of the experience count α_{ij} indicates the weight of the domain expert knowledge compared to the size of the observed data set $\mathcal{D}$. The experience count may be considered as a case count for *virtual cases*. For instance, if $\alpha_{ij} = N$, then $\mathcal{D}$, and the expert knowledge carry the same weight in the estimation. The experience table over a variable X_i and its parent variables $\text{pa}(X_i)$ indicates the experience related to the child distribution for each possible configuration of $\text{pa}(X_i)$.

In the case of expert knowledge, the E-step does not change, whereas the M-step becomes

$$\theta_{ijk}^* = \frac{\alpha_{ijk} + \mathbb{E}_\Theta(N_{ijk})}{\sum_{k=1}^{\|X_i\|} (\alpha_{ijk} + \mathbb{E}_\Theta(N_{ijk}))},$$

where $\alpha_{ijk} = P(X_i = k | \text{pa}(X_i) = j)\alpha_{ij}$ is the initial count for $(X_i, \text{pa}(X_i)) = (k, j)$. Thus, the M-step is changed to take the expert knowledge into account.

Fig. 8.33 Two equivalent
DAGs over X and Y

 a   **b**

Table 8.17 Data frequencies
for estimating the
distributions of X and Y

	X	
Y	x_0	x_1
y_0	0	2	2
y_1	4	2	6
y_2	6	3	9
	10	7	17

Example 8.22 (Penalized EM: Coin Tossing).   Consider again the problem of predicting the result of a coin toss. Assume we have reason to believe that the coin is not fair. Instead of assuming a uniform prior distribution on the parameters, we will assume a nonuniform prior on X, for example, we assume the parameter assignment is $P(X) = (0.75, 0.25)$ with an experience count of $\alpha = 5$. This will serve as the initial parameter assignment.

The estimated distribution is $P(X) = (0.75, 0.25)$ with an experience count $\alpha = 15$. This distribution is the result of only two iterations of the EM algorithm with $\delta = 0.0001$ and

$$l_1(\Theta) = l_2(\Theta) = -4.49868.$$

The parameter estimation is completed after only two iterations of the EM algorithm as described above.

Taking the observations on the coin and the additional knowledge of the coin into account, we predict tails to be three times as likely as heads.   □

The penalized EM algorithm is useful for combining expert domain knowledge and data in parameter estimation. It is, however, important to be careful when using the penalized EM algorithm as illustrated by the following example.

Example 8.23 (Penalized EM: Experience Counts).   Assume we need to estimate the conditional probability distributions of a network with two dependent variables X and Y, where $\|X = 2\|$ and $\|Y\| = 3$. To model the dependency between variables X and Y, we have either the network in Fig. 8.33a or the network in Fig. 8.33b.

From the point of view of modeling the joint probability distribution over variables X and Y, the choice of network does not matter as the two models are equivalent. Assume the complete database of cases to be used in the parameter estimation has data frequencies as shown in Table 8.17. The database consists of 17 complete cases, that is, no missing values. Notice that the configuration (x_0, y_0) does not appear in the database.

The probability of the configuration (x_0, y_0) will be zero in both networks if the EM algorithm is used for the estimation. The penalized EM algorithm may be used

X	Y			
x_0	y_0	0.0177	0.0216	0.0093
x_0	y_1	0.2298	0.2262	0.2315
x_0	y_2	0.3359	0.3370	0.3426
x_1	y_0	0.1215	0.1080	0.1204
x_1	y_1	0.1215	0.1257	0.1204
x_1	y_2	0.1736	0.1815	0.1759

Table 8.18 Joint probability distributions over X and Y

to avoid zero probabilities in the joint probability distribution. The trick is to assign a positive probability distribution to the initial parameter values and a nonzero value to the experience counts. The values of the experience counts should be chosen with care, though.

Assume we decide to assign a uniform probability distribution to the parameters, for example, $P(X) = (1/2, 1/2)$, $P(Y \mid X = x_0) = (1/3, 1/3, 1/3)$, and $P(Y \mid X = x_1) = (1/3, 1/3, 1/3)$ for the network in Fig. 8.33a. To avoid a zero probability for the configuration (x_0, y_0), we assign experience counts as $\alpha_X = 1$, $\alpha_{Y0} = 1$ and $\alpha_{Y1} = 1$. The resulting joint probability distribution over X and Y is shown in the second column of Table 8.18.

If we use the same approach on the network in Fig. 8.33b, then the resulting joint probability distribution over X and Y is shown in the third column of Table 8.18.

It is clear from Table 8.18 that the approach taken does not produce the same results for two equivalent models given complete data and a uniform prior. The problem is the value assigned to the experience counts.

Instead of assigning the value 1 to all experience counts, we assign the value $1/\|\mathrm{pa}(X)\|$ to each parent configuration of X and value $1/\|\mathrm{pa}(Y)\|$ to each parent configuration of Y. This approach will ensure that the two equivalent networks in Fig. 8.33 represent the same joint probability distribution over X and Y after EM learning. With this approach, the result is the same for both models, and it is shown in the last column of Table 8.18. $\qquad\square$

8.6 Sequential Parameter Learning

Sequential parameter learning or parameter adaptation is the task of sequentially updating the parameters of the conditional probability distributions of a Bayesian network when the structure and an initial specification of the conditional probability distributions are given in advance. We consider a Bayesian approach to sequential parameter learning (Spiegelhalter & Lauritzen 1990, Cowell et al. 1999).

In sequential learning, experience is extended to include both quantitative expert knowledge and past cases (e.g. from EM learning). Thus, the result of EM learning may be considered as input for sequential learning.

Let X_i be a variable with n states. Then, the prior belief in the parameter vector $\Theta_{ij} = (\theta_{ij1}, \dots, \theta_{ijn})$, that is, the conditional probability distribution

Fig. 8.34 Retrieval and
dissemination of experience

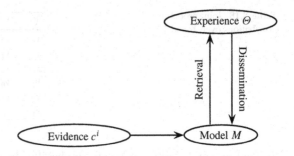

of X_i given its parents $\text{pa}(X_i) = j$, is specified as an n-dimensional Dirichlet distribution $\mathcal{D}(\alpha_{ij1}, \ldots, \alpha_{ijn})$. This distribution is represented using a single
experience count $\alpha_{ij} = 1/\|\text{pa}(X_i)\|$ (*equivalent sample size*)and the initial distri-
bution $P(X_i | \text{pa}(X_i) = j)$. The experience count α_{ijk} for a particular state k of X_i
given $\text{pa}(X_i) = j$ is $\alpha_{ijk} = \alpha_{ij} P(X_i = k | \text{pa}(X_i) = j)$. This setting is similar to
the setting of the EM algorithm.

Parameter adaptation proceeds by updating the experience associated with the
parameters and subsequently updating the parameters to reflect the new experience.
The process of updating the experience associated with a distribution is referred to
as *retrieval of experience*. *Dissemination of experience* is the process of calculating
prior conditional probability distributions for the variables in the Bayesian network
given the experience, and it proceeds by setting each parameter equal to the mean of
the corresponding updated Dirichlet distribution, that is, θ_{ijk}^*, as shown below. See
Fig. 8.34 for a graphical representation of dissemination and retrieval of experience.

After a complete observation $(X_i, \text{pa}(X_i)) = (k, j)$, the posterior belief in the
distribution is updated as $\alpha_{ijk}^* = \alpha_{ijk} + 1$ and $\alpha_{ijl}^* = \alpha_{ijl}$ for $l \neq k$. After an
incomplete observation, the resulting weighted sum of Dirichlet distributions over
the parameters is approximated with a single Dirichlet distribution with the same
means and sum of variances as the mixture. The approximation is used in order to
avoid the combinatorial explosion, which would otherwise occur when subsequent
incomplete observations are made. For each i, j, k, the updated value θ_{ijk}^* of each
parameter θ_{ijk} is

$$\theta_{ijk}^* = \frac{\alpha_{ijk} + P(X_i = k, \text{pa}(X_i) = j \mid \varepsilon) + \theta_{ijk}(1 - P(\text{pa}(X_i) = j \mid \varepsilon))}{\alpha_{ij} + 1}.$$

The updated equivalent sample size α_{ij}^* is a function of the means and the sum
of variances, see, for example, Cowell et al. (1999) for details on computing the
updated α_{ij}^*. Notice that $P(X_i = k, \text{pa}(X_i) = j \mid \varepsilon)$ and $P(\text{pa}(X_i) = j | \varepsilon)$ are
readily available after a propagation of evidence.

In order to reduce the influence of past and possibly outdated information, an
optional feature of fading is provided. Fading proceeds by reducing the experience
count before the retrieval of experience takes place. The experience count α_{ij} is
faded by a factor of $0 < \lambda_{ij} \leq 1$ typically close to one according to $p_{ij} = $

Table 8.19 Experience counts for B, L, and S before and after adaptation

	α_S	$\alpha_{L \mid S=\text{no}}$	$\alpha_{L \mid S=\text{yes}}$	$\alpha_{B \mid S=\text{no}}$	$\alpha_{B \mid S=\text{yes}}$
Before	10,000	4,970.88	5,029.12	4,970.88	5,029.12
After	9,001	4,472.71	5,029.12	4,473.73	5,029.12

$P(\text{pa}(X_i) = j)$ such that $\alpha_{ij}^* = \alpha_{ij}((1 - p_{ij}) + \lambda_{ij} p_{ij})$. Notice that experience counts corresponding to parent configurations, which are inconsistent with the evidence, are unchanged. The fading factors of a variable X_i are specified in a separate table including one fading factor for each configuration of $\text{pa}(X_i)$.

Example 8.24 (Sequential Parameter Learning). Consider again the Asia example (Example 4.2 on page 73). Assume we have evidence $\varepsilon = \{S = \text{no}, A = \text{yes}, D = \text{yes}\}$ on a patient, that is, a nonsmoking patient with dyspnoea who has recently been to Asia. The evidence is entered and propagated followed by an adaptation of parameters. Table 8.19 shows the experience counts for L, B, and S before (i.e., after EM learning using 10,000 randomly generated cases) and after the adaptation with fading factor of $\lambda = 0.999$ for each distribution.

Note that since S is an observed variable without parents, the experience count α_S for $P(S)$ will converge to $\frac{1}{\lambda} = 1,001$ if $S = \text{no}$ is observed multiple times. □

Sequential updating may be applied to the parameters of conditional probability distributions in mixed Bayesian networks and in influence diagrams when all decisions have been instantiated.

8.7 Summary

In this chapter, we have considered data-driven modeling as the process of inducing a Bayesian network from data and domain expert knowledge. We have considered this as a two-step learning process where the first step is to induce the structure of the graph of the Bayesian network, whereas the second step is to estimate the parameters of the conditional probability distributions induced by the graphical structure of the Bayesian network.

We have considered both a constraint-based and a search and score-based approach to learning the structure of a Bayesian network from data. For the constraint-based approach, we have described in some detail the steps of the PC algorithm. The PC algorithm is based on performing a sequence of statistical hypothesis tests for (conditional) independence. Based on the set of CDIRs derived by the test, the skeleton, the colliders, and derived directions of the graph are identified. For the search and score-based approach, we have described in some detail the steps of a hill climbing algorithm using the BIC score to measure the goodness of a candidate structure. We have also in detail described search and score-based approaches for learning structure restricted models such as the naive

Bayes models, Chow–Liu trees, and tree-augmented naive Bayes models. The naive Bayes and the tree-augmented Naive Bayes models, in particular, have shown high performance on classification tasks.

Since the result of the PC algorithm (i.e., the PDAG) is rather sensitive to errors in the CDIRs, the notion of a necessary path condition for the absence of an edge in the skeleton of the induced graph is introduced. The necessary path condition produces the NPC algorithm, which has also been described in some detail.

The graph resulting from structure learning defines the set of conditional probability distributions of the Bayesian network. The parameters of this set of distributions may be set manually, but more often the parameters of the distributions will be estimated from the same database of cases as used by the structure learning algorithm. We have described the EM algorithm and the penalized EM algorithm for estimating the parameters of a conditional probability distribution from data.

Finally, we have described a Bayesian approach for adaptation of parameters as the model is used.

In the next part of this book, we consider different methods for analyzing probabilistic networks. This includes methods for conflict analysis, sensitivity analysis, and value of information analysis.

Exercises

Exercise 8.1. Let variables A, C, F, S, and T represent angina, cold, fever, spots in throat, and sore throat, respectively. Assume that the following set of conditional pairwise independence statements have resulted from performing statistical independence tests on this set of variables:

$$\mathcal{M}_{\perp\!\!\!\perp} = \left\{ A \perp\!\!\!\perp C \,|\, \{\}, C \perp\!\!\!\perp S \,|\, \{\}, F \perp\!\!\!\perp S \,|\, \{A\}, T \perp\!\!\!\perp S \,|\, \{A\}, T \perp\!\!\!\perp F \,|\, \{A, C\} \right\}.$$

For each independence statement $X \perp\!\!\!\perp Y \,|\, S_{XY}$, the conditioning set S_{XY} is minimal.

(a) Identify the skeleton of the graph of the network.
(b) Identify colliders.
(c) Identify derived directions.
(d) Identify remaining undirected edges according to your interpretation of the problem domain.

Exercise 8.2. Generate 10,000 sample cases from the Asia network shown in Fig. 8.35 (see Example 4.2 on page 73 for more details).

The Asia network consists of the three hypothesis variables Bronchitis, Cancer, and Tuberculosis. The risk factors are Smoking and a recent visit to Asia, while the symptoms of the network are X_ray and Dyspnoea. The risk factors and symptoms are the possible observations a physician can make on a patient.

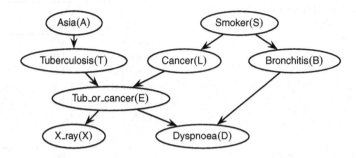

Fig. 8.35 A graph specifying the independence and dependence relations of the Asia example

Fig. 8.36 The angina network

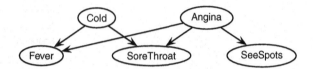

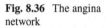

Table 8.20 Data frequencies for the class variable

Class	n
edible	5,333
poisonous	4,667

(a) Use a software package for learning the equivalence class of DAGs representing the data generated. Specify expert knowledge on the structure of the DAG as constraints.

(b) Resolve ambiguous regions.

(c) Complete orientation of the DAG.

(d) Specify expert knowledge on the distributions of the model and estimate the parameters from the data generated.

Exercise 8.3. Consider the network in Fig. 8.36. Assume tat the variables are all binary and that the network is the result of learning from a complete database with $1,000$ cases and let $l(\Theta) = -964$.

(a) Compute the AIC score.

(b) Compute the BIC score.

(c) Compare the AIC and BIC scores. Explain the difference.

Exercise 8.4. Assume we plan to pick up mushrooms to prepare for a nice dinner. In the process, we want to classify each mushroom as either edible or poisonous. We want to construct a network for classifying each mushroom based on a database of mushrooms. Assume Tables 8.20–8.23 specify the data frequencies of a database we found on the Internet.

Table 8.21 Data frequencies
for the class variable
and Population

	Class	
Population	edible	poisonous
abundant	587	0
clustered	389	75
numerous	480	0
scattered	1,173	420
several	1,440	3,384
solitary	1,264	788

Table 8.22 Data frequencies
for the class variable
and Cap_Shape

	Class	
Cap_Shape	edible	poisonous
bell	478	56
conical	0	5
convex	2,453	2,034
flat	2,098	1,853
knobbed	267	719
sunken	37	0

Table 8.23 Data frequencies
for the class variable
and Odor

	Class	
Odor	edible	poisonous
almond	475	0
anise	475	0
creosote	0	229
fishy	0	685
foul	0	2,567
musty	0	57
none	4,383	141
pungent	0	303
spicy	0	685

Table 8.24 Data frequencies for A, D, S, and X

	$A = $ no	$A = $ yes	$D = $ no	$D = $ yes	$S = $ no	$S = $ yes
$D = $ no	5,675	42				
$D = $ yes	4,241	42				
$S = $ no	4,921	40	3,422	1,539		
$S = $ yes	4,995	44	2,295	2,744		
$X = $ no	8,825	74	5,311	3,588	4,621	4,278
$X = $ yes	1,091	10	406	695	340	761

Exercise 8.5. Construct a maximal-weight spanning tree using the data in
Table 8.24.

(a) Construct a Naive Bayes model for classifying mushrooms based on the data in
Tables 8.20–8.23.
(b) What is the probability of a mushroom with no odor being edible?

Part III
Model Analysis

Chapter 9
Conflict Analysis

It is difficult or even impossible to construct models covering all aspects of (complex) problem domains of interest. A model is therefore most often an approximation of a problem domain that is designed to be applied according to the assumptions as determined by the background condition or context of the model. If a model is used under circumstances not consistent with the background condition, the results will in general be unreliable. The evidence need not be inconsistent with the model in order for the results to be unreliable. It may be that evidence is simply in conflict with the model. This implies that the model in relation to the evidence may be weak, and therefore the results may be unreliable.

Evidence-driven conflict analysis is used to detect possible conflicts in the evidence or between the evidence and the model. If a possible conflict is detected, we should alert the user that the model given the evidence may be weak or even misleading. In this way, conflict analysis can also be used for model revision.

Hypothesis-driven conflict analysis is used to identify findings acting in favor of or against a hypothesis. If the evidence set consists of a large number of findings, it may be crucial to identify which individual findings act in favor of or against a hypothesis.

In this chapter we use the Asia example to illustrate the concepts of evidence and hypothesis-driven conflict analysis in Bayesian networks.

Example 9.1 (Conflict Analysis). As an example, we apply conflict analysis to the Asia example shown in Fig. 9.1 (see Example 4.2 on page 73 for more details).

Assume we see a smoking patient with no shortness of breath and a negative X-ray result, that is, the initial set of evidence is $\varepsilon = \{\text{Dyspnoea} = \text{no}, \text{Smoker} = \text{yes}, \text{X\_ray} = \text{no}\}$. In the remainder of this section, we write $\varepsilon = \{\varepsilon_D, \varepsilon_S, \varepsilon_X\}$ for short.

From our knowledge about the problem domain and the assumptions of the model, we would say that the findings are in conflict. The patient visited the chest clinic, but she does not have any of the symptoms even though she is a smoker. A propagation of the evidence produces posterior probability distributions over the

U.B. Kjærulff and A.L. Madsen, *Bayesian Networks and Influence Diagrams: A Guide to Construction and Analysis*, ISS 22, DOI 10.1007/978-1-4614-5104-4_9, © Springer Science+Business Media New York 2013

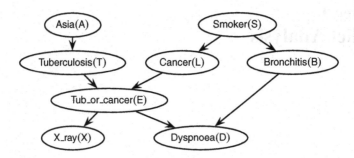

Fig. 9.1 A graph specifying the independence and dependence relations of the Asia example

diseases $P(B = \text{yes}|\varepsilon) = 0.25$, $P(L = \text{yes}|\varepsilon) = 0.089$, and $P(T = \text{yes}|\varepsilon) = 0.009$. This does not disclose a possible conflict, but only indicates that the most likely disease the patient is suffering from is bronchitis. Using the model only, we cannot distinguish between flawed evidence, a case not covered by the model, or a rare case.   □

In Sect. 9.1, we describe evidence-driven conflict analysis. A *conflict measure* designed to be efficient to compute and to give indications of possible conflicts in evidence is introduced. Possible conflicts should be detected and traced or explained as rare cases that the model may not cover well. In Sect. 9.2, we describe hypothesis-driven conflict analysis. A *cost-of-omission measure* is introduced. This measure is useful for relating the impact of findings on a hypothesis variable. In addition, we describe how to investigate the impact of a finding on the probability of a hypothesis.

9.1   Evidence-Driven Conflict Analysis

The objective in evidence-driven conflict analysis is to detect possible conflicts in a set of evidence. As a tool for detecting possible conflicts, we want a conflict measure which is easy to calculate and which gives a reliable indication of a possible conflict.

9.1.1   Conflict Measure

In order to detect a possible conflict, we need to compare the results obtained from our best model with the results obtained from an alternative model. This model is referred to as a *straw model*. A straw model is a simple and computationally efficient model used as an alternative model in the detection of possible conflicts.

In the design of the conflict measure, we make the assumption that for the *normally* (according to the model) behaving evidence, it is the case that $P(\varepsilon_i | \varepsilon_j) > P(\varepsilon_i)$

where ε_i and ε_j are two pieces of evidence. Based on this assumption and since $P(\varepsilon_i, \varepsilon_j) = P(\varepsilon_i | \varepsilon_j) P(\varepsilon_j)$, the conflict measure is designed to indicate a possible conflict when the joint probability of the evidence is less than the product of the probabilities of the individual pieces of evidence given the model. We thus assume that there is a possible conflict between two pieces of evidence ε_i and ε_j, if

$$\frac{P(\varepsilon_i) P(\varepsilon_j)}{P(\varepsilon_i, \varepsilon_j)} > 1 \Leftrightarrow \log \frac{P(\varepsilon_i) P(\varepsilon_j)}{P(\varepsilon_i, \varepsilon_j)} > 0,$$

that is, ε_i and ε_j are negatively correlated. Thus, we define the conflict measure as

$$\text{conf}(\varepsilon) = \log \frac{P(\varepsilon_i) P(\varepsilon_j)}{p(\varepsilon)},$$

where $\varepsilon = \{\varepsilon_i, \varepsilon_j\}$. Notice that we are comparing the joint probability of the evidence with a model where the observed variables are independent (the straw model).

The main assumption is that pieces of evidence are positively correlated such that $P(\varepsilon) > \prod_{i=1}^{n} P(\varepsilon_i)$. With this assumption, the general conflict measure is defined as

$$\text{conf}(\varepsilon) = \text{conf}(\{\varepsilon_1, \ldots, \varepsilon_n\}) = \log \frac{\prod_{i=1}^{n} P(\varepsilon_i)}{p(\varepsilon)}.$$

This implies that a positive value of the conflict measure $\text{conf}(\varepsilon)$ indicates a possible conflict. Notice that the conflict measure is easy to calculate once the evidence ε has been propagated in a junction tree representation of the model. The marginal probabilities $P(\varepsilon_i)$ are available once the junction tree has been initialized, and the probability of the evidence $P(\varepsilon)$ is available as a by-product of message passing, see Sect. 5.1 for details on propagation of evidence.

Another way to look at the definition of the conflict measure is the following. In the general case, two pieces of evidence ε_i and ε_j are either:

- Positively correlated, that is, $P(\varepsilon_i | \varepsilon_j) > P(\varepsilon_i)$
- Negatively correlated, that is, $P(\varepsilon_i | \varepsilon_j) < P(\varepsilon_i)$
- Independent, that is, $P(\varepsilon_i | \varepsilon_j) = P(\varepsilon_i)$

Given these three options, we choose to assume that two pieces of evidence ε_i and ε_j are positively correlated.

Example 9.2 (Conflict Detection).   Returning to Example 9.1, the evidence is $\varepsilon = \{\varepsilon_D, \varepsilon_S, \varepsilon_X\}$. We compute the conflict measure to be

$$\text{conf}(\varepsilon) = \text{conf}(\{D = \text{no}, S = \text{yes}, X = \text{no}\})$$
$$= \log \frac{P(D = \text{no}) P(S = \text{yes}) P(X = \text{no})}{P(\varepsilon)}$$

$$= \log \frac{0.56 * 0.5 * 0.89}{0.2} = 0.22$$

$$> 0.$$

Thus, $\mathrm{conf}(\varepsilon)$ indicates a possible conflict in ε. $\qquad\qquad\qquad\square$

9.1.2 Tracing Conflicts

Once a possible conflict has been detected, the origin of the conflict should be determined such that it can be presented to the analyst or user. Tracing the source of a conflict amounts to computing the conflict measure $\mathrm{conf}(\varepsilon')$ for (all) subsets $\varepsilon' \subseteq \varepsilon$.

Tracing the conflict to all subsets of the evidence is a computationally complex problem as the number of subsets increases exponentially with the size of the evidence set. It is not always possible or meaningful to assume monotonicity with respect to conflict in subsets, that is, no subset $\varepsilon'' \subseteq \varepsilon'$ with $\mathrm{conf}(\varepsilon'') > 0$ exists for ε' with $\mathrm{conf}(\varepsilon') \leq 0$. That is, the monotonicity assumption states that if ε' is not in conflict, then no subset of ε' is in conflict.

Example 9.3 (Conflict Traced).   In Example 9.2, a possible conflict was identified, but not traced, that is, located. That is, after the conflict measure has been found to indicate a possible conflict, the source of the conflict should be traced. This is achieved by computing the conflict measure for different subsets ε' of ε.

With three pieces of evidence, there are three pairs of evidence to consider. The pair $\varepsilon_{DS} = \{\varepsilon_D, \varepsilon_S\}$ has conflict measure

$$\mathrm{conf}(\varepsilon_D, \varepsilon_S) = \log \frac{P(D = \text{no}) P(S = \text{yes})}{P(D = \text{no}, S = \text{yes})}$$

$$= \log \frac{0.56 * 0.5}{0.22} = 0.24,$$

the pair $\varepsilon_{DX} = \{\varepsilon_D, \varepsilon_X\}$ has conflict measure

$$\mathrm{conf}(\varepsilon_D, \varepsilon_X) = \log \frac{P(D = \text{no}) P(X = \text{no})}{P(D = \text{no}, X = \text{no})}$$

$$= \log \frac{0.56 * 0.89}{0.52} = -0.04,$$

and the pair $\varepsilon_{SX} = \{\varepsilon_S, \varepsilon_X\}$ has conflict measure

$$\mathrm{conf}(\varepsilon_S, \varepsilon_X) = \log \frac{P(S = \text{yes}) P(X = \text{no})}{P(S = \text{yes}, X = \text{no})}$$

$$= \log \frac{0.5 * 0.89}{0.47} = -0.06.$$

The (partial) conflicts show that there is a conflict in the pair $\varepsilon_{DS} = \{\varepsilon_D, \varepsilon_S\}$ while there are no conflicts in the pairs $\varepsilon_{DX} = \{\varepsilon_D, \varepsilon_X\}$ and $\varepsilon_{SX} = \{\varepsilon_S, \varepsilon_X\}$. Hence, the source of the global conflict in $\text{conf}(\varepsilon)$ can be traced to the partial conflict between ε_D and ε_S. The finding that the patient is a smoker is in conflict with the finding that the patient is not suffering from dyspnoea (under the assumptions of the model). □

Let ε_i and ε_j be a partitioning of the evidence ε into two disjoint subsets such that $\varepsilon = \varepsilon_i \cup \varepsilon_j$ is the evidence under consideration. The *global conflict* $\text{conf}(\varepsilon)$ can be computed from *local* $\text{conf}(\{\varepsilon_i, \varepsilon_j\})$ and *partial conflicts* $\text{conf}(\varepsilon_i)$ and $\text{conf}(\varepsilon_j)$

$$\text{conf}(\varepsilon) = \text{conf}(\{\varepsilon_i, \varepsilon_j\}) + \text{conf}(\varepsilon_i) + \text{conf}(\varepsilon_j).$$

This property holds in general, and it may be used as a tool for tracing conflicts.

Example 9.4 (Partial Conflicts). In the example, we have three subsets with partial conflicts computed in Example 9.3 $\text{conf}(\varepsilon_D, \varepsilon_S) = 0.24$, $\text{conf}(\varepsilon_{DX}) = \text{conf}(\{\varepsilon_D, \varepsilon_X\}) = -0.04$, and $\text{conf}(\varepsilon_{DS}) = \text{conf}(\{\varepsilon_S, \varepsilon_X\}) = -0.06$. The local conflict between $\{\varepsilon_D, \varepsilon_S\}$ and ε_X is

$$\text{conf}(\{\varepsilon_D, \varepsilon_S\}, \varepsilon_X) = \log \frac{P(D = \text{no}, S = \text{yes}) P(X = \text{no})}{P(\varepsilon)}$$

$$= \log \frac{0.22 * 0.89}{0.2} = -0.02.$$

The global conflict can be computed from local and partial conflicts as

$$\text{conf}(\{\varepsilon_D, \varepsilon_S, \varepsilon_X\}) = \text{conf}(\{\varepsilon_D, \varepsilon_S\}) + \text{conf}(\{\{\varepsilon_D, \varepsilon_S\}, \varepsilon_X\})$$

$$= 0.24 + (-0.02) = 0.22.$$

We notice that the finding ε_X reduces the global conflict slightly. □

9.1.3 Conflict Resolution

Typical evidence from a *rare case* may indicate a possible conflict (a rare case is identified as a finding or set of evidence with low (prior) probability). Let $\varepsilon = \{\varepsilon_1, \ldots, \varepsilon_n\}$ be findings for which the conflict measure indicates a possible conflict, that is, $\text{conf}(\varepsilon) > 0$. Also, let h be a hypothesis which could explain the findings (i.e., $\text{conf}(\varepsilon \cup \{h\}) \leq 0$). That is, if we also know the hypothesis h to be true, then we will not expect a conflict. We compute

$$\text{conf}(\varepsilon \cup \{h\}) = \text{conf}(\varepsilon) + \log \frac{P(h)}{P(h \mid \varepsilon)}.$$

If $\text{conf}(\varepsilon) \leq \log \frac{P(h \mid \varepsilon)}{P(h)}$, then h can explain away the conflict where $\frac{P(h \mid \varepsilon)}{P(h)}$ is the *normalized likelihood*.

Table 9.1 The
log-normalized likelihood of
each possible instantiation of
each variable in Example 9.1

Hypothesis	$P(h)$	$P(h\mid\varepsilon)$	$\log\frac{P(h\mid\varepsilon)}{P(h)}$
Asia = no	0.99	0.99	0
Asia = yes	0.01	0.01	0
Bronchitis = no	0.55	0.75	0.31
Bronchitis = yes	0.45	0.25	−0.59
Cancer = no	0.545	0.1	1.7
Cancer = yes	0.455	0.9	0.68
Tuberculosis = no	0.99	0.999	0.01
Tuberculosis = yes	0.01	0.001	−2.3
Tub_or_cancer = no	0.995	0.9999	0.005
Tub_or_cancer = yes	0.005	0.0001	−3.9

Example 9.5 (Conflict Resolution). Table 9.1 shows the log-normalized likelihood
of each possible instantiation of each variable in Example 9.1. From this table, it is
clear that there are five possible explanations of the conflict (some of which have a
low log-normalized likelihood).

For instance, the posterior probability of the patient not having bronchitis
is $P(B = \text{no}\mid\varepsilon) = 0.75$ while the prior probability is $P(B = \text{no}) = 0.55$. We
compute the logarithm of the normalized likelihood

$$\log\frac{P(h\mid\varepsilon)}{P(h)} = \log\frac{0.75}{0.55} = 0.31.$$

From this, we compute the conflict measure under the hypothesis of the patient not
having bronchitis

$$\text{conf}(\varepsilon \cup \{h\}) = \text{conf}(\varepsilon) + \log\frac{P(h)}{P(h\mid\varepsilon)}$$

$$= 0.22 - 0.31 < 0.$$

Thus, the conflict may be explained away as the rare case where the patient
is not suffering from bronchitis given the symptoms and risk factors (similarly
for Cancer). In general, if the normalized likelihood is greater than the conflict,
then we may have a rare case.   □

The above method for detecting conflicts may fail. This will happen if the
assumption of positively correlated pieces of evidence does not hold. The above
approach can be combined with other methods such as those reported in Kim &
Valtorta (1995) and Laskey (1991). These methods are based on using a more
complex straw model for the comparison (i.e., evidence variables are not assumed
independent) and a different measure to detect conflicts. The advantage of the above
approach is that it is computationally efficient.

The above method for detecting conflicts was introduced by Andersen et al.
(1989), and the method is described in Jensen (1996) and Jensen (2001).

9.2 Hypothesis-Driven Conflict Analysis

In hypothesis-driven conflict analysis, the impact of a finding on the probability of a hypothesis is investigated. In order to be able to relate the impacts of different findings on the probability of the hypothesis given evidence, a measure is needed. This will allow us to identify pieces of evidence that conflict with the impact of the entire set of evidence.

9.2.1 Cost-of-Omission Measure

The main purpose of hypothesis-driven conflict analysis is to identify pieces of evidence with an impact on the evidence that conflicts with the impact of the entire set of evidence. In order to perform this investigation, a measure providing a numerical value specifying the *cost* of omitting a single piece of evidence is required. Suermondt (1992) defines the *cost-of-omission* $c(P(X \mid \varepsilon), P(X \mid \varepsilon \setminus \{\varepsilon_i\}))$ of ε_i as

$$c(P(X \mid \varepsilon), P(X \mid \varepsilon \setminus \{\varepsilon_i\})) = \sum_{x \in \mathrm{dom}(X)} P(x \mid \varepsilon) \log \frac{P(x \mid \varepsilon)}{P(x \mid \varepsilon \setminus \{\varepsilon_i\})}. \qquad (9.1)$$

The above equation is undefined for values $P(x \mid \varepsilon) = 0$ and $P(x \mid \varepsilon \setminus \{\varepsilon_i\}) = 0$. For these two cases, we define cost-of-omission to be 0 and infinity, respectively.

9.2.2 Evidence with Conflict Impact

Let H be a hypothesis variable with states $\mathrm{dom}(H) = (h_1, \ldots, h_n)$ and let $\varepsilon = \{\varepsilon_1, \ldots, \varepsilon_m\}$ be the set of evidence. The *impact* of a finding $\varepsilon_i \in \varepsilon$ on a hypothesis $h \in \mathrm{dom}(H)$ is determined by computing and comparing the prior probability of the hypothesis $P(h)$, the posterior probability of the hypothesis given all evidence $P(h \mid \varepsilon)$, and the posterior probability of the hypothesis given all evidence except the finding under consideration $P(h \mid \varepsilon \setminus \varepsilon_i)$. By comparing these three probabilities, we can identify findings that have a conflicting impact on the probability of the hypothesis compared with the impact of the entire set of evidence.

Example 9.6 (Evidence with Conflict Impact). In the Asia example, we assume evidence $\varepsilon = \{\mathsf{Dyspnoea} = \mathsf{no}, \mathsf{Smoker} = \mathsf{yes}, \mathsf{X\_ray} = \mathsf{no}\} = \{\varepsilon_D, \varepsilon_S, \varepsilon_X\}$. Assume further that $B = \mathsf{Bronchitis}$ is the hypothesis variable under consideration.

Figure 9.2 shows the impact of the finding ε_S on the hypothesis $B = \mathsf{no}$ while Fig. 9.3 shows the impact of the finding ε_S on the hypothesis $B = \mathsf{yes}$. From Fig. 9.2, it is clear that the finding ε_S acts against the hypothesis $B = \mathsf{no}$.

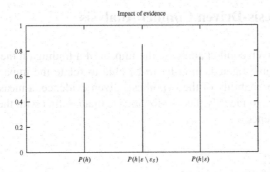

Fig. 9.2 The impact of finding ε_S on the hypothesis $h : B = \text{no}$

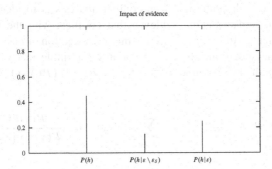

Fig. 9.3 The impact of finding ε_S on the hypothesis $h : B = \text{yes}$

The probability of the hypothesis is higher when the finding ε_S is excluded than when it is included. The posterior is higher than the prior. This implies that the combined effect of the evidence acts in favor of the hypothesis.

Similarly, from Fig. 9.3, it is clear that the finding ε_S acts in favor of the hypothesis $B = \text{yes}$. The probability of the hypothesis is higher when the finding ε_S is included than when it is excluded, but not as high as the prior though. The posterior is lower than the prior. This implies that the combined effect of the evidence acts against the hypothesis.

The numbers in the two graphs are pairwise complementary since B is binary.□

When considering the impact of findings given a large set of evidence, it may be an advantage to use a cost-of-omission threshold to focus on findings with a cost-of-omission greater than the threshold.

Example 9.7 (Example 9.6, cont.). In Example 9.6 the cost-of-omission of finding ε_S is 0.03, that is,

$$c(P(B \,|\, \varepsilon), P(B \,|\, \varepsilon \setminus \{\varepsilon_S\})) = 0.03.$$

□

Further information on hypothesis-driven conflict analysis can be found in Suermondt (1992).

9.3 Summary

In this chapter we have considered evidence and hypothesis-driven conflict analysis in Bayesian networks.

The objective of evidence-driven conflict analysis is to detect possible conflicts in the evidence. To support this analysis, we have defined a conflict measure that is simple to compute. The conflict measure is computed based on an alternative and much simpler model (the straw model). The conflict measure is defined such that a positive value is an indication that a conflict may be present in the evidence. Once a possible conflict is detected, we try to trace and resolve the conflict. We say that a hypothesis may resolve the conflict if the log of the normalized likelihood of the hypothesis is greater than the conflict. Furthermore, a positive conflict measure may originate from a rare case.

The objective of hypothesis-driven conflict analysis is to investigate the impact of a single piece of evidence on the probability of a hypothesis compared with the impact of all the evidence. To support this investigation, we have defined a cost-of-omission measure. The cost-of-omission measure is used to measure the difference between including and excluding the selected piece of evidence on the probability of the hypothesis given evidence. In hypothesis-driven conflict analysis, we relate the prior probability of the hypothesis to the probability of the hypothesis given the entire set of evidence and the probability of the hypothesis given the entire set of evidence except the selected piece of evidence. This enables us to determine whether or not a single piece of evidence conflicts with the remaining set of evidence with respect to the probability of the hypothesis.

In Chap. 10, we consider sensitivity analysis. Evidence sensitivity analysis is to determine the sensitivity of the posterior probability of a hypothesis relative to observations made. Parameter sensitivity is to determine the sensitivity of the posterior probability of a hypothesis relative to parameters of the model.

Exercises

Exercise 9.1. From Example 2.4 on page 25, we know that Dr Watson makes frequent calls to Mr. Holmes regarding the burglar alarm; however, till now the cause of activation of the alarm has been small earthquakes or a big truck passing by near the house. Every time Mr. Holmes rushes home just to find that everything is in order; so now Mr. Holmes is installing a seismometer in his house with a direct line to his office. In this exercise, we assume $P(B) = P(E) = (0.1, 0.9)$.

Fig. 9.4 Mr. Holmes installs
a seismometer with a direct
line to his office

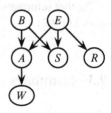

Table 9.2 The conditional probability distributions $P(W \mid A)$ and $P(S \mid B, E)$

W	A = no	A = yes		B	E	S no	some	large
no	0.99	0.65		no	no	0.97	0.02	0.01
yes	0.01	0.35		no	yes	0.01	0.97	0.02
				yes	no	0.01	0.02	0.97
				yes	yes	0	0.03	0.97

Table 9.3 The conditional
probability distribution
$P(A \mid B, E, F)$

E	B	F	A no	yes
no	no	no	0.99	0.01
no	no	yes	0.01	0.99
no	yes	no	0.01	0.99
no	yes	yes	0	1
yes	no	no	0.01	0.99
yes	no	yes	0	1
yes	yes	no	0	1
yes	yes	yes	0	1

The revised model in Fig. 9.4 captures the situation where Mr. Holmes has installed a seismometer in his house with a direct line to the office. Assume S has states reflecting no, some, and large vibrations in the house. The conditional probability distributions $P(W \mid A)$ and $P(S \mid B, E)$ are shown in Table 9.2.

(a) One afternoon, Dr. Watson calls again and announces that Mr. Holmes' alarm has gone off. Mr. Holmes checks the seismometer; it is in state *no*.

Are the two observations in conflict?

(b) Mr. Holmes looks out his window. It rains heavily. When it rains heavily, a flood is likely to occur. Extend the model in Fig. 9.4 to capture these events when the prior of rain is 0.01 and rain causes a flood in one out of ten cases. The conditional probability table $P(A \mid B, E, F)$ where F represents flood is shown in Table 9.3.

Are the observations in conflict in this model and what is the value of the conflict measure?

(c) Is there a potential explanation of the value of the conflict measure?

(d) What is the partial conflict of each subset of the observations and what conclusion can be derived from the partial conflicts?

Exercise 9.2. Consider again the model with a seismometer of Mr. Holmes in Exercise 9.1.

(a) Perform hypothesis-driven conflict analyses with respect to both Burglar = yes and Earthquake = yes.

(b) What is the cost-of-omission for ε_R with respect to Burglar and Earthquake?

Chapter 10
Sensitivity Analysis

We construct probabilistic networks to support and solve problems of belief update and decision making under uncertainty. In problems of belief update the posterior probability of a single hypothesis variable is sometimes of interest. When the evidence set consists of a large number of findings or even when it consists of only a small number of findings questions concerning the impact of subsets of the evidence on the hypothesis or a competing hypothesis emerge.

Evidence sensitivity analysis may, for instance, give answers to questions like *what are the minimum and maximum beliefs produced by observing a variable?*, *which evidence acts in favor of or against a hypothesis?*, *which evidence discriminates one hypothesis from an alternative hypothesis?*, and *what if a certain observed variable had been observed to a value different from the actual value?* Knowing the answers to these and similar questions may help to explain and understand the conclusions reached by the model as a result of probabilistic inference. It will also help to understand the impact of subsets of the evidence on a certain hypothesis and alternative hypotheses.

Evidence sensitivity analysis is not the only possible kind of sensitivity analysis that can be performed on a probabilistic network. Parameter sensitivity analysis is another type of sensitivity analysis that is supported by probabilistic networks. We focus on parameter sensitivity analysis in discrete Bayesian networks. The parameters considered in parameter sensitivity analysis are the entries of the conditional probability distributions specified in the Bayesian network. The analysis is performed relative to a hypothesis and a given set of evidence. It has been shown that there is a (surprisingly) simple correlation between the probability of a set of evidence and an entry of a conditional probability distribution. The probability of the evidence is a linear function of the parameter. This knowledge can be exploited to determine the functional relation between the probability of a hypothesis given a subset of evidence and a parameter of a conditional probability distribution.

Parameter sensitivity analysis is particularly useful for identifying parameters of a probabilistic network that have a large or small impact on the probability of a hypothesis given evidence. When knowledge elicitation resources are limited,

U.B. Kjærulff and A.L. Madsen, *Bayesian Networks and Influence Diagrams: A Guide to Construction and Analysis*, ISS 22, DOI 10.1007/978-1-4614-5104-4_10,
© Springer Science+Business Media New York 2013

parameter sensitivity analysis is a useful tool for identifying and focusing resources on the parameters that are most influential on the posterior probability of a hypothesis given evidence. That is, parameter sensitivity analysis can be used in an attempt to focus knowledge elicitation resources in the model construction process.

In Sect. 10.1, we introduce evidence sensitivity analysis. A distance measure designed to measure the impact of evidence on the probability of a hypothesis is introduced. In the following subsections, we consider identifying minimum and maximum beliefs in a hypothesis given various subsets of the evidence, the impact of different evidence subsets on a hypothesis, how subsets of the evidence discriminates between a pair of competing hypotheses, what-if analysis, and the impact of findings on a hypothesis variable. In Sect. 10.2 (one-way), parameter sensitivity analysis is introduced, and in Sect. 10.3, two-way parameter sensitivity analysis is considered while Sect. 10.4 introduces parameter tuning.

10.1 Evidence Sensitivity Analysis

Evidence sensitivity analysis (SE analysis) is the analysis of how sensitive the results of a belief update (propagation of evidence) is to variations in the set of evidence (observations, likelihood, etc.).

Consider the situation where a decision maker has to make a decision based on the probability distribution of a hypothesis variable. It could, for instance, be a physician deciding on a treatment of a patient given the probability distribution of a disease variable. Prior to deciding on a treatment, the physician may have the option to investigate the impact of the collected information on the posterior distribution of the hypothesis variable, that is, given a set of findings and a hypothesis, which sets of findings are in favor of, against, or irrelevant for the hypothesis, which sets of findings discriminate the hypothesis from an alternative hypothesis, what if a variable had been observed to a different value than the one observed, etc. These questions can be answered by SE analysis. Given a Bayesian network model and a hypothesis variable, the task is to determine how sensitive the belief in the hypothesis variable is to variations in the evidence. We consider one-step look-ahead hypothesis-driven SE analysis on discrete random variables.

Example 10.1 (Evidence Sensitivity Analysis). As an example, we consider SE analysis on the Asia example shown in Fig. 10.1 (see Example 4.2 on page 74 for more details). The hypothesis variable is Bronchitis (B), and the initial set of evidence is $\varepsilon = \{\varepsilon_S, \varepsilon_D\} = \{S = \text{no}, D = \text{yes}\}$. That is, we are considering whether or not the patient is suffering from bronchitis after observing that the patient does not smoke (Smoker = no) and has shortness of breath (Dyspnoea = yes).

This example is used in the following sections to illustrate concepts of SE analysis. $\square$

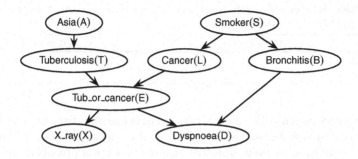

Fig. 10.1 A graph specifying the independence and dependence relations of the Asia example

10.1.1 Distance and Cost-of-Omission Measures

The main purpose of hypothesis-driven SE analysis is to investigate how changes in the set of evidence impact the probability of a hypothesis. In order to perform this investigation, two distance measures are required. Each distance measure will provide a numerical value specifying the *distance* between either two probabilities or two probability distributions.

Let X be a hypothesis variable with state space $\mathrm{dom}(X) = (x_1, \ldots, x_n)$ and let $\varepsilon = \{\varepsilon_1, \ldots, \varepsilon_m\}$ be a set of evidence (findings). We let $\varepsilon_Y \in \varepsilon$ denote the finding on variable $Y \in \mathfrak{X}(\varepsilon)$.

The distance $d(p, q)$ between two probabilities p and q is defined, for $p \neq 0$, as

$$d(p, q) = \left| \frac{q}{p} - 1 \right|.$$

This measure is, for instance, useful for measuring the distance between the probability $P(x \mid \varepsilon)$ of hypothesis x given evidence ε and the probability $P(x \mid \varepsilon \setminus \{\varepsilon_i\})$ of hypothesis x given evidence $\varepsilon \setminus \{\varepsilon_i\}$, that is, the set of evidence where ε_i is excluded from ε.

A pair of probabilities p and q are said to be *almost equal* when their distance $d(p, q)$ is below a predefined threshold δ, that is, $d(p, q) < \delta$.

The *cost-of-omission* $c(P(X \mid \varepsilon), P(X \mid \varepsilon \setminus \{\varepsilon_i\}))$ of ε_i was defined in Sect. 9.2.1 on page 297 as

$$c(P(X \mid \varepsilon), P(X \mid \varepsilon \setminus \{\varepsilon_i\})) = \sum_{x \in \mathrm{dom}(X)} P(x \mid \varepsilon) \log \left(\frac{P(x \mid \varepsilon)}{P(x \mid \varepsilon \setminus \{\varepsilon_i\})} \right).$$

Notice the difference between the distance measure and the cost-of-omission measure. The distance measure evaluates the distance between probability values whereas the cost-of-omission measure evaluates the distance between two posterior

Table 10.1 Sensitivity of the posterior probability distribution of the hypothesis variable B to findings on A

b	$\min_a P(B = b\,|\,\varepsilon, a)$	$P(B = b\,|\,\varepsilon)$	$\max_a P(B = b\,|\,\varepsilon, a)$
no	0.228	0.228	0.236
yes	0.764	0.772	0.772

probability distributions relative to omitting a certain finding ε_i from the evidence ε. The cost-of-omission measure is a special case of the more general cross entropy distance (or Kullback–Leibler distance) measure between a probability distribution P and an approximation P' of P:

$$
D_{KL}(P, P') = \sum_{x \in \text{dom}(X)} P(x) \log \left(\frac{P(x)}{P'(x)} \right).
$$

The cost-of-omission measure is, for instance, useful for measuring the distance between the posterior probability distribution $P(X\,|\,\varepsilon)$ of hypothesis variable X given evidence ε and the posterior probability distribution $P(X\,|\,\varepsilon \setminus \{\varepsilon_i\})$ of hypothesis variable X given evidence $\varepsilon \setminus \{\varepsilon_i\}$, that is, the set of evidence where ε_i is excluded from ε.

In the following sections, the distance measures defined above are used to introduce different concepts related to SE analysis.

10.1.2   Identify Minimum and Maximum Beliefs

As part of performing SE analysis, we may be interested in knowing the minimum and maximum values of the posterior belief for each possible state $x \in \text{dom}(X)$ of the hypothesis variable X given all possible observations on a given variable $Y \notin \mathcal{X}(\varepsilon)$, that is, what are the minimum and maximum values of $P(x\,|\,\varepsilon, y)$ as a function of $y \in \text{dom}(Y)$.

The minimum $\min_{y \in \text{dom}(Y)} P(x\,|\,\varepsilon, y)$ and maximum $\max_{y \in \text{dom}(Y)} P(x\,|\,\varepsilon, y)$ values of the posterior belief are determined by entering and propagating each state y of the information variable Y. This analysis requires one belief update for each state of variable Y.

This analysis identifies the range of the posterior belief in a hypothesis as a function of possible observations on an unobserved variable. This may help to determine the impact of a possible observation on the probability of the hypothesis.

Example 10.2 (Example 10.1, cont.).   Table 10.1 shows the sensitivity of the posterior probability distribution $P(B\,|\,\varepsilon, a)$ of the hypothesis variable B relative to instantiations of the unobserved variable Asia (A).

For each state $b \in \text{dom}(B)$ of B, the minimum posterior belief $\min_a P(B = b\,|\,\varepsilon, a)$, the current belief $P(B = b\,|\,\varepsilon)$, and the maximum posterior belief

ε′			NL
ε_S	ε_X	ε_D	1.256
ε_S	ε_X	–	0.667
ε_S	–	ε_D	1.675
ε_S	–	–	0.667
–	ε_X	ε_D	1.515
–	ε_X	–	1.125
–	–	ε_D	1.853
–	–	–	1

Table 10.2 Normalized likelihood of hypothesis h_B given all subsets of the evidence ε

$\min_a P(B = b|\varepsilon, a)$ are shown. From the table, it is clear that an observation on A would produce insignificant variations in the posterior belief in any state of B. □

10.1.3 Impact of Evidence Subsets

Investigation of the impact of different subsets of the evidence ε on each state $x \in$ dom(X) of the hypothesis variable X is a useful part of SE analysis. Investigating the impact of different subsets of the evidence on states of the hypothesis may help to determine subsets of the evidence acting in favor of or against each possible hypothesis.

The impact of a subset of the evidence $\varepsilon' \subseteq \varepsilon$ on a state x of the hypothesis variable X is determined by computing the *normalized likelihood NL* of the hypothesis x given evidence ε', that is,

$$NL = \frac{P(\varepsilon'|x)}{P(\varepsilon')} = \frac{P(\varepsilon', x)/P(x)}{P(\varepsilon')} = \frac{P(x|\varepsilon')P(\varepsilon')/P(x)}{P(\varepsilon')} = \frac{P(x|\varepsilon')}{P(x)},$$

where we assume $P(\varepsilon') > 0$ and $P(x) > 0$. This fraction is computed by entering and propagating ε'. Therefore, this analysis requires one belief update for each subset ε' of the evidence ε.

Each normalized likelihood is a measure of the impact of a subset of evidence on the hypothesis. By comparing the normalized likelihoods of different subsets of the evidence, we compare the impacts of the subsets of evidence on the hypothesis.

Example 10.3 (Example 10.2, cont.).   Assume that we observe the patient to have a positive X-ray result $X =$ yes, such that the set of evidence is $\varepsilon = \{\varepsilon_S, \varepsilon_D, \varepsilon_X\} = \{S =$ no, $D =$ yes, $X =$ yes$\}$. Table 10.2 shows the normalized likelihood of the hypothesis $h_B : B =$ yes given the evidence $\varepsilon = \{\varepsilon_S, \varepsilon_D, \varepsilon_X\}$.

From Table 10.2, it is clear that the finding ε_D on D acts in favor of the hypothesis h_B. On the other hand, the evidence ε_S acts slightly against the hypothesis h_B while ε_X is irrelevant, against, and in favor of h_B depending on the remaining evidence. □

Table 10.3 Discrimination
between hypothesis h_B and
the alternative hypothesis h_L

ε'			B
ε_S	ε_X	ε_D	0.281
ε_S	ε_X	$-$	0.258
ε_S	$-$	ε_D	3.869
ε_S	$-$	$-$	3.667
$-$	ε_X	ε_D	0.134
$-$	ε_X	$-$	0.127
$-$	$-$	ε_D	0.992
$-$	$-$	$-$	1

10.1.4 Discrimination of Competing Hypotheses

A central question considered by SE analysis is how different subsets of the evidence discriminate between competing hypotheses. The challenge is to compare the impact of subsets of the evidence on competing hypotheses.

We consider the discrimination between two different hypotheses represented as states of two different variables. Thus, let X be the hypothesis variable of interest and let Y be an alternative hypothesis variable where $X \neq Y$.

In Sect. 3.5, we describe how the discrimination of a pair of competing hypotheses $x \in \text{dom}(X)$ and $y \in \text{dom}(Y)$ may be based on the calculation of *Bayes' factor B* (or *Bayesian likelihood ratio*)for all subsets $\varepsilon' \subseteq \varepsilon$ of a set of evidence ε:

$$B = \frac{\text{posterior odds ratio}}{\text{prior odds ratio}} = \frac{P(x|\varepsilon')/P(y|\varepsilon')}{P(x)/P(y)} = \frac{P(\varepsilon'|x)}{P(\varepsilon'|y)} = \frac{L(x|\varepsilon')}{L(y|\varepsilon')}, \quad (10.1)$$

where we assume $P(x) > 0$, $P(y) > 0$, and $P(\varepsilon') > 0$. Bayes' factor is the ratio of the likelihoods of hypothesis x and y given the evidence ε'.

From (10.1), we see that:

$B > 1$ if the evidence ε' provides more support for x than for y.

$B < 1$ if the evidence ε' provides less support for x than for y.

$B = 1$ if the evidence ε' does not provide useful information for differentiating between x and y.

This analysis requires one belief update for each subset $\varepsilon' \subseteq \varepsilon$.

Example 10.4 (Example 10.3, cont.). Assume that h_L : Cancer $=$ yes is an alternative hypothesis to the hypothesis h_B. Table 10.3 shows Bayes' factor for the hypothesis h_B and the alternative hypothesis h_L.

From Table 10.3, it is clear that subsets $\{\varepsilon_S\}$ and $\{\varepsilon_S, \varepsilon_D\}$ act in favor of the hypothesis h_B when compared with the alternative hypothesis h_L. On the other hand, the remaining subsets act slightly against the hypothesis h_B when compared with the alternative hypothesis h_L. □

Table 10.4 What-if analysis on findings

Y	y'	$P(B = \text{yes}\,\vert\,\varepsilon \setminus \{\varepsilon_Y\}, y')$
D	no	0.092
D	yes	0.565
S	no	0.565
S	yes	0.714
X	no	0.774
X	yes	0.565

10.1.5 What-If Analysis

In *what-if* analysis, the type of question considered is the following. What if the finding on a discrete random variable $Y \in X(\varepsilon)$ had been the observation $Y = y'$ instead of $Y = y$ (represented by the finding $\varepsilon_Y \in \varepsilon$, where ε is the set of evidence)? We consider a hypothesis-driven approach to what-if SE analysis.

Hypothesis-driven what-if analysis is performed by computing the posterior probability distribution $P(X \,\vert\, \varepsilon \setminus \{\varepsilon_Y\}, y')$ of the hypothesis variable X for each possible state $y' \neq y$ of the observed variable Y.

The posterior probability distribution $P(X \,\vert\, \varepsilon \setminus \{\varepsilon_Y\}, y')$ specifies the impact of each possible instantiation of Y on the posterior distribution of the hypothesis variable X. The analysis requires one belief update for each $y' \in \text{dom}(Y)$. Notice that ε_Y need not be an instantiation of Y, that is, ε_Y may be soft evidence.

Example 10.5 (Example 10.4, cont.). For each finding variable Y, we may consider the impact of each possible observation $Y = y$. Table 10.4 shows the posterior belief in the hypothesis $B = \text{yes}$ given the evidence where the finding y is substituted with each possible state $y' \in \text{dom}(Y)$.

From Table 10.4, it is clear that changing the observation on D from yes to no has a significant impact of the posterior belief in the hypothesis $B = \text{yes}$. On the other hand, the posterior belief in the hypothesis $B = \text{yes}$ has a lower sensitivity to observations on S and X.

Since each finding (i.e., ε_S, ε_X, or ε_D) is an instantiation, one row for each observed variable Y corresponds to the posterior belief in the hypothesis $B = \text{yes}$. That is, $P(B = \text{yes}\,\vert\,\varepsilon) = 0.565$ is represented three times in the table. □

10.1.6 Impact of Findings

Let $X = x$ be the hypothesis of interest where X is the hypothesis variable and let ε be the entire set of evidence. The impact of each finding $\varepsilon_Y \in \varepsilon$ on the probability of x is determined by computing and comparing the prior probability of the hypothesis, $P(x)$, the posterior probability of the hypothesis given the entire set of evidence, $P(x\,\vert\,\varepsilon)$, and the posterior probability of the hypothesis given the entire set of evidence except the finding ε_Y, $P(x\,\vert\,\varepsilon \setminus \{\varepsilon_Y\})$.

Table 10.5 Findings impact analysis

ε_Y	$P(B = \text{yes})$	$P(B = \text{yes}\,|\,\varepsilon \setminus \{\varepsilon_Y\})$	$P(B = \text{yes}\,|\,\varepsilon)$
ε_S	0.45	0.682	0.565
ε_X	0.45	0.754	0.565
ε_D	0.45	0.3	0.565

To relate the impact of individual findings on the probability of the hypothesis, we define the notions of an important finding, a redundant finding, and an irrelevant finding:

- A finding $\varepsilon_Y \in \varepsilon$ is *important* when the difference between the probability $q = P(x\,|\,\varepsilon \setminus \{\varepsilon_Y\})$ of the hypothesis given the entire set of evidence except ε_Y and the probability $p = P(x\,|\,\varepsilon)$ of the hypothesis given the entire set of evidence is *too large*; that is, the probabilities p and q are not *almost equal* $(d(p,q) \geq \delta)$.
- A finding $\varepsilon_Y \in \varepsilon$ is *redundant* when $q = P(x\,|\,\varepsilon \setminus \{\varepsilon_Y\})$ is *almost equal* to $p = P(x\,|\,\varepsilon)$; that is, $d(p,q) < \delta$.
- A finding $\varepsilon_Y \in \varepsilon$ is *irrelevant* when $q = P(x\,|\,\varepsilon' \setminus \{\varepsilon_Y\})$ is *almost equal* to $p = P(x\,|\,\varepsilon')$ for all subsets ε'; that is, $d(p,q) < \delta$ for all subsets ε'. That is, the finding ε_Y is redundant for all subsets of the evidence.

The term *almost equal* is defined based on the distance measure introduced in Sect. 10.1.1. Similarly, a sufficient set of evidence is defined as:

- A subset of evidence $\varepsilon' \subseteq \varepsilon$, for example, the entire set of evidence ε except a certain finding ε_Y, is *sufficient* when $q = P(x\,|\,\varepsilon')$ is *almost equal* to $p = P(x\,|\,\varepsilon)$; that is, $d(p,q) < \delta$.

The impact of each finding ε_Y may be considered for each state or a certain state of the hypothesis variable X. Sufficiency δ_s and importance δ_i thresholds should be specified by the user.

Example 10.6 (Example 10.5, cont.). We may be interested in considering the impact of each finding $\varepsilon_Y \in \varepsilon$ on the probability of the hypothesis $B = \text{yes}$ by comparing $P(B = \text{yes})$, $P(B = \text{yes}\,|\,\varepsilon)$, and $P(B = \text{yes}\,|\,\varepsilon \setminus \{\varepsilon_Y\})$. Table 10.5 shows the prior belief in the hypothesis $B = \text{yes}$, the posterior belief given all evidence $P(B = \text{yes}\,|\,\varepsilon)$, and the posterior belief given all evidence except a single finding $P(B = \text{yes}\,|\,\varepsilon \setminus \{\varepsilon_Y\})$.

From Table 10.5, we make the following observations:

- The finding ε_S is important. At sufficiency threshold $\delta_s = 0.02$, the finding ε_S is redundant, and $\varepsilon \setminus \{\varepsilon_S\}$ is sufficient. At cost-of-omission threshold $\delta_o = 0.03$, the evidence ε_S would not be included in the analysis.
- The finding ε_X is important.
- The finding ε_D is important.

The subset $\varepsilon' = \{\varepsilon_S, \varepsilon_X, \varepsilon_D\} = \varepsilon$ is, of course, also sufficient. The analysis is performed using threshold values $\delta_o = 0.0001$, $\delta_s = 0.02$, and $\delta_i = 0.05$.

Table 10.6 shows the cost-of-omission $c(P(B\,|\,\varepsilon), P(B\,|\,\varepsilon \setminus \{\varepsilon_Y\}))$ and the distance $d(P(B = \text{yes}\,|\,\varepsilon), P(B = \text{yes}\,|\,\varepsilon \setminus \{\varepsilon_Y\}))$ for each finding ε_Y.   □

Table 10.6 Cost-of-omission and distance in posterior beliefs of the hypothesis for each finding

ε_Y	$c(P(B\mid\varepsilon), P(B\mid\varepsilon\setminus\{\varepsilon_Y\}))$	$d(P(B = \text{yes}\mid\varepsilon), P(B = \text{yes}\mid\varepsilon\setminus\{\varepsilon_Y\}))$
ε_S	0.03	0.171
ε_X	0.085	0.25
ε_D	0.151	0.884

10.2 Parameter Sensitivity Analysis

Parameter sensitivity analysis (SP analysis) is the analysis of how sensitive the results of a belief update (propagation of evidence) is to variations in the value of a parameter of the model. The parameters of a model are the entries of the conditional probability distributions.

Consider the situation where company management allocates resources to research and development projects based on an estimation of projects' successes. The success of a project depends on the ability of the project management to obtain certain environmental permissions from the authorities. This and other properties of the domain are modeled as a Bayesian network. As part of the model construction, the knowledge engineers (and the management of the company) would like to assess how sensitive the conclusion of the model (i.e., the probability of project success) is to the prior probability of a specific environmental permission being obtained. Parameter sensitivity analysis is designed to answer such questions. Given a Bayesian network model, a hypothesis, and a set of evidence, the task is to determine the sensitivity of the posterior belief in the hypothesis to variations in the value of an assessed parameter.

We consider scenario-based SP analysis on discrete random variables with respect to changes in a single parameter value where a scenario S is defined as a vector consisting of a hypothesis variable H, a state of the hypothesis variable h, and a set of evidence ε, that is, $S = (H, h, \varepsilon)$.

Example 10.7 (Parameter Sensitivity Analysis). Apple Jack, see Example 4.1 on page 71, is interested in computing the probability of his finest apple tree being sick. His hypothesis is that the tree is sick, that is, the hypothesis is Sick = yes. The evidence available to support the reasoning of Apple Jack is that the tree is losing its leaves. Thus, the scenario under consideration is $S = (\text{Sick}, \text{yes}, \{\text{Loses} = \text{yes}\})$. Given the Bayesian network shown in Fig. 4.1 on page 72 and the conditional probability distributions $P(\text{Sick}) = (0.95, 0.05)$, $P(\text{dry}) = (0.9, 0.1)$, and $P(\text{Loses}\mid\text{Sick}, \text{Dry})$ as specified in Table 4.1 on page 73, the posterior distribution of the hypothesis given the observation, $P(\text{Sick} = \text{yes}\mid\varepsilon)$, is 31.62 %.

How sensitive is this posterior probability to small variations in the quantification of the model? For instance, how would the posterior probability of the hypothesis change if the prior probability of Dry = yes decreases from 0.1 to 0.075? Setting the prior probability of Dry = yes to 0.075, the posterior distribution of the hypothesis given the observation, $P(\text{Sick} = \text{yes}\mid\varepsilon)$, is 36.64 %.

This shows that the posterior probability of the hypothesis Sick = yes (and other events) changes when the prior on Dry = yes is changed to 0.075 (from 0.1). The posterior probability of Sick = yes increases from 31.52 % to 36.64 %. This seems intuitive as the prior of one cause decreases and the posterior of another cause increases given the observed effect. However, what if the prior had been changed to a different value? Is it necessary to compute the posterior probability of the hypothesis for all possible values of $P(\text{Dry} = \text{yes})$, that is, the parameter we are investigating?

The aforementioned questions may be answered more efficiently using parameter sensitivity analysis. $\square$

10.2.1   Sensitivity Function

Parameter sensitivity analysis is based on the observation that the probability of the evidence is a linear function of any single parameter in the model, that is, any entry of any conditional probability distribution (Castillo, Gutiérrez & Hadi 1997, Coupé & van der Gaag 1998). That is, $y = P(\varepsilon)$ as a function of a conditional probability $t = P(X = x \mid \text{pa}(X) = z)$ has the simple form $y = \alpha \cdot t + \beta$ where $\alpha, \beta \in \mathbb{R}$. This implies that the conditional probability of a hypothesis h given evidence ε as a function of a parameter t has the form

$$f(t) = P(h \mid \varepsilon)(t) = \frac{P(h, \varepsilon)(t)}{P(\varepsilon)(t)} = \frac{\alpha \cdot t + \beta}{\gamma \cdot t + \delta},$$

where $\alpha, \beta, \gamma, \delta \in \mathbb{R}$. Hence, the posterior probability $P(h \mid \varepsilon)$ is a quotient of two linear functions of the parameter t.

The function $f(t)$ is known as the *sensitivity function*. The coefficients of the sensitivity function are determined separately for its numerator and denominator functions. The coefficients of a linear function can be determined from two values of the function for two different values of the parameter. We can compute the value of the function for two different values of the parameter by propagating the evidence twice (once for each of the two parameter values). When determining the coefficients of the sensitivity function, we use proportional scaling to change the remaining related parameters such that they keep the original proportion. This implies that when we change the parameter value for t, the remaining probability values for the corresponding parent configuration (i.e., $P(X = x' \mid \text{pa}(X) = z)$ for all $x \neq x'$) are scaled proportionally (Chan & Darwiche 2002a). We need to scale the values such that the values sum to one; that is, $\sum_x P(x \mid \text{pa}(X)) = 1$. Assume $P(X \mid \text{pa}(X) = z) = (p_1, \dots, p_n)$ is the initial assessment of the probability of X given $\text{pa}(X) = z$ and p_i is the parameter value under consideration. Proportional scaling on p_j for $j \neq i$ when changing p_i to p_i* amounts to computing

$$p_j^* = \frac{p_j(1 - p_i^*)}{\sum_{j \neq i} p_j},$$

where $(p_1^*, \dots, p_n^*)$ is the updated probability of X given $\text{pa}(X) = z$.

We assume that each parameter can be varied independently of other parameters and that each parameter is non-extreme, that is, it can be varied in an open interval around its initial assessment.

Example 10.8 (Proportional Scaling). Assume that the variable Dry has three states, no, dry, and Very dry, with a prior distribution $P(\text{Dry}) = (0.9, 0.08, 0.02)$. Let $S = (\text{Dry}, \text{no}, \{\text{Loses} = \text{yes}\})$ be the scenario under consideration. If we want to investigate the impact of adjusting the parameter $P(\text{Dry} = \text{no}) = 0.9$ to 0.875, then it is necessary to adjust the values of the other two parameters such that all three adjusted parameters sum to one. This is achieved by proportional scaling such that the adjusted prior distribution becomes

$$P(\text{Dry}) = \left(0.875, \frac{0.08 \cdot (1 - 0.875)}{0.08 + 0.02}, \frac{0.02 \cdot (1 - 0.875)}{0.08 + 0.02}\right)$$

$$= (0.875, 0.1, 0.025).$$

When a variable has only two states, a change in the value of one parameter must induce a similar (but opposite) change in the other parameter. $\square$

Example 10.9 (Sensitivity Function). Let $S = (\text{Sick}, \text{yes}, \varepsilon = \{\text{Loses} = \text{yes}\})$ be the scenario under consideration. Hence, the hypothesis under investigation is Sick = yes while the parameter in focus is $t = P(\text{Dry} = \text{yes})$. The sensitivity function $f(t)$ where t is the parameter for $P(\text{Dry} = \text{yes})$ is

$$f(t) = P(\text{Sick} = \text{yes} \mid \text{Loses} = \text{yes})$$

$$= \frac{P(\text{Sick} = \text{yes}, \text{Loses} = \text{yes})}{P(\text{Loses} = \text{yes})}$$

$$= \frac{\alpha \cdot t + \beta}{\gamma \cdot t + \delta}$$

$$= \frac{0.0025 \cdot t + 0.045}{0.791 \cdot t + 0.064}$$

The coefficients of denominator and numerator functions are determined separately. Both functions are linear in the parameter t. Thus, the coefficients of each function can be determined by propagating evidence for two different parameter values. For instance, the coefficients γ and δ of the denominator can be determined as

$$\gamma = \frac{P(\text{Loses} = \text{yes})(t_1) - P(\text{Loses} = \text{yes})(t_0)}{t_1 - t_0} = \frac{0.2222 - 0.1431}{0.2 - 0.1} = 0.791$$

$$\delta = P(\text{Loses} = \text{yes})(t_0) - \gamma \cdot t_0 = 0.1431 - 0.791 \cdot 0.1 = 0.064,$$

where $t_0 = 0.1$ and $t_1 = 0.2$ are two different values of the parameter t.

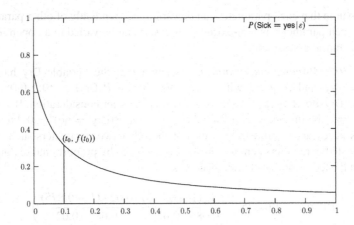

Fig. 10.2 The graph of the sensitivity function $f(t) = P(\text{Sick} = \text{yes} | \text{Loses} = \text{yes})$ as a function of $t = P(\text{Dry} = \text{yes})$

The graph of the sensitivity function $f(t)$ for all possible values of t, that is, values of t between zero and one, is plotted in Fig. 10.2.

Figure 10.2 shows that the minimum value of the probability of the hypothesis is 0.0556 for $t = 1$ while the maximum value of the probability of the hypothesis is 0.7031 for $t = 0$. Thus, no matter what value of t is specified $P(\text{Sick} = \text{yes} | \varepsilon)$ is between 0.0556 and 0.7031. In addition, it is clear from Fig. 10.2 that the posterior probability of the hypothesis is more sensitive to small variations in the parameter value when the initial parameter value is in the range from 0 to, say, 0.25 than when the initial parameter is in the range from 0.25 to 1. $\qquad\square$

Performing two full propagations of the evidence for each parameter value may be inefficient if the number of parameters is large. Jensen (2001) describes a modeling technique for computing the coefficients of the linear function based on introducing an auxiliary variable (for each parameter inspected). By introducing an auxiliary variable, it is possible to reduce the number of messages to be passed in the junction tree representation. Kjærulff & van der Gaag (2000) describe an approach for making sensitivity analysis computationally efficient while Madsen (2005) describes a propagation method that makes it possible to compute the coefficients for all parameters from a single propagation in a junction tree representation.

10.2.2   Sensitivity Value

The partial derivative $f'(t) = \partial P(h | \varepsilon) / \partial t$ of the sensitivity function $f(t)$ with respect to t expresses how much $P(h | \varepsilon)(t)$ changes as a function of t given small variations in the initial assessment. The partial derivative $f'(t)$ of the sensitivity function on t is

$$f'(t) = \frac{\alpha \cdot \delta - \beta \cdot \gamma}{(\gamma \cdot t + \delta)^2}.$$

The partial derivative reflects how the posterior probability of the hypothesis h changes with changes in the parameter t under evidence scenario ε. Wang, Rish & Ma (2002) define parameter sensitivity $S(t \mid h, \varepsilon)$ as $f'(t) = \partial P(h \mid \varepsilon)/\partial t$. The sensitivity value of a parameter is defined as $|S(t \mid h, \varepsilon)|$, that is, the absolute value of the derivative of the sensitivity function at the initial assessment of the parameter (Laskey 1993). The sign of $S(t \mid h, \varepsilon)$ indicates whether the probability of the hypothesis is increasing or decreasing in t.

The *sensitivity value* can be used as a guide to identify parameter assessments where small variations in the value may have the largest impact on the posterior probability of the hypothesis given the evidence. In general, a parameter is of interest when the sensitivity value is greater than zero. Parameter sensitivity analysis enables us to identify the most important parameter assessments in the Bayesian network. Let t' and t'' be two parameters. If t' has a higher sensitivity value than t'', then t' will intuitively induce a larger change on the probability of interest than t'' given the same variation in the parameter assessments.

When considering sensitivity analysis with respect to multiple evidence sets, we may choose to weigh the parameter sensitivity with the probability of the evidence, that is, $P(\varepsilon) \cdot S(t \mid h, \varepsilon)$ as different evidence scenarios may have different probabilities. Parameter importance defines the importance of a parameter across multiple evidence scenarios ε and multiple hypotheses h. Wang et al. (2002) define parameter importance $I(t)$ as

$$I(t) = \frac{1}{mn} \sum_{h,\varepsilon} S(t \mid h, \varepsilon) = \frac{1}{mn} \sum_{h,\varepsilon} \frac{\partial P(h \mid \varepsilon)}{\partial t},$$

where m is the number of hypotheses and n is the number of evidence scenarios. The importance value of a parameter is defined as $|I(t_0)|$, that is, the absolute value of the parameter importance function at the initial assessment of the parameter.

Example 10.10 (Sensitivity Value). Let $S = (\text{Dry}, \text{yes}, \{\text{Loses} = \text{yes}\})$ be the scenario under consideration. The initial value of the parameter of interest $t = P(\text{Dry} = \text{yes})$ is $t_0 = 0.1$. This implies that the sensitivity value of the parameter t is $f'(t_0) = 17$.

On the other hand, the conditional probability $P(\text{Loses} = \text{yes} \mid \text{Dry} = \text{yes}, \text{Sick} = \text{yes})$ has sensitivity function

$$f(t) = \frac{0.005 \cdot t + 0.13835}{0.005 \cdot t + 0.0405}.$$

The initial parameter assessment is $t_0 = 0.95$, and the sensitivity value is 0.24. □

10.2.3 Admissible Deviation

Parameter sensitivity values may be used to focus the knowledge elicitation resources in the model construction process. The sensitivity function, its derivative, and the sensitivity value are not sufficient tools for analyzing how the change in a parameter t may change the most likely state of a hypothesis variable H.

We extend the basic sensitivity analysis method with the calculation of an interval within which the parameter under investigation can be varied without changing the most likely value of the hypothesis variable of interest. Let H be the hypothesis variable of interest, ε the evidence scenario under consideration, and t the parameter under investigation with initial assessment t_0. The admissible deviation of t from t_0 is a pair of real numbers (r, s) such that t can be varied from $\max(0, t_0 - r)$ to $\min(1, t_0 + s)$ without changing the most likely state of H, that is, $\arg \max P(h | \varepsilon)$ is unchanged by the deviation of t from t_0. The values of (r, s) should be the largest such numbers for which the property is satisfied (van der Gaag & Renooij 2001). The admissible deviation interval is $[\min(0, t_0 - r); \max(1, t_0 + s)]$. Notice that the interval specifying the admissible deviation is in general not symmetric around the value t_0, that is, in general r is not equal to s.

Some parameters may take any value without changing the most likely state of H. This implies that the value of the parameter can be varied over the entire range $[0; 1]$ without changing the most likely state of the hypothesis. In this case the admissible deviation interval is specified as (∞, ∞).

Example 10.11 (Admissible Deviation).   Assume that Apple Jack is interested in determining whether or not his apple tree is sick. The hypothesis variable of interest is Sick, the evidence scenario is Loses = yes, and the parameter under investigation is $t = P(\text{Dry} = \text{yes})$. How much can the parameter t be varied without inducing a change in the most likely state of Sick? Apple Jack wants to know the admissible deviation of t.

The sensitivity function $f(t)$ for Sick = no is

$$f(t) = \frac{0.7885 \cdot t + 0.019}{0.791 \cdot t + 0.064}.$$

The sensitivity functions for Sick = no and Sick = yes are shown in Fig. 10.3.

For $t = 0.033$, we have $P(\text{Sick} = \text{yes} | \varepsilon) = P(\text{Sick} = \text{no} | \varepsilon) = 0.5$. Since the hypothesis variable Sick is binary, the two states have equal probability when they both have probability 0.5. Assuming that $t_0 = 0.1$, the admissible deviation of t is the pair $(-0.0967, \infty)$, that is, the value of the parameter t can be varied from $0.033 = 0.1 - 0.0967$ to 1 without changing the hypothesis with highest probability. $\Box$

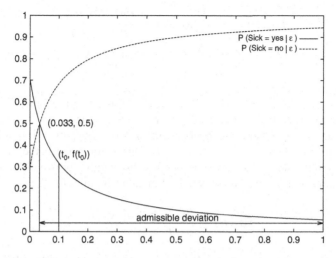

Fig. 10.3 The graph of the sensitivity functions for Sick = yes and Sick = no

10.3 Two-Way Parameter Sensitivity Analysis

In Sect. 10.2, we considered one-way parameter sensitivity analysis, that is, how sensitive the results of a belief update are to variations in the value of a single parameter of the model. In this section, we consider the case of two-way parameter sensitivity analysis where the values of two parameters of the model are varied jointly.

Two-way parameter sensitivity analysis is considered under the same assumptions as the one-way parameter sensitivity analysis considered in Sect. 10.2.

10.3.1 Sensitivity Function

As mentioned in Sect. 10.2, one-way parameter sensitivity analysis is based on the observation that the probability of the evidence is a linear function of any single parameter in the model, that is, any entry of any conditional probability distribution. In the case of two-way parameter sensitivity analysis, the probability of the evidence is a multilinear function of any two parameters in the model, i.e., any two entries of any one or two conditional probability distributions. That is, $y = P(\varepsilon)$ as a function of two conditional probabilities $t_1 = P(X_1 = x_1 | \mathrm{pa}(X_1) = z_1)$ and $t_2 = P(X_2 = x_2 | \mathrm{pa}(X_2) = z_2)$ has the slightly more complex form of a multilinear function $y = \alpha \cdot t_1 \cdot t_2 + \beta \cdot t_1 + \gamma \cdot t_2 + \delta$ where $\alpha, \beta, \gamma, \delta \in \mathbb{R}$. This implies that the conditional probability of a hypothesis h given evidence ε as a function of two parameters t_1 and t_2 has the form

$$f(t_1, t_2) = P(h \mid \varepsilon)(t_1, t_2) = \frac{P(h, \varepsilon)(t_1, t_2)}{P(\varepsilon)(t_1, t_2)}$$

$$= \frac{\alpha_1 \cdot t_1 \cdot t_2 + \beta_1 \cdot t_1 + \gamma_1 \cdot t_2 + \delta_1}{\alpha_2 \cdot t_1 \cdot t_2 + \beta_2 \cdot t_1 + \gamma_2 \cdot t_2 + \delta_2},$$

where $\alpha_1, \alpha_2, \beta_1, \beta_2, \gamma_1, \gamma_2, \delta_1, \delta_2 \in \mathbb{R}$. Hence, the posterior probability $P(h \mid \varepsilon)$ is a quotient of two multilinear functions of the parameters t_1 and t_2.

If the parameters t_1 and t_2 are independent, then $y = \alpha \cdot t_1 \cdot t_2 + \beta \cdot t_1 + \gamma \cdot t_2 + \delta$ simplifies to $y = \beta \cdot t_1 + \gamma \cdot t_2 + \delta$ as $\alpha = 0$. This is, for instance, the case for two parameters in the same conditional probability distribution for two different parent configurations. In this case, the sensitivity function simplifies to

$$f(t_1, t_2) = \frac{\beta_1 \cdot t_1 + \gamma_1 \cdot t_2 + \delta_1}{\beta_2 \cdot t_1 + \gamma_2 \cdot t_2 + \delta_2},$$

where $\beta_1, \beta_2, \gamma_1, \gamma_2, \delta_1, \delta_2 \in \mathbb{R}$. Hence, the posterior probability $P(h \mid \varepsilon)$ is a quotient of two multilinear functions of the parameters t_1 and t_2.

Example 10.12 (Two-Way Sensitivity Analysis). As mentioned in Sect. 10.2, Apple Jack, see Example 4.1 on page 71, is interested in computing the probability of his finest apple tree being sick. His hypothesis is that the tree is sick, that is, the hypothesis is Sick = yes. The evidence available to support the reasoning of Apple Jack is that the tree is losing its leaves. Thus, the scenario under consideration is $S = $ (Sick, yes, {Loses = yes}). Given the Bayesian network shown in Fig. 4.1 on page 72 and the conditional probability distributions $P(\text{Sick}) = (0.95, 0.05)$, $P(\text{Dry}) = (0.9, 0.1)$, and $P(\text{Loses} \mid \text{Sick}, \text{Dry})$ as specified in Table 4.1 on page 73, the posterior distribution of the hypothesis given the observation, $P(\text{Sick} = \text{yes} \mid \varepsilon)$, is 31.62 %.

In Sect. 10.2, we considered how sensitive this posterior probability is to small variations in a single parameter of the model. The sensitivity function $f(t)$ for parameter $t = P(\text{Dry} = \text{yes})$ was computed as

$$f(t) = \frac{0.0025 \cdot t + 0.045}{0.791 \cdot t + 0.064}.$$

Here we consider how sensitive this posterior probability is to small variations in a pair of parameters of the model. Let $t_1 = P(\text{Dry} = \text{yes})$ and $t_2 = P(\text{Loses} = \text{yes} \mid \text{Dry} = \text{yes}, \text{Sick} = \text{yes})$ be the two parameters in focus. The sensitivity function $f(t_1, t_2)$ is computed as

$$f(t_1, t_2) = \frac{0.05 \cdot t_1 \cdot t_2 - 0.045 \cdot t_1 + 0 \cdot t_2 + 0.045}{0.05 \cdot t_1 \cdot t_2 + 0.7435 \cdot t_1 + 0 \cdot t_2 + 0.064}.$$

For each multilinear function, the coefficients can be calculated by entering and propagating two different (non-extreme) values for each parameter (resulting in

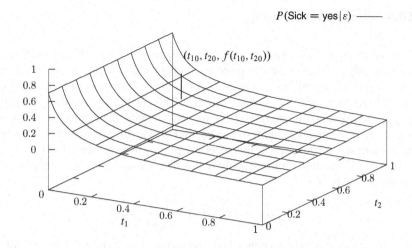

$P(\text{Sick} = \text{yes} \,|\, \varepsilon)$ ———

Fig. 10.4 The graph of the sensitivity function $f(t_1, t_2) = P(\text{Sick} = \text{yes} \,|\, \text{Loses} = \text{yes})$ as a function of $t_1 = P(\text{Dry} = \text{yes})$ and $t_2 = P(\text{Loses} = \text{yes} \,|\, \text{Dry} = \text{yes}, \text{Sick} = \text{yes})$, where $t_0 = (t_{1_0} = 0.1, t_{2_0} = 0.95)$ are the initial values of the two parameters in focus

four propagations) and solving the resulting four linear equations. That is, these calculations are performed for both the numerator and the denominator.

The graph of the sensitivity function $f(t_1, t_2)$ for all possible values of t_1 and t_2, that is, values of t_1 and t_2 between zero and one, is plotted in Fig. 10.4.

Figure 10.4 shows that the minimum value of the probability of the hypothesis is 0 for $t_1 = 1$ and $t_2 = 0$ while the maximum value of the probability of the hypothesis is 0.7031 for $t_1 = 0$ and $t_2 = 1$ (t_2 can take any value for $t_1 = 0$). Thus, no matter what values of t_1 and t_2 are specified $P(\text{Sick} = \text{yes} \,|\, \varepsilon)$ is between 0 and 0.7031. In addition, it is clear from Fig. 10.4 that the posterior probability of the hypothesis is sensitive to small variations in the initial parameter values of t_1 and t_2. Given the initial parameter assessments, the posterior probability seems most sensitive to variations in t_1.   □

Notice that the parameters under consideration must not be 0 or 1 because the method used to compute the constants in the sensitivity function will fail in these cases.

In principle, it is possible to perform an n-way sensitivity parameter analysis on any n parameters of the model. The computations involved in determining the coefficients of the sensitivity function become more and more complex as n increases. For practical purposes, it is recommended to consider one-way or at most two-way parameter sensitivity analysis only.

10.4   Parameter Tuning

In some cases, we may want to tune the parameters of a Bayesian network model by imposing constraints on the posterior probabilities computed. For instance, it may be that we would like to impose a constraint of the posterior probability of a hypothesis under a specific evidence scenario. This can be achieved using properties of parameter sensitivity analysis. Here we consider how the results related to one-way parameter sensitivity analysis can be used to tune the parameters of a Bayesian network model. Major parts of this section are taken from Jensen (2010).

Recall the equation for the sensitivity function in the case of one-way parameter sensitivity analysis

$$f(t) = P(h|\varepsilon)(t) = \frac{\alpha \cdot t + \beta}{\gamma \cdot t + \delta}.$$

The sensitivity function describes how the posterior probability $P(h|\varepsilon)(t)$ of the hypothesis h given ε changes as a function of the parameter t. Using this function, we can compute the values of t that satisfy a given constraint on the posterior probability. Chan & Darwiche (2002b) consider three different types of constraints

$$P(h|\varepsilon) \geq \tau \; (Simple)$$

$$P(h_1|\varepsilon) - P(h_2|\varepsilon) \geq \tau \; (Difference)$$

$$P(h_1|\varepsilon)/P(h_2|\varepsilon) \geq \tau \; (Ratio)$$

where τ is a constant and $P(h_1|\varepsilon)$ and $P(h_2|\varepsilon)$ are posterior probabilities for different hypotheses h_1 and h_2 given the same evidence ε. The hypotheses h_1 and h_2 may or may not be states of the same variable. Hence, we assume two sensitivity functions $f_{h_1}(t)$ and $f_{h_2}(t)$ such that

$$f_{h_1}(t) = \frac{\alpha_1 \cdot t + \beta_1}{\gamma \cdot t + \delta}$$

$$f_{h_2}(t) = \frac{\alpha_2 \cdot t + \beta_2}{\gamma \cdot t + \delta}.$$

Since the evidence scenario is the same, the denominators of $f_{h_1}(t)$ and $f_{h_2}(t)$ are identical. For the *Simple* constraint $P(h|\varepsilon) \geq \tau$, we can write

$$P(h|\varepsilon) = \frac{\alpha \cdot t + \beta}{\gamma \cdot t + \delta} \geq \tau$$

$$\Longleftrightarrow (\alpha - \gamma \cdot \tau) \cdot t \geq -\beta + \delta \cdot \tau.$$

This means that

$$t \geq \frac{-\beta + \delta \cdot \tau}{\alpha - \gamma \cdot \tau}, \text{ if } \alpha - \gamma \cdot \tau > 0,$$

$$t \leq \frac{-\beta + \delta \cdot \tau}{\alpha - \gamma \cdot \tau}, \text{ if } \alpha - \gamma \cdot \tau < 0.$$

That is, if the value of $\alpha - \gamma \cdot \tau$ is positive, then a lower bound for t is obtained whereas an upper bound is obtained if the value is negative. For the *Difference* constraint $P(h_1|\varepsilon) - P(h_2|\varepsilon) \geq \tau$, we can write

$$P(h_1|\varepsilon) - P(h_2|\varepsilon) = \frac{\alpha_1 \cdot t + \beta_1}{\gamma \cdot t + \delta} - \frac{\alpha_2 \cdot t + \beta_2}{\gamma \cdot t + \delta} \geq \tau$$

$$\Longleftrightarrow (\alpha_1 - \alpha_2 - \gamma \cdot \tau) \cdot t \geq -\beta_1 + \beta_2 + \gamma \cdot \tau.$$

This means that

$$t \geq \frac{-\beta_1 + \beta_2 + \gamma \cdot \tau}{\alpha_1 - \alpha_2 - \gamma}, \text{ if } \alpha_1 - \alpha_2 - \gamma \cdot \tau > 0,$$

$$t \leq \frac{-\beta_1 + \beta_2 + \gamma \cdot \tau}{\alpha_1 - \alpha_2 - \gamma}, \text{ if } \alpha_1 - \alpha_2 - \gamma \cdot \tau < 0.$$

That is, if the value of $\alpha_1 - \alpha_2 - \gamma \cdot \tau$ is positive, then a lower bound for t is obtained whereas an upper bound is obtained if the value is negative. For the *Ratio* constraint $P(h_1|\varepsilon)/P(h_2|\varepsilon) \geq \tau$, we can write

$$P(h_1|\varepsilon)/P(h_2|\varepsilon) = \frac{\alpha_1 \cdot t + \beta_1}{\gamma \cdot t + \delta} \Big/ \frac{\alpha_2 \cdot t + \beta_2}{\gamma \cdot t + \delta} \geq \tau$$

$$\Longleftrightarrow (\alpha_1 - \alpha_2 \cdot \tau) \cdot t \geq -\beta_1 + \beta_2 \cdot \tau.$$

This means that

$$t \geq \frac{-\beta_1 + \beta_2 \cdot \tau}{\alpha_1 - \alpha_2 \cdot \tau}, \text{ if } \alpha_1 - \alpha_2 \cdot \tau > 0,$$

$$t \leq \frac{-\beta_1 + \beta_2 \cdot \tau}{\alpha_1 - \alpha_2 \cdot \tau}, \text{ if } \alpha_1 - \alpha_2 \cdot \tau < 0.$$

That is, if the value of $\alpha_1 - \alpha_2 \cdot \tau$ is positive, then a lower bound for t is obtained whereas an upper bound is obtained if the value is negative.

In each case we assume that the parameter t under consideration has an impact on both h_1 and h_2. If this is not the case, then the equations above will simplify accordingly. The equations can be used to find the parameter value of t that satisfy the chosen constraint. The next example illustrates how the equations for the *Simple*

Table 10.7 Parameter values that satisfy the constraint $P(\text{Sick} = \text{yes}|\varepsilon) \geq 0.35$

Parameter	Current value	Suggested value
$P(\text{Dry} = \text{no})$	0.9	≥ 0.9177
$P(\text{Dry} = \text{yes})$	0.1	≤ 0.0823
$P(\text{Sick} = \text{no})$	0.95	≤ 0.9422
$P(\text{Sick} = \text{yes})$	0.05	≥ 0.0578

constraint can be used to adjust the value of a parameter of the model in order to satisfy the constraint.

Example 10.13 (Parameter Tuning). Consider again Apple Jack from Example 4.1 on page 71, and assume as in Example 10.12 that the scenario under consideration is $S = (\text{Sick}, \text{yes}, \{\text{Loses} = \text{yes}\})$. Apple Jack is interested in computing the probability of his finest apple tree being sick. His hypothesis is that the tree is sick, that is, the hypothesis is $\text{Sick} = \text{yes}$. Given the initial parameterization of the Bayesian network, the posterior distribution of the hypothesis given the observation, $P(\text{Sick} = \text{yes}|\varepsilon)$, is 0.3162. Assume we want to adjust the model such that $P(\text{Sick} = \text{yes}|\varepsilon) \geq 0.35$. The question is how can we adjust the parameters of the model such that this constraint is satisfied.

Let t be the parameter under consideration for adjustment in order to satisfy the constraint. In Example 10.9, the sensitivity function $f(t)$ for parameter $t = P(\text{Dry} = \text{yes})$ was computed as

$$f(t) = \frac{0.0025 \cdot t + 0.045}{0.791 \cdot t + 0.064}.$$

Using the equation above, we can compute the value for each parameter where the constraint is satisfied. For the parameter $t = P(\text{Dry} = \text{yes})$, we may compute

$$f(t) = \frac{0.0025 \cdot t + 0.045}{0.791 \cdot t + 0.064} \geq 0.35$$

$$\Longleftrightarrow \quad (0.0025 - 0.791 \cdot 0.35) \cdot t \geq -0.045 + 0.064 \cdot 0.35$$

$$\Longleftrightarrow \quad t \leq 0.0823.$$

Hence, if $t \leq 0.0823$, then the constraint is satisfied. A similar computation can be performed for each of the other parameters that can be adjusted.

Table 10.7 shows examples of parameter values that satisfy the constraint $P(\text{Sick} = \text{yes}|\varepsilon) \geq 0.35$. Which parameter to adjust will depend on the circumstances, but in the example, it is clear that the adjustment producing the smallest absolute change in a parameter value is to change the prior on $P(\text{Sick})$. □

10.5 Summary

In this chapter we have considered evidence and parameter sensitivity analysis in Bayesian networks.

The objective of evidence sensitivity analysis is to investigate how sensitive the result of a belief update is to variations in the set of evidence. To support this analysis, we have defined two distance measures designed to provide a numerical value specifying the distance between either two probabilities or two probability distributions. Based on the distance measures, we have described five different types of evidence sensitivity analysis: identifying minimum and maximum beliefs, impact of evidence subsets, discrimination of competing hypotheses, what-if analysis, and impact of findings. We described how parameter sensitivity analysis can be used for parameter tuning and how to perform two-way parameter sensitivity analysis.

The objective of parameter sensitivity analysis is to investigate how sensitive the result of a belief update is to variations in a parameter of the model. It has been shown that there is a simple functional relation between the probability of a set of evidence and an entry of a conditional probability table, that is, a parameter. The probability of the evidence is a simple linear function of the parameter. This insight may be used to perform parameter sensitivity analysis on the posterior probability of a hypothesis given a set of evidence. Parameter sensitivity values may be used to focus the knowledge elicitation resources in the model construction process.

In Chap. 11, we consider value of information analysis. Value of information analysis is to compute the value of potential new observations.

Exercises

Exercise 10.1. In the morning when Mr. Holmes leaves his house, he realizes that his grass is wet. He wonders whether it has rained during the night or whether he has forgotten to turn off his sprinkler. He looks at the grass of his neighbors, Dr. Watson and Mrs. Gibbon. Both lawns are dry, and he concludes that he must have forgotten to turn off his sprinkler. (This problem was also discussed in Exercise 6.3 on page 188.)

The structure of a network for modeling the above scenario is shown in Fig. 10.5.

The prior probabilities are $P(\text{Rain} = \text{no}) = P(\text{Sprinkler} = \text{no}) = 0.9$ while the conditional probability distributions are shown in Tables 10.8–10.10.

The hypothesis under consideration is Sprinkler = yes.

(a) Identify the set of evidence.
(b) What is the impact of subsets of the evidence on the hypothesis?
(c) What subsets of the evidence discriminate the hypothesis from the alternative hypothesis Rain = yes?
(d) How does the posterior distribution of the hypothesis change with changes in the observed state for each evidence variable?

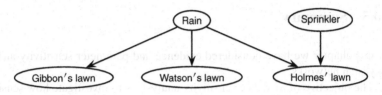

Fig. 10.5 The wet grass network

Table 10.8 The conditional probability distribution $P(\text{Holmes' lawn} \mid \text{Rain, Sprinkler})$

		Holmes' lawn	
Rain	Sprinkler	dry	wet
no	no	1	0
no	yes	0.1	0.9
yes	no	0.01	0.99
yes	yes	0	1

Table 10.9 The conditional probability distribution $P(\text{Gibbon's lawn} \mid \text{Rain})$

	Gibbon's lawn	
Rain	dry	wet
no	0.9	0.1
yes	0.01	0.99

Table 10.10 The conditional probability distribution $P(\text{Watson's lawn} \mid \text{Rain})$

	Watson's lawn	
Rain	dry	wet
no	0.9	0.1
yes	0.01	0.99

(e) What is the impact of each individual piece of evidence on the posterior distribution of the hypothesis?

Exercise 10.2. Consider the Asia network shown in Fig. 10.6 (see Example 4.2 on page 73 for more details).

The Asia network consists of the three hypothesis variables Bronchitis, Cancer, and Tuberculosis. The risk factors are Smoking and a recent visit to Asia while the symptoms of the network are X_ray and Dyspnoea. The risk factors and symptoms are the possible observations a physician can make on a patient.

Assume the physician is diagnosing a smoking patient with dyspnea who has recently been to Asia. The hypothesis under consideration is Bronchitis = yes.

(a) Identify the set of evidence.
(b) What is the impact of subsets of the evidence on the hypothesis?
(c) What subsets of the evidence discriminate the hypothesis from the alternative hypothesis Cancer = yes?
(d) How does the posterior distribution of the hypothesis change with changes in the observed state for each evidence variable?
(e) What is the impact of each individual piece of evidence on the posterior distribution of the hypothesis?

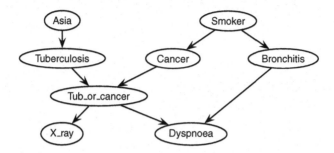

Fig. 10.6 A graph specifying the independence and dependence relations of the Asia example

Exercise 10.3. One in a thousand people has a prevalence for a particular heart disease. There is a test to detect this disease. The test is 100 % accurate for people who have the disease and is 95 % accurate for those who do not (this means that 5 % of people who do not have the disease will be wrongly diagnosed as having it).

(a) If a randomly selected person tests positive, what is the probability that the person actually has the heart disease?
(b) Compute the sensitivity function $f(t) = P(\text{Heart\_Disease} = \text{yes}|\text{Test} = \text{yes})(t)$ where $t = P(\text{Heart\_Disease} = \text{yes})$.
(c) Compute the sensitivity value for $t_0 = 0.001$.
(d) Identify the admissible deviation of t.

Exercise 10.4. Let us consider parameter sensitivity analysis in the wet grass network (cf. Exercise 10.1).

(a) Compute the sensitivity function $f(t) = P(\text{Sprinkler} = \text{yes}|\text{Holmes' lawn} = \text{wet}, \text{Watson's lawn} = \text{dry})(t)$ where $t = P(\text{Rain} = \text{yes})$.
(b) Compute the sensitivity value for $t_0 = 0.2$.
(c) Compute the sensitivity function $f(t) = P(\text{Rain} = \text{yes}|\text{Holmes' lawn} = \text{wet}, \text{Watson's lawn} = \text{dry})(t)$ where $t = P(\text{Rain} = \text{yes})$.
(d) Identify the admissible deviation of t when the hypothesis is Sprinkler = yes and the alternative hypothesis is Rain = yes.

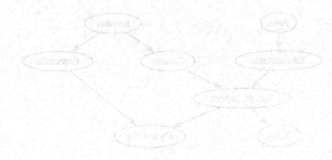

Chapter 11
Value of Information Analysis

Probabilistic networks are constructed to support belief update and decision making under uncertainty. A common solution to a belief update problem is the posterior probability distribution over a hypothesis variable given a set of evidence. Similarly, the solution to a decision-making problem is an optimal decision given a set of evidence. When faced with a belief update or decision-making problem, we may have the option to consult additional information sources for further information that may improve the solution. Value of information analysis is a tool for analyzing the potential usefulness of additional information before the information source is consulted.

We consider a greedy hypothesis-driven approach to value of information (VOI) analysis. At any time, at most one additional information source may be consulted in the search for additional information. In the case of a belief update problem, we assume that the posterior probability distribution of a certain hypothesis variable is of interest and that a set of evidence is available. The task of value of information analysis is to determine the value of information from different information sources, that is, the value of making additional observations before accepting the posterior distribution of the hypothesis as the solution to belief update. On the other hand, in the case of a decision-making problem, we assume that we are about to make a certain decision based on a set of observations on its relevant past. Again, the task of value of information analysis is to consider the value of information from different information sources, that is, the value of making additional observations before making a decision based on the current expected utility function over the decision options available.

In Sect. 11.1, we describe value of information analysis in Bayesian networks. The value of information analysis in Bayesian networks is based on an information theoretic approach using concepts such as entropy, mutual information, and information gain. Entropy and mutual information are introduced as *information measures* specifying the information gain by observing a variable. These information measures are easy to compute using probabilistic inference. In Sect. 11.2, we describe value of information analysis in influence diagrams where the change in expected utility is used as the information measure. In both sections, we consider

U.B. Kjærulff and A.L. Madsen, *Bayesian Networks and Influence Diagrams: A Guide to Construction and Analysis*, ISS 22, DOI 10.1007/978-1-4614-5104-4_11,

a greedy hypothesis-driven approach to value of information analysis where we assume at most a single information source may be consulted in a quest for additional information.

11.1   VOI Analysis in Bayesian Networks

Consider the situation where a decision maker has to make a decision based on the probability distribution of a hypothesis variable. It could, for instance, be a physician deciding on a treatment of a patient given the probability distribution of a disease variable. For instance, if the probability of the patient suffering from the disease is above a certain threshold, then the patient should be treated immediately. Prior to deciding on a treatment, the physician may have the option to gather additional information about the patient such as performing a test or asking a certain question. Given a range of options, what option should the physician choose next? That is, which of the given options will (on average) produce the most information? These questions can be answered by a value of information analysis.

Given a Bayesian network model and a hypothesis variable, the task is to identify the variable which is most informative with respect to the hypothesis variable. Hence, we consider a greedy hypothesis-driven value of information analysis in Bayesian networks.

11.1.1   Entropy and Mutual Information

The main reason for acquiring additional information is to decrease the uncertainty about the hypothesis under consideration. The selection of the variable to observe next (e.g., the question to ask next) can be based on the notion of *entropy*. Entropy is a measure of how much the probability mass is scattered over the states of a variable (the degree of chaos in the distribution of the variable), see Cover & Thomas (1991). As such, entropy is a measure of randomness. The more random a variable is, the higher its entropy will be.

Let X be a discrete random variable with n states $x_1, \ldots, x_n$ and probability distribution $P(X)$; then the entropy of X is defined as

$$H(X) = -\mathbb{E}_{P(X)}[\log P(X)]$$
$$= -\sum_X P(X) \log P(X)$$
$$\geq 0.$$

The maximum entropy, $\log(n)$, is achieved when the probability distribution, $P(X)$, is uniform while the minimum entropy, 0, is achieved when all the probability mass is located on a single state. Thus, $H(X) \in [0, \log(n)]$.

Since entropy can be used as a measure of the uncertainty in the distribution of a variable, we can determine how the entropy of a variable changes as observations are made. In particular, we can identify the most informative observation.

If Y is a random variable, then the entropy of X given an observation on Y is

$$H(X \mid Y) = -\mathbb{E}_{P(X,Y)}[\log P(X \mid Y)]$$

$$= -\sum_Y P(Y) \sum_X P(X \mid Y) \log P(X \mid Y)$$

$$= H(X) - I(X, Y),$$

where $I(X, Y)$ is the *mutual information* (also known as *cross entropy*)of X and Y. The *conditional entropy* $H(X \mid Y)$ is a measure of the uncertainty of X given an observation on Y, while the mutual information $I(X, Y)$ is a measure of the information shared by X and Y (i.e., the reduction in entropy from observing Y). If X is the variable of interest, then $I(X, Y)$ is a measure of the value of observing Y. The mutual information is computed as

$$I(X, Y) = H(X) - H(X \mid Y)$$

$$= H(Y) - H(Y \mid X)$$

$$= \sum_Y P(Y) \sum_X P(X \mid Y) \log \frac{P(X, Y)}{P(X)P(Y)}.$$

In principle, $I(X, Y)$ is a measure of the distance between $P(X)P(Y)$ and $P(X, Y)$. The *conditional mutual information* given a set of evidence ε is computed by conditioning the probability distributions on the available evidence ε:

$$I(X, Y \mid \varepsilon) = \sum_Y P(Y \mid \varepsilon) \sum_X P(X \mid Y, \varepsilon) \log \frac{P(X, Y \mid \varepsilon)}{P(X \mid \varepsilon)P(Y \mid \varepsilon)}.$$

We compute $I(X, Y \mid \varepsilon)$ for each possible observation Y. The next variable to observe is the variable Y that has the highest nonzero mutual information with X (i.e., $I(X, Y \mid \varepsilon)$), if any.

The probabilities needed for the computation of mutual information are readily computed by message passing in a junction tree representation of the model.

11.1.2 *Hypothesis-Driven Value of Information Analysis*

Value of information analysis is the task of estimating the value of additional information. When considering hypothesis-driven value of information analysis in Bayesian networks, we need to define a *value function* in order to determine the value of an information scenario. Entropy can be used as a value function.

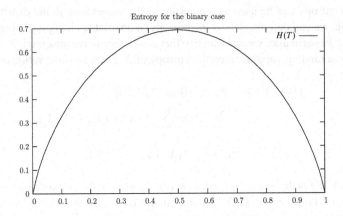

Fig. 11.1 The entropy of T

In a hypothesis-driven value of information analysis, the value of an information scenario is defined in terms of the probability distribution of the hypothesis variable. If T is the hypothesis variable and entropy is used as the value function, then the value function is defined as

$$V(T) = -H(T) = \sum_T P(T) \log(P(T)).$$

The reason for using the negation of the entropy is best illustrated using an example. Consider a binary hypothesis variable T with states false and true. Hence, the distribution of T is fully specified by a single parameter p; that is, $P(T = \text{false}, T = \text{true}) = (p, 1 - p)$. Figure 11.1 illustrates the entropy as a function of p while Fig. 11.2 illustrates the negation of the entropy as a function of p.

As can be seen from Fig. 11.1, the entropy takes on its maximum value for the uniform distribution and its minimum value for the extreme cases ($p = 0$ and $p=1$). Since the value function should take on its maximum value at the extreme cases and the minimum value in the uniform case, the negation of the entropy is used as the value function as illustrated in Fig. 11.2.

The value of the information scenario after observing a variable X is

$$V(T \mid X) = -(H(T) - I(X, T)).$$

Thus, greedy hypothesis-driven value of information analysis in Bayesian networks amounts to computing the value of the initial information scenario $V(T)$ and the value of information scenarios where a variable X is observed, that is, $V(T \mid X)$. The task is to identify the variable that increases the value of information the most. The most informative variable to observe is the variable with the highest mutual information with the hypothesis variable.

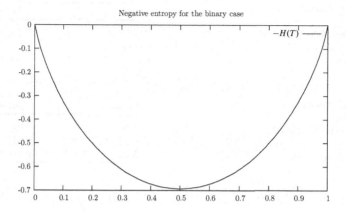

Fig. 11.2 The negation of the entropy of T

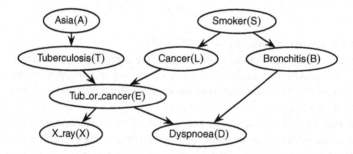

Fig. 11.3 A graph specifying the independence and dependence relations of the Asia example

Example 11.1 (VOI Analysis in a Bayesian Network). As an example we consider a greedy hypothesis-driven value of information analysis on the Asia example shown in Fig. 11.3. The hypothesis variable is Bronchitis (B), and the initial set of evidence is $\varepsilon = \emptyset$. That is, we are considering whether or not the patient is suffering from bronchitis.

Given the network of Fig. 11.3, the hypothesis variable Bronchitis, and the initial set of evidence $\varepsilon = \{\varepsilon_D\} = \{D = \text{yes}\}$, we want to determine the most valuable observation. We may compute the value of the initial information scenario as

$$V(\text{Bronchitis}) = -H(\text{Bronchitis})$$

$$= - \sum_{x \in \{\text{no,yes}\}} P(\text{Bronchitis} = x) \log P(\text{Bronchitis} = x)$$

$$= -0.69,$$

Table 11.1 Mutual
information
between Bronchitis and other
variables given no
observations

Variable name (X)	I(Bronchitis, X)
Dyspnoea	0.25
Smoker	0.05
X_ray	0.0008
Asia	0

Table 11.2 Mutual
information
between Bronchitis and other
variables
given Dyspnoea = yes

Variable name (X)	I(Bronchitis, X)
X_ray	0.014
Smoker	0.0129
Asia	0.0002

where

$$P(\text{Bronchitis} = \text{yes}) = 1 - P(\text{Bronchitis} = \text{no})$$
$$= 0.45;$$

that is, $P(\text{Bronchitis}) = (\text{yes}, \text{no}) = (0.55, 0.45)$.

To identify the most informative observation, we compute the mutual information between the hypothesis variable and each of the other variables in the model. Table 11.1 specifies the mutual information between Bronchitis and each of the other (unobserved) variables.

Notice that one of the variables has a mutual information measure of value zero. A mutual information measure of zero specifies that the two variables are independent (this can be easily verified applying d-separation).

From Table 11.1, it is clear that the most informative variable is Dyspnoea. Thus, we choose to observe this variable. Assume we observe the patient to suffer from dyspnoea; that is, Dyspnoea = yes. The value of the new information scenario can be computed as described above:

$$V(\text{Bronchitis} | \text{Dyspnoea} = \text{yes}) = -H(\text{Bronchitis} | \text{Dyspnoea} = \text{yes})$$
$$= -0.45,$$

where $P(\text{Bronchitis} | \text{Dyspnoea} = \text{yes}) = (\text{yes}, \text{no}) = (0.834, 0.166)$. Once the Dyspnoea variable has been observed to be in state yes, we may be satisfied with the certainty in the hypothesis or we may search for the next observation to make.

Table 11.2 shows the mutual information between Bronchitis and each of the remaining unobserved variables when Dyspnoea is observed to be in state yes. The variable with the highest mutual information score is X_ray.

If the variable with the highest mutual information score is unobservable, then we proceed to the variable with the second highest score. Notice that the mutual information scores change as observations are made. Often both the score and the order of the variables will change with observations.

Variables with a score of zero should not be observed as they will not add any information to the analysis. Notice that additional information cannot decrease the value of an information scenario. □

11.2 VOI Analysis in Influence Diagrams

The value of information is a core element of decision analysis. We perform decision analysis using influence diagram representations of decision problems. The structure of an influence diagram $N = (X, G, P, U)$ specifies a partial order on observations relative to the order of decisions

$$\mathfrak{I}_0 \prec D_1 \prec \mathfrak{I}_1 \prec \cdots \prec D_n \prec \mathfrak{I}_n.$$

Value of information analysis in influence diagrams considers the impact of changing the partial order of observations relative to decisions.

Assume D_j is the next decision to be made and let ε be the set of observations and decisions made up to decision D_j. Initially, the basis for making decision D_j is the expected utility function $\mathrm{EU}(D_j \mid \varepsilon)$ over the options encoded by D_j.

Let $X_i \in \mathfrak{I}_k$ where $k \geq j$ such that $X_j \notin \mathcal{F}(D_j)$ be a discrete random variable with n states $x_1, \ldots, x_n$; that is, X_i is a variable observed after D_j or never observed such that X_i is not a descendant of D_j. Assume $X_i = x$ is observed prior to making decision D_j. The revised basis for making decision D_j is the expected utility function $\mathrm{EU}(D_j \mid \varepsilon, x)$. Prior to observing the state of X_i, the probability distribution of X_i is $P(X_i \mid \varepsilon)$. Thus, we can compute the expected utility of the optimal decision at D_j after X_i is observed $\mathrm{EUO}(X_i, D_j \mid \varepsilon)$ to be

$$\mathrm{EUO}(X_i, D_j \mid \varepsilon) = \sum_{X_i} P(X_i \mid \varepsilon) \max_{D_j} \mathrm{EU}(D_j \mid \varepsilon, X_i).$$

This value should be compared with the expected utility $\max_{D_j} \mathrm{EU}(D_j \mid \varepsilon)$ of the optimal decision at D_j without the observation on X_i. The value $\mathrm{VOI}(X_i, D_j \mid \varepsilon)$ of observing X_i before decision D_j is

$$\mathrm{VOI}(X_i, D_j \mid \varepsilon) = \mathrm{EUO}(X_i, D_j \mid \varepsilon) - \max_{D_j} \mathrm{EU}(D_j \mid \varepsilon).$$

Example 11.2 (VOI Analysis in Influence Diagrams). Appendicitis may cause fever, pain, or both. If a patient has appendicitis, then the patient will have an increased count of white blood cells in addition to fever and pain. Assume that fever and pain are observed.

When a patient potentially has appendicitis, the physician may choose to operate right away or wait for the result of a blood test. The question considered is whether or not the result of the blood test provides sufficient value.

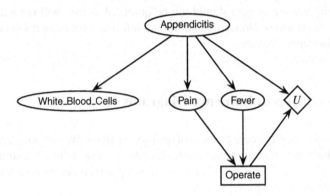

Fig. 11.4 A graph representing the appendicitis example

Table 11.3 The conditional probability distribution $P(\text{Fever}\,|\,$ Appendicitis)

	Appendicitis = no	Appendicitis = yes
Fever = no	0.5	0.02
Fever = yes	0.5	0.98

Table 11.4 The conditional probability distribution $P(\text{Pain}\,|\,\text{Appendicitis})$

	Appendicitis = no	Appendicitis = yes
Pain = no	0.4	0.05
Pain = yes	0.6	0.95

Table 11.5 The conditional probability distribution $P(\text{White\_Blood\_Cells}\,|\,$ Appendicitis)

	Appendicitis = no	Appendicitis = yes
White_Blood_Cells = no	0.95	0.01
White_Blood_Cells = yes	0.05	0.99

Table 11.6 The utility function $U(\text{Appendicitis, Operate})$

Appendicitis	Operate	
no	now	5
no	wait	−5
yes	now	−10
yes	wait	10

Figure 11.4 shows a graphical representation of the decision problem where we assume fever and pain are observed while the blood test result is not (yet) observed.

To compute the value of information on White_Blood_Cells, the model has to be quantified. Let $P(\text{Appendicitis} = \text{no}, \text{Appendicitis} = \text{yes}) = (0.85, 0.15)$ and the remaining conditional probability distributions be given as specified in Tables 11.3–11.5. Table 11.6 shows the utility function $U(\text{Appendicitis, Operate})$.

Assume that the physician observes the patient to suffer from pain and fever; that is, $\varepsilon = \{$Fever $=$ yes, Pain $=$ yes$\}$. With the above quantification we compute the expected utility function over Operate to be EU(Operate$|$Fever $=$ yes, Pain $=$ yes$)$ $=$ $(0.31, -0.31)$. We compute the expected utility of the optimal decision at Operate after White_Blood_Cells is observed to be

$$\text{EUO(White\_Blood\_Cells, Operate}|\varepsilon)$$

$$= \sum_{\text{White\_Blood\_Cells}} P(\text{White\_Blood\_Cells}|\varepsilon)$$

$$\max_{\text{Operate}} \text{EU(Operate}|\varepsilon, \text{White\_Blood\_Cells})$$

$$= 6.375.$$

The value of observing White_Blood_Cells before decision Operate is

$$\text{VOI(White\_Blood\_Cells, Operate}|\varepsilon)$$

$$= \text{EUO(White\_Blood\_Cells, Operate}|\varepsilon) - \max_{\text{Operate}} \text{EU(Operate}|\varepsilon)$$

$$= 6.375 - 0.31 = 6.065.$$

Thus, the physician should wait for the result of the blood test.   □

Instead of considering which observation to make next, if any, at a decision in the middle of the decision process, we may consider how the expected utility of the optimal strategy $\hat{\Delta}$ changes as the partial order of observations is altered. Let EU$(\hat{\Delta})$ be the expected utility of the original formulation of the decision problem $\mathcal{N}$ and let EU$(\hat{\Delta}^*)$ be the expected utility of the revised formulation of the decision problem $\mathcal{N}^*$ where a variable $X \in \mathcal{I}_j$ in $\mathcal{N}$ and $X \in \mathcal{I}_k$ in $\mathcal{N}^*$ such that $j > k$ and $X \notin \mathcal{F}(D_{k-1})$. The value of observing X before decision $k-1$ instead of before decision $j-1$ is EU$(\hat{\Delta}^*) -$ EU$(\hat{\Delta})$.

Example 11.3 (Reconsidering the Decision Problem in Example 11.2). Let us compute the expected utility EU$(\hat{\Delta})$ of the optimal strategy $\hat{\Delta}$ for the information scenario where White_Blood_Cells is not observed before making any observation to be (where we use the first letter of each variable name to shorten the presentation)

$$\text{EU}(\hat{\Delta}) = \sum_{P} \sum_{F} \max_{O} \sum_{W} \sum_{A} P(A)P(F|A)P(P|A)P(W|A)U(O, A)$$

$$= 2.99.$$

Similarly, we compute the expected utility EU$^*(\hat{\Delta}^*)$ of the optimal strategy $\hat{\Delta}^*$ for the information scenario where White_Blood_Cells is observed to be

$$\text{EU}^*(\hat{\Delta}^*) = \sum_W \sum_P \sum_F \max_O \sum_A P(A)P(F\mid A)P(P\mid A)P(W\mid A)U(O,A)$$

$$= 5.45.$$

Based on the above analysis, the physician should wait for the result of the blood test. The value of observing White_Blood_Cells prior to the decision is

$$\text{VOI}(\text{White\_Blood\_Cells}, \text{Operate}\mid\varepsilon) = \text{EU}^*(\hat{\Delta}^*) - \text{EU}(\hat{\Delta})$$

$$= 5.45 - 2.99$$

$$= 2.46,$$

where ε is the set of observations on Fever and Pain.   □

11.3   Summary

In this chapter we have considered value of information analysis in Bayesian networks and influence diagrams in two separate sections. In both cases, we have described a greedy approach to value of information analysis, that is, value of information analysis performed under the assumption that we may at most consult one additional information source in the search for further information before accepting the posterior distribution of the hypothesis variable or making the decision. In order to perform value of information analysis in Bayesian networks, we have defined entropy and mutual information as information measures. Entropy is a measure of how much the probability mass is scattered over the states of the hypothesis variable. In the evaluation of possible information sources, we identify the possible observation that reduces the entropy of the hypothesis variable the most. This will be the variable with the highest mutual information with the hypothesis variable.

In the case of value of information analysis in influence diagrams, expected utility is used as the information measure. In the evaluation of possible information sources, we identify the possible observation that increases the expected utility of the decision the most. This variable is identified by computing the expected utility of the decision given that the variable is observed prior to the decision. This expected utility is computed for each possible observation and compared with the expected utility of the decision given no additional information.

Exercises

Exercise 11.1. Consider the Asia network shown in Fig. 11.5 (see Example 4.2 on page 73 for more details).

The Asia network consists of the three hypothesis variables Bronchitis, Cancer, and Tuberculosis. The risk factors are Smoking and a recent visit to Asia while the symptoms of the network are X_ray and Dyspnoea. The risk factors and symptoms are the possible observations a physician can make on a patient.

(a) What is the entropy of the prior distribution on each of the diseases?
(b) What is the most informative observation with respect to each of the diseases?
(c) What is the most informative observation with respect to each of the diseases if the patient is a smoker suffering from dyspnoea?

Exercise 11.2. A used car salesman offers all potential customers a test performed on the car they are interested in buying. The test should reveal whether the car has either no defects or one (or more) defect; the prior probability that a car has one or more defects is 0.3. There are two possible tests: $Test_1$ has three possible outcomes, namely, no defects, defects, and inconclusive. If the car does not have any defects, then the probabilities for these test results are 0.8, 0.05, and 0.15, respectively. On the other hand, if the car has defects, then the probabilities for the test results are 0.05, 0.75, and 0.2. For $Test_2$, there are only two possible outcomes (no defects and defects). If the car does not have any defects, then the probabilities for the test results are 0.8 and 0.2, respectively, and if the car has defects, then the probabilities are 0.25 and 0.75.

(a) Construct a Bayesian network (both structure and probabilities) representing the relations between the two tests and the state of the car.
(b) Calculate the probabilities $P(\text{StateOfCar}|\text{Test}_1)$ and $P(\text{Test}_1)$.
(c) Perform a value of information analysis on both $Test_1$ and $Test_2$ with respect to StateOfCar.

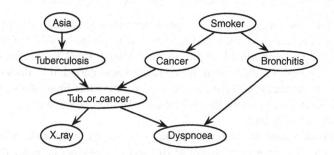

Fig. 11.5 A graph specifying the independence and dependence relations of the Asia example

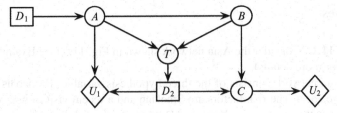

Fig. 11.6 An influence diagram with two decisions

Exercise 11.3. Assume we are given the influence diagram in Fig. 11.6.

$P(A\,\vert\,D_1)$	d_{11} d_{12}
a_1	0.3 0.6
a_2	0.7 0.4

$P(B\,\vert\,A)$	a_1 a_2
b_1	0.1 0.8
b_2	0.9 0.2

$P(T\,\vert\,A,B)$	a_1	a_2
b_1	$(0,1)$	$(0.2,0.8)$
b_2	$(0.6,0.4)$	$(0.8,0.2)$

$P(C\,\vert\,B,D_2)$	b_1	b_2
d_{21}	$(0.9,0.1)$	$(0.5,0.5)$
d_{22}	$(0.5,0.5)$	$(0.9,0.1)$

$U_1(A,D_2)$	a_1 a_2
d_{21}	10 0
d_{22}	0 -6

$U_2(C) = (20,0)$

(a) Compute the solution to this decision problem (i.e., compute the strategy maximizing the expected utility).

(b) Describe the impact of the information on variable T.

(c) Repeat parts (a) and (b) when $P(A\,\vert\,D_1 = d_{12}) = (0.4, 0.6)$ and $U_2(C) = (20, 100)$.

(d) Assume the informational link between T and D_2 is not present in the influence diagram. Compute $\mathrm{VOI}(T)$ when $D_1 = d_{12}$ for both quantifications.

Exercise 11.4. Assume that Frank wakes up one morning feeling ill. Frank thinks that he may have caught the flu, and he now has to decide whether to go to the pharmacy to buy some medicine (at the cost of € 150). If Frank has the flu, then the medicine will relieve his discomfort during the sickness period; if he does not have the flu, then the medicine will have no effect. Assuming that Frank does not suffer from the discomfort caused by a flu, then he can take some additional overtime work which will be worth € 2000.

Before Frank decides to go to the pharmacy, he can try to get more information by buying a thermometer (at a cost of € 10.) and test whether he has a fever; the thermometer is very precise and will indicate a fever if and only if Frank actually has a fever.

Table 11.7 The joint			Flu	
probability distribution $P($Fever, Flu$)$			no	yes
	Fever	no	0.89298	0.00095
		yes	0.09702	0.00905

(a) Perform a myopic value of information analysis for the decision problem above and calculate the expected profit of performing the test (i.e., buying the thermometer at a cost of € 10 and taking the temperature). Calculate the required probabilities from the joint probability table (over the variables Flu and Fever) specified in Table 11.7.

Quick Reference to Model Construction

In this quick reference, we have listed those modeling methods, recommendations, and tricks that appear in Chaps. 6 and 7. For each item, we give reference to the page in the book, where the item is described in greater detail, as well as listing some associated "dos" and "don'ts."

Identifying the Variables (page 149).   There are two kinds of variables: *Chance variables* model events of the problem domain, which are not under control of the decision maker. *Decision variables* represent decisions taken by the decision maker. Variables can be *discrete* or *continuous*. Discrete variables take values from a finite set of possible values. Continuous variables (page 75) take values from the set of real numbers. For discrete variables:

Do:

- Make sure that the variables are well defined (page 149) such that it can pass the clarity test (page 151); that is, each variable consists of an exhaustive and mutually exclusive set of states, can pass the uniqueness test, and each state is semantically absolutely clear.
- Categorize the variables as *problem variables*, *background information variables*, *symptom variables* or *mediating variables* (page 152).

Don't:

- Define states of a variable that are not mutually exclusive.
- Define a set of states that is not complete (i.e., the set does not exhaustively define the space of possible values of the variable).
- (Normally) define a state of a variable that is mutually exclusive with a state of a single other variable.
- Define a state of a variable that is not semantically absolutely clear.

Identifying the Links (page 154).   It is the links of a probabilistic network that determine the factorization of the joint probability distribution defined over the variables represented in the network; that is, they determine the conditional

U.B. Kjærulff and A.L. Madsen, *Bayesian Networks and Influence Diagrams: A Guide to Construction and Analysis*, ISS 22, DOI 10.1007/978-1-4614-5104-4, © Springer Science+Business Media New York 2013

independence properties of the network and hence the complexity of inference. Thus, if the links give rise to independence properties that only approximately or not at all match reality, erroneous inferences can be made, or if there are independences among the domain variables that are not represented in the model, inference may become unnecessarily complex or even intractable.

Therefore, identifying a "correct" set of links is important for obtaining a good balance between correctness and speed of the inference process.

Do:

- Make sure links are directed from cause to effect, for example, from a background variable to a problem variable or from a problem variable to a symptom variable (see Figs. 6.3 and 6.20).
- If the relations among a (sub)set of variables can be appropriately characterized through one of the five idioms on page 156, use the appropriate idiom to define the links between the variables. See the flow chart on page 162 for a guide on how to select the right idiom.
- Try to keep the number of parents of each (discrete) variable as low as possible. If the number of parents of a variable exceeds two, consider using *parent divorcing*, *temporal transformation*, or an *independence of causal influence* model to reduce the burden of eliciting the conditional probability distributions and (potentially) the complexity of inference (see below).
- If the relation between a pair of variables is not naturally described using a directed link, consider using the trick described below to specify an *undirected dependence relation*.

Don't:

- Add a directed link from effect to cause, for example, from a symptom variable to a problem variable. This is a typical error, probably resulting from thinking in terms of "input" and "output" (i.e., from observation to "consequence"). Not only does adding a link opposite the causal direction typically lead to erroneous inference and wrong (conditional) independence assumptions, but it also makes elicitation of conditional probabilities awkward.
- Generally, a variable shouldn't have many parent variables. As mentioned above, this can make the inference and the burden of eliciting the conditional probability distributions unnecessarily heavy.
- Add a direct edge (X, Y) between two variables X and Y when X has only an indirect impact on Y, for example, there is a third variable Z such that $X \rightarrow Z \rightarrow Y$.

Eliciting the numbers (page 176). The task of eliciting the (conditional) probabilities of a probabilistic network can be quite huge and may involve several domain experts. Therefore, as the structure of the network dictates the factorization of the joint probability distribution of the model and hence the structure and the domain of each factor [i.e., (conditional) probability distribution] to be specified, care must be taken to ensure that the network structure (i.e., variables and links) is as correct as possible.

Do:

- Make sure the structure of the network has been carefully verified before elicitation of (conditional) probabilities commences.
- Consider making assumptions to allow use of, for example, parent divorcing, independence of causal influence, or Noisy-OR to ease the elicitation process.
- Consider using a qualitative approach to assess subjective Probabilities, for example, the *probability wheel* (Section page 178) or verbal statements (page 178).
- Consider using indirect assessment of a subjective probability (i.e., parameter value) or a subjective utility value through a simple gamble-based approach (Sect. 6.5.1, page 177 and Sect. 6.5.2, page 179, respectively).
- Consider performing elicitation of (conditional) probabilities in a three-step process, where only rough estimates of the probabilities are provided in the first step. In the second step, perform sensitivity analysis (see Chap. 10) to identify probabilities that need to be specified with care. Then, in the third step, provide more precise estimates of those probabilities. Probably, the second and the third steps need to be repeated a number of times, as the careful assessment of the critical probabilities in the third step might reveal new critical probabilities in the reiterated second step.
- Use algorithmic/mathematical expressions to specify (conditional) probability tables (CPTs) and utility tables (UTs) whenever possible, as this may greatly simplify the elicitation task (Sect. 6.5.3, page 180).

Don't:

- Start eliciting probabilities and utilities before the structure of the model has been properly verified.
- Elicit conditional probabilities of a variable with many parents, unless *parent divorcing*, *temporal transformation*, or *independence of causal influence* model are not applicable.
- Elicit parameter values [i.e., (conditional) probabilities or utilities] manually if the CPT or UT can be specified using an algorithmic/mathematical expression. Specification as well as maintenance becomes much easier if the table has been specified via an expression.

Parent divorcing (page 192).   An effect (e.g., a symptom) can have many causes. In a probabilistic network representation, this can be modeled with a node, say v, (representing the effect) with one parent node for each cause. As the number of entries in the CPT representing $P(X_v \mid X_{\mathrm{pa}(v)})$ grows exponentially with the number of parents, one should try to exploit or make assumptions about independences that allow a factorization of $P(X_v \mid X_{\mathrm{pa}(v)})$ so as to minimize the parameter elicitation effort. Such independences are represented in a probabilistic network by introducing additional (mediating) variables representing contributions to the effect from a subset of the cause variables. The process of introducing such mediating variables is called *parent divorcing*.

Do:

- Whenever a node has more than two parents, consider using parent divorcing to reduce the number of parents per node to two, potentially by recursive application of the technique.
- Be sure not to violate crucial dependences among the parent variables.
- If parent divorcing does not apply, consider using some other technique like temporal transformation or independence of causal influence (see below).

Don't:

- Assume that parent divorcing is a universally applicable trick to reduce the size of the CPTs of a probabilistic network.
- Let the number of states of an added mediating variable be greater than or equal to the product of the number of states of its parent variables.

Temporal Transformation (page 196).   Instead of combining causes pairwise as in parent divorcing, the influence of causes on the effect variable is taken into account one cause at a time in their causal or temporal order.

Temporal transformation can be used to implement independence of causal influence (see below).

Do:

- Consider using this technique if the impacts from a number of cause variables act cumulatively on a common effect variable.
- Make sure that the "temporal" ordering of the cause variables is respected.
- Make sure the states of the effect variable are ordered and represent different levels of abnormality and that there is a *designated state* indicating absence of abnormality.
- Similarly, make sure that each of the causes has a state corresponding to "no impact" on the effect variable.

Don't:

- Make a cause variable representing the impact of an act seeking to eliminate the effect appear before the end of the "temporal" ordering, that is, the variable must appear as the last in the ordering.

Structural and Functional Uncertainty (page 197).   If it is difficult or impossible to specify a set of independence and dependence assumptions, maybe because they change or are unknown, a trick can be applied to model this structural uncertainty. Similarly, uncertain functional dependence relation between a variable and (a subset of) its parents can be modeled structurally.

Do:

- Consider using this technique if you are uncertain if one or the other of two variables should be parent of a third variable (e.g., in paternity models).
- Consider using this technique if the functional relationship that governs the conditional probability distribution of a variable given its parents can assume different forms depending on circumstances.

Don't:

- Assume the two competing parent variables in structural uncertainty modeling to have different domains or even states that are ordered differently.

Undirected dependence relations (page 201). When capturing a set of dependence relations between variables using a DAG, it is not unusual to encounter the problem of how to (most efficiently) represent a dependence relation which by nature is undirected.

Do:

- Use this technique to implement constraints among variables, for example, a value of one variable prohibits another variable from assuming a certain value.

Don't:

- Generally, use this technique to implement joint probability distributions.

Bidirectional Relations (page 204). A problem similar to implementing undirected dependence relations is the problem of representing bidirectional relations, that is, when a pair of variables, say X_u and X_v, are dependent and it is not evident which direction the link between them should have.

Do:

- If neither $u \to v$ nor $u \leftarrow v$ is obvious, consider adding a common cause variable, say X_w, such that $w \to u$ and $w \to v$.
- If u and v already have a common parent, say w, and $X_u \not\perp\!\!\!\perp X_v \mid X_w$, then consider introducing a new common cause, say X_z, such that $X_u \perp\!\!\!\perp X_v \mid X_z$, and let w be a parent of z (see Example 7.6 on page 205).

Don't:

- Implement a bidirectional relation using the undirected dependence relations technique (see above).

Naive Bayes Model (page 206). A naive Bayes model is characterized by a structure where all variables but one are children (attribute variables) of a single class variable and the attribute variables are conditionally independent given the class variable. This makes the model construction very easy and the inference very efficient. The model is popular for solving classification problems.

Do:

- Consider using the naive Bayes model if you are faced with a classification problem or (equivalently) a diagnostic problem with a single variable representing a set of mutually exclusive diagnoses with symptoms that are conditionally independent given the diagnosis.

Don't:

- Use the naive Bayes model if two or more problems (e.g., classes in a classification problem or diagnoses in a diagnostic problem) can coexist (i.e., are not mutually exclusive).
- Use the naive Bayes model if the attribute variables (e.g., features in a classification problem or symptoms in a diagnostic problem) are (strongly) conditionally dependent given the class or diagnosis variable.
- Use the naive Bayes model if the probability model is supposed to be accurate, as the joint probability distribution of the naive Bayes model tends to sometimes not be well calibrated.

Measurement Uncertainty (page 209).   The value observed for variables in a probabilistic network is often subject to uncertainty, for example, because of noise or inaccuracies of sensors. Then an extra node should be added to the network, representing the actual observed value. In addition, it should be considered to add a node representing the source of the uncertainty.

Do:

- Consider using this technique if the observed value of some observable variable is subject to uncertainty.
- If known, consider explicitly modeling the source of the uncertainty by adding a variable representing the uncertainty mechanism (e.g., known accuracy specifications of a sensor).

Don't:

- Use explicit representation of measurement/observation accuracy if the measurement uncertainty can be naturally encoded in the conditional probability distribution of the observation given the true value.

Expert opinions (page 211).   Domain experts might sometimes disagree on the values of the probability parameters (i.e., the (conditional) probabilities) of a model. Such disagreement may profitably be represented explicitly in the model structure.

Do:

- If experts disagree on the (conditional) probability distribution of some variable, add an extra variable with states representing the disagreeing experts, and let the variable be a parent of the variable for which there is disagreement on its (conditional) probability distribution.

- Encode the relative trust in the experts in the probability distribution of the new variable.

Don't:

- Let one "experts" node be parent of many CPTs. Use separate "experts" nodes, that is, one for each relevant CPT.
- Leave "experts" nodes in the final model. That is, avoid explicit distributions on experts; instead, use node absorption to eliminate "experts" nodes (see below).

Node Absorption (page 213).   An unobserved variable with no children has no impact on the probability distributions of the other variables in the network. Hence, such (barren) variables can simply be removed from the network. An unobserved variable with children can be turned into a barren variable through arc reversal(s) and subsequently removed. Such an operation is called *node absorption* and can be useful for removing *nuisance variables*.

Do:

- Use this technique to remove nuisance variables (Definition 5.3, page 115), for example, intermediate auxiliary variables like "experts" nodes, to improve efficiency of inference.

Don't:

- Absorb a node unless there is no chance it will be observed and its presence in the model is unimportant to understand the model and capture the (conditional) independence properties of the model.

Set Value by Intervention (page 214).   An important distinction should be made between a passive observation of the state of a variable and an active action forcing a variable to be in a certain state. Under the assumption of a causal ordering (see Sect. 4.2), a passive observation of a variable impacts the beliefs of the ancestors of the variable, whereas an active action enforcing a certain state on a variable does not. We refer to this type of active action as *intervention*.

Do:

- Use this technique if you wish to model an active action forcing a variable to be in a certain state.

Don't:

- Use this technique if the assumption of a causal ordering is not fulfilled.

Independence of causal influence (page 216).   Parent variables interacting independently on a common child variable (called the effect variable) are said to exhibit *independence of causal influence*. This can be modeled structurally in such a way that the complexity of eliciting the probability parameters reduces from exponential to linear in the number of parents.

Do:

- See *temporal transformation* above.
- If the variables involved are Boolean, consider using the noisy-OR model.

Don't:

- Use on variables with states that cannot be ordered.
- Use when there is high synergy among causes.

Mixture of Gaussian Distributions (page 221). An arbitrary continuous distribution of a variable can be approximated by a mixture of Gaussian distributions.

Do:

- Add a discrete parent variable of the continuous variable, let the number of states of the discrete variable determine the number of components of the mixture distribution, and let the probabilities of the states of the discrete variable determine the weights of the individual components.

Don't:

- Overdo it. Mixtures of Gaussian distributions are expensive in terms of complexity of inference.
- Use a mixture of Gaussian distributions if a discrete/interval variable is sufficient.
- Use mixture of Gaussian distributions if the variable is supposed to have child variables.

Test Decisions (page 224). A decision maker may be faced with the option to perform a test. This option is represented by a binary decision variable with states representing whether or not the test is performed. If the test is not performed, the random variable representing the test outcome will remain unobserved, and hence, the informational link to a subsequent decision variable will be violated, unless special modeling tricks are employed.

Do:

- Use this technique if you are faced with the problem of modeling a test decision (i.e., a decision that determines whether or not the value of a particular random variable (potentially) becomes known, where this value has an impact on subsequent decisions).
- Consider introducing a utility function associated with the test decision if there is a cost associated with performing the test.

Don't:

- Consider a test decision as any other decision.

Missing Informational Links (page 227). Correct specification of informational links in influence diagrams is crucial for making correct inferences (i.e., solution of the decision problem). However, if the decision problem involves only a single decision, informational links can be avoided provided the influence diagram is solved (i.e., propagation of evidence is performed) each time new evidence becomes available. Using this kind of "on-line" solution (as opposed to "off-line" solution, where the decision policy specifies an optimal decision once and for all for each configuration of the parents of the decision) may substantially increase the efficiency of inference. Moreover, the influence diagram may appear less cluttered.

Do:

- Consider using this trick if your influence diagram includes only a single decision variable.
- If informational links are excluded, solve the decision problem (i.e., propagate evidence) after entering the evidence

Don't:

- Avoid informational links if your influence diagram contains more than one decision variable.
- Rely on the decision policy after entering evidence if the influence diagram contains no informational link.

Missing Observations (page 229). An informational link (X, D) from a random variable X to a decision variable D of an influence diagram signifies that a value of X is observed before decision D is made. Sometimes, however, the value of X may be unavailable, in which case (X, D) should be absent. This modeling dilemma can be solved by introducing an auxiliary variable between X and D.

Do:

- Use this technique if an observation of a random variable of an influence diagram may be missing.

Don't:

- Use this technique unless you have to.
- Use this technique if there is only a single decision variable.

Hypothesis of highest probability (page 231). In some decision scenarios, it can be useful to define a decision variable that represents a decision to select a decision option corresponding to the state of a random variable with highest probability (e.g., in medical diagnosis).

Do:

- Consider using this technique if a decision variable is supposed to represent a decision option to select a state of a random variable or a set of random variables with highest probability.

- If there is only one decision variable, consider leaving out informational links (see *missing informational links* above).

Don't:

- Possibly, use this technique if the influence diagram already contains one or more decision variables, and the addition of the new decision variable either makes solution of the decision problem very complex or makes the model very cluttered.

Constraints on Decisions (page 233).   There may be constraints on the configurations of two or more decision variables, that is, not all combinations of decision options across the variables make sense or are legal. In such cases, a random variable must be included, which contains the space of acceptable combinations.

Do:

- If an influence diagram contains decision variables where not all combinations of decision options are legal, introduce a constraining binary random variable as a child of the decision variables.

Don't:

- Use this technique unless you have to — use large negative utility instead.

List of Examples

2.1	Sample Variable Domains	20
2.2	Cartesian Product and Projecting Variable Domains	21
2.3	Evidence Function	24
2.4	Burglary or Earthquake (Pearl 1988)	25
2.5	The Power of DAGs	30
2.6	Burglary or Earthquake, page 25	33
2.7	d-Separation	34
2.8	Directed Global Markov Criterion	35
3.1	Burglary or Earthquake, page 25	40
3.2	Probability of Mutually Exclusive Events	41
3.3	Burglary or Earthquake, page 25	42
3.4	Balls in An Urn	42
3.5	Balls in An Urn, page 42	43
3.6	Balls in An Urn, page 42	45
3.7	Burglary or Earthquake, page 25	46
3.8	Normalization	48
3.9	Example 3.8, cont.	49
3.10	Combination and Marginalization	50
3.11	Distributive Law	52
3.12	Barren Variables	53
3.13	Burglary or Earthquake, page 25	55
3.14	Arc Reversal	57
3.15	Balls in An Urn, page 42	59
3.16	Conditional Independence	60
3.17	Independence and DAGs	61
3.18	No D-maps	61
3.19	Chain Decomposition and DAGs	63
3.20	Example 3.19, cont.	64
4.1	Apple Jack (Madsen et al. 1998)	71
4.2	Chest Clinic (Lauritzen & Spiegelhalter 1988)	73
4.3	CLG Bayesian Network	77

U.B. Kjærulff and A.L. Madsen, *Bayesian Networks and Influence Diagrams: A Guide to Construction and Analysis*, ISS 22, DOI 10.1007/978-1-4614-5104-4,
© Springer Science+Business Media New York 2013

4.4   Adapted from Lauritzen (1992a) 78
4.5   Oil Wildcatter (Raiffa 1968) 82
4.6   Partial Order of Information Set 83
4.7   Oil Wildcatter Strategy 84
4.8   Apple Jack 85
4.9   Jensen et al. (1994) 88
4.10   Guessing Game (Madsen & Jensen 2005) 91
4.11   Oil Wildcatter (Madsen & Jensen 2005)..................................... 92
4.12   LIMID 94
4.13   Breeding Pigs (Lauritzen & Nilsson 2001) 94
4.14   Object-Oriented Probabilistic Network 98
4.15   Apple Jack's Garden 98
4.16   Instance Tree 101
4.17   Apple Jack's Finest Tree.. 103
4.18   Breeding Pigs 104
5.1   Prior Probability (Apple Jack) 112
5.2   Posterior Probability (Apple Jack) 114
5.3   Barren Variables and Nuisance Variables................................... 115
5.4   Arc Reversal 117
5.5   Burglary or Earthquake, page 25.. 119
5.6   Chest Clinic 121
5.7   Association of CPTs to Cliques 122
5.8   Cluster Trees Vs. Junction Trees 123
5.9   Density Function.. 126
5.10   Density Function vs. Weak Marginal 126
5.11   Oil Wildcatter 129
5.12   Oil Wildcatter 132
5.13   Marketing Budget (Madsen & Jensen 2005) 133
5.14   Decomposition of Influence Diagrams (Nielsen 2001) 135
5.15   Solving LIMID: Breeding Pigs 138
5.16   Probability of Future Decisions 138
5.17   Unsoluble LIMID: Breeding Pigs.. 140
6.1   Set of States of a Variable 149
6.2   Variables of a Probabilistic Network...................................... 150
6.3   Clarity Test: Doors... 151
6.4   Classification 153
6.5   Insemination (Jensen 2001) 153
6.6   Doors, cont. 155
6.7   Classification, cont. 156
6.8   Definitional/Synthesis Idiom (Neil et al. 2000) 157
6.9   Cause–Consequence Idiom (Neil et al. 2000) 157
6.10   Measurement Idiom (Neil et al. 2000) 158
6.11   Induction Idiom (Neil et al. 2000) 158
6.12   Reconciliation Idiom (Neil et al. 2000)................................... 158
6.13   Model Verification 174

6.14	Probability Wheel	177
6.15	Number of People (HUGIN 2006)	181
6.16	Fair or Fake Die (HUGIN 2006)	181
6.17	Discretizing a Variable (HUGIN 2006)	182
7.1	Parent Divorcing	192
7.2	Paternity	198
7.3	Functional Uncertainty	199
7.4	Functional Uncertainty: Guessing Game	200
7.5	Washing Socks (Jensen 1996)	202
7.6	Insemination (Jensen 1996)	205
7.7	Classification of Mushrooms	207
7.8	Explicit Representation of Uncertainty	209
7.9	Implicit Representation of Uncertainty	210
7.10	Expert Opinions: Chest Clinic	211
7.11	Node Absorption: Expert Opinions	213
7.12	Set Value by Intervention	216
7.13	Noisy-OR: Sore Throat	219
7.14	Mixture of Gaussian Distributions	222
7.15	Oil Wildcatter (Raiffa 1968)	224
7.16	Aspirin (Jensen 1996)	225
7.17	Missing Informational Links	227
7.18	Missing Observations: Oil Wildcatter	230
7.19	Hypothesis of Highest Probability	232
7.20	Constraints on Decisions	234
8.1	Data Cases	239
8.2	Equivalent Models	241
8.3	Equivalence Class	241
8.4	Statistical Test	244
8.5	Independence Tests	246
8.6	Example 8.5, cont.	247
8.7	Example 8.6, cont.	248
8.8	Example 8.7, cont.	249
8.9	Example 8.8, cont.	250
8.10	Necessary Path Condition	252
8.11	Condition Graph	253
8.12	Strongly Connected Component	254
8.13	Skeleton (PC Algorithm)	255
8.14	Overfitting	259
8.15	Log-Likelihood vs. BIC	262
8.16	Hill Climbing	264
8.17	Naive Bayes Model	265
8.18	Chow–Liu Tree	267
8.19	Tree-Augmented Naive Bayes	269
8.20	Maximum Likelihood Estimate	278
8.21	Toss of a Coin	280

8.22   Penalized EM: Coin Tossing 282
8.23   Penalized EM: Experience Counts 282
8.24   Sequential Parameter Learning 285
9.1   Conflict Analysis 291
9.2   Conflict Detection 293
9.3   Conflict Traced 294
9.4   Partial Conflicts 295
9.5   Conflict Resolution 296
9.6   Evidence with Conflict Impact 297
9.7   Example 9.6, cont. 298
10.1   Evidence Sensitivity Analysis 304
10.2   Example 10.1, cont. 306
10.3   Example 10.2, cont. 307
10.4   Example 10.3, cont. 308
10.5   Example 10.4, cont. 309
10.6   Example 10.5, cont. 310
10.7   Parameter Sensitivity Analysis 311
10.8   Proportional Scaling 313
10.9   Sensitivity Function 313
10.10   Sensitivity Value 315
10.11   Admissible Deviation 316
10.12   Two-Way Sensitivity Analysis 318
10.13   Parameter Tuning 322
11.1   VOI Analysis in a Bayesian Network 331
11.2   VOI Analysis in Influence Diagrams 333
11.3   Reconsidering the Decision Problem in Example 11.2 335

List of Figures

Fig. 1.1 Graphical representation of rules "if C_1 then E_1" and
"if C_1 and C_2 then E_2" 8
Fig. 1.2 Bayesian network for the "car would not start" problem 11
Fig. 1.3 Influence diagram for the "car would not start" problem 13

Fig. 2.1 (a) An acyclic, directed graph (DAG). (b) Moralized graph 19
Fig. 2.2 (a) Hard evidence on X. (b) Soft (or hard) evidence on X 24
Fig. 2.3 Structure of a probabilistic network model for the
"Burglary or Earthquake" story of Example 2.4 on page 25. 26
Fig. 2.4 Serial connection (causal chain) with no hard evidence
on Alarm. Evidence on Burglary will affect our belief
about the state of Watson_calls and vice versa 27
Fig. 2.5 Serial connection (causal chain) with hard evidence on
Alarm. Evidence on Burglary will have no affect on our
belief about the state of Watson_calls and vice versa 27
Fig. 2.6 Diverging connection with no evidence on Earthquake.
Evidence on Alarm will affect our belief about the state
of Radio_news and vice versa 28
Fig. 2.7 Diverging connection with hard evidence on
Earthquake. Evidence on Alarm will not affect our
belief about the state of Radio_news and vice versa 28
Fig. 2.8 Converging connection with no evidence on Alarm or
any of its descendants. Information about Burglary will
not affect our belief about the state of Earthquake and
vice versa 29
Fig. 2.9 Converging connection with (possibly soft) evidence
on Alarm or any of its descendants. Information
about Burglary will affect our belief about the state of
Earthquake and vice versa 29
Fig. 2.10 Flow of information in the "Burglary or Earthquake"
network with evidence on Alarm 30

U.B. Kjærulff and A.L. Madsen, *Bayesian Networks and Influence Diagrams: A Guide
to Construction and Analysis*, ISS 22, DOI 10.1007/978-1-4614-5104-4,
© Springer Science+Business Media New York 2013

Fig. 2.11 Flow of information in the "Burglary or Earthquake"
network with additional evidence on Radio after
evidence on Alarm... 31

Fig. 2.12 Wrong model for the "Burglary or Earthquake"
story of Example 2.4 on page 25, where the links
are directed from effects to causes, leading to faulty
statements of (conditional) dependence and independence 32

Fig. 2.13 Sample DAG with a few sample dependence
(d-connected) and independence (d-separated) statements........... 34

Fig. 2.14 (a) A DAG, $\mathcal{G}$. (b) $\mathcal{G}$ with subsets A, B, and S
indicated, where the variables in S are assumed to be
observed. (c) The subgraph induced by the ancestral
set of $A \cup B \cup S$. (d) The moral graph of the DAG of part (c)...... 35

Fig. 3.1 Graphical representation of $P(X \mid Y_1, \ldots, Y_n)$ 46

Fig. 3.2 (a) Model for $P(X, Y, Z)$. (b) Equivalent model when
Z is barren... 54

Fig. 3.3 Two equivalent models that can be obtained from each
other through arc reversal 56

Fig. 3.4 (a) Model for $P(X, Y, Z)$. (b) Equivalent model
obtained by reversing $Y \rightarrow Z$. (c) Equivalent model
provided Y is barren. (d) Equivalent model obtained
by reversing $X \rightarrow Z$ 57

Fig. 3.5 Graphical representations of $X \perp\!\!\!\perp_P Y \mid Z$,
representing, respectively, (3.24)–(3.26) 61

Fig. 3.6 (a) Preliminary skeleton for the independence
statements of Example 3.18. (b) DAG representing
CID_1 and CID_3. (c) DAG representing CID_1 and CID_2 62

Fig. 3.7 (a) DAG corresponding to (3.28). (b) DAG
corresponding to (3.29) 63

Fig. 3.8 A perfect map of a distribution P; see Example 3.20 64

Fig. 4.1 The Apple Jack network 72

Fig. 4.2 A graph specifying the independence and dependence
relations of the Asia example 74

Fig. 4.3 CLG Bayesian network with X_1 discrete and X_2 and X_3
continuous 77

Fig. 4.4 CLG Bayesian network for credit account management 78

Fig. 4.5 The oil wildcatter network 82

Fig. 4.6 We model the system at two different points in time
(before and after a decision) by replicating the structure............. 86

Fig. 4.7 Addition of a decision variable for treatment to the
Bayesian network in Fig. 4.6 86

Fig. 4.8 A complete qualitative representation of the influence
diagram used for decision making in Apple Jack's orchard.......... 87

Fig. 4.9   A simplified influence diagram for the decision problem of Apple Jack 88

Fig. 4.10   An influence diagram representing the sequence of decisions D_1, D_2, D_3, D_4 89

Fig. 4.11   The influence diagram of Fig. 4.10 with no-forgetting links 90

Fig. 4.12   A CLQG influence diagram for a simple guessing game 91

Fig. 4.13   A revised version of the oil wildcatter problem 92

Fig. 4.14   A LIMID representation of a decision scenario with two unordered decisions 94

Fig. 4.15   Three test-and-treat cycles are performed prior to selling a pig.. 95

Fig. 4.16 $\mathcal{M}$ is an instance of a network class $C_{\mathcal{M}}$ within another network class $C_{\mathcal{N}}$ 96

Fig. 4.17   The apple tree network class... 99

Fig. 4.18   The apple garden network consisting of three instantiations of the apple tree network 99

Fig. 4.19   An instance tree 101

Fig. 4.20   The structure of a static network model............................ 102

Fig. 4.21   The structure of a dynamic model with n time-slices 103

Fig. 4.22   A model with four time-slices 104

Fig. 4.23   The test-and-treat cycle of the breeding pigs network in Fig. 4.15 .. 105

Fig. 4.24   The breeding pigs network as a time-sliced OOPN 105

Fig. 4.25   The stud farm pedigree.. 109

Fig. 5.1   The relevant networks for computing (a) $P(T|\varepsilon)$ and $P(L|\varepsilon)$ and (b) $P(B|\varepsilon)$... 116

Fig. 5.2   An illustration of reversal of the arc (w, v)........................... 117

Fig. 5.3   Computing $P(L)$ by arc reversal 117

Fig. 5.4   A graphical illustration of the process of eliminating Y from $\phi(X_1, X_2, Y)$ and X_1 from $\phi(X_1, X_2, X_3, X_4)$, where the ovals represent the domain of a potential before elimination and rectangles represent the domain of a potential after elimination.. 119

Fig. 5.5   The Burglary or Earthquake network 119

Fig. 5.6   A graphical illustration of the process of computing $P(A)$ in (5.6), where the ovals represent the domain of a potential before elimination and rectangles represent the domain of a potential after elimination 120

Fig. 5.7   A graphical illustration of the process of computing $P(W)$ in (5.7), where the ovals represent the domain of a potential before elimination and rectangles represent the domain of a potential after elimination 120

Fig. 5.8   A junction tree representation $\mathcal{T}$ for the chest clinic network 121

Fig. 5.9   When C_j has absorbed information from its other neighbors, C_i can absorb from C_j 122

Fig. 5.10   A junction tree representation $\mathcal{T}$ of the Bayesian
             network depicted in Fig. 5.5 on page 119 123
Fig. 5.11   The undirected graph corresponding to Figs. 5.6, 5.7, and 5.10 124
Fig. 5.12   Message passing in $\mathcal{T}$ 124
Fig. 5.13   The density function for X_3... 127
Fig. 5.14   The density function $f(X_3)$ for X_3 and its weak marginal $g(X_3)$ ... 127
Fig. 5.15   Optimization of price given marketing budget size.................. 133
Fig. 5.16   Expected utility as a function of price and marketing budget 134
Fig. 5.17   The DAG induced by the subset of requisite
             observations and relevant variables for D_4 136
Fig. 5.18   The DAG induced by the subset of requisite
             observations and relevant variables for D_3 136
Fig. 5.19   The DAG induced by the subset of requisite
             observations and relevant variables for D_2 136
Fig. 5.20   A graph specifying the independence and dependence
             relations of the Asia example 142

Fig. 6.1   A model for determining pregnancy state........................... 154
Fig. 6.2   A refined model for determining the pregnancy state,
             reflecting the fact that both tests are indications of
             the hormonal state, which in turn is an indication of
             pregnancy state 154
Fig. 6.3   Typical overall causal structure of a probabilistic network........... 155
Fig. 6.4   Causal structure for the prize problem 156
Fig. 6.5   Structure of the model for the classification problem
             of Example 6.4 156
Fig. 6.6   Sample instantiation of the definitional/synthesis
             idiom (Neil et al. 2000) 157
Fig. 6.7   Sample instantiation of the cause–consequence
             idiom (Neil et al. 2000) 157
Fig. 6.8   Sample instantiation of the measurement idiom
             (Neil et al. 2000) 158
Fig. 6.9   Sample instantiation of the induction idiom (Neil et al. 2000)....... 158
Fig. 6.10   Sample instantiation of the reconciliation idiom
             (Neil et al. 2000) 159
Fig. 6.11   The five basic kinds of idioms defined by Neil et al. (2000) 161
Fig. 6.12   Choosing the right idiom (Neil et al. 2000) 162
Fig. 6.13   Structure of potential causes of bronchitis 165
Fig. 6.14   Structure of potential causes of bronchitis and lung cancer 168
Fig. 6.15   Structure of potential causes of tuberculosis......................... 168
Fig. 6.16   Alternative structure of potential causes of tuberculosis 169
Fig. 6.17   Structure of potential symptoms of bronchitis and pneumonia 171
Fig. 6.18   Structure of potential symptoms shared by bronchitis,
             pneumonia, and lung cancer 172
Fig. 6.19   Complete structure of the extended chest clinic model.............. 172

Fig. 6.20   Typical expanded overall causal structure of a
            probabilistic network.. 174
Fig. 6.21   Two competing models for Example 6.13 175
Fig. 6.22   Probability wheel with three pie wedges corresponding
            to three states of a variable representing the increase
            in the average global temperature over the next 100
            years. The relative area occupied by a particular pie
            wedge represents the (conditional) probability for the
            associated state (indicated by the numbers next to the
            pie wedges) 178
Fig. 6.23   Mapping of verbal statements of probability to probabilities 178
Fig. 6.24   A model for the fake die problem 182
Fig. 6.25   Model development is an activity that iteratively passes
            through design, implementation, test, and analysis
            phases until model tests no longer uncover undesired
            behavior of the model 186
Fig. 6.26   Structure of a Bayesian network for the "burglary or
            earthquake" inference problem 189

Fig. 7.1   (a) X_1, X_2, and X_3 are direct parents of Y. (b) X_3 is a
            direct parent of Y, while the combined influence of X_1
            and X_2 is mediated through I 192
Fig. 7.2   Parent divorcing applied to the distribution for
            $Y = X_1 \vee X_3 \vee X_3$.. 193
Fig. 7.3   Parent divorcing in general 194
Fig. 7.4   Modeling a temporal order of the impacts
            of $X_1, \ldots, X_4$ on Y 196
Fig. 7.5   (a) Should A or B be the parent of Y? (b) Modeling
            structure and functionality uncertainty 197
Fig. 7.6   Either George or Henry is the true sire of Ann...................... 199
Fig. 7.7   Either $Y = X_1 \vee X_2$ or $Y = X_1 \wedge X_2$............................... 199
Fig. 7.8   Functional uncertainty on the height of a person 200
Fig. 7.9   Functional uncertainty on the height of a person 201
Fig. 7.10   (a) A functional relation $f(X_1, X_2, X_3)$ is to be
            enforced. (b) A constraint over X_1, X_2, and X_3 is
            enforced by instantiating C to on 202
Fig. 7.11   The constraint over $S_1, \ldots, S_4$ is enforced by
            instantiating C to on 203
Fig. 7.12   (a) How should the bidirectional correlation
            between X_1 and X_2 be captured? (b) A mediating
            variable Y between X_1 and X_2 captures the
            bidirectional relation 204

Fig. 7.13   (a) How should the bidirectional correlation
             between BT and UT be captured? (b) The bidirectional
             correlation between BT and UT is captured by the
             mediating variable HS 205
Fig. 7.14   The structure of the naive Bayes model 206
Fig. 7.15   A naive Bayes model for classifying mushrooms 207
Fig. 7.16   The observed value of a phenomenon is a function of
             the accuracy of the measurement and the actual value
             of the measured phenomenon.. 209
Fig. 7.17   The measured temperature is a function of the quality
             of the thermometer and the actual temperature 209
Fig. 7.18   The observations on color and pattern are imperfect 210
Fig. 7.19   The variable Experts has one state for each expert 212
Fig. 7.20   A graph specifying the independence and dependence
             relations of the Asia example 212
Fig. 7.21   Modeling the option of setting a value of B by intervention......... 215
Fig. 7.22   Taking an aspirin forces the fever to a certain level.
             Subsequent observations on Sleepy should not change
             our belief in Flu.. 216
Fig. 7.23   The causal influence of C_i on E is independent of the
             causal influence of C_j on E (for $i \neq j$) 217
Fig. 7.24   Two model structures for capturing independence of
             causal influence 217
Fig. 7.25   One inhibitor probability for each parent C_i of the effect E 218
Fig. 7.26   Sore throat may be caused by both angina and cold.................. 219
Fig. 7.27   Approximation of a continuous distribution using the
             MoGs modeling technique... 222
Fig. 7.28   A two-component mixture approximation of the
             Gamma(2, 2) distribution 223
Fig. 7.29   The test result is only available after a test is performed 224
Fig. 7.30   Prior to deciding on whether or not to take an aspirin,
             we may measure the temperature..................................... 226
Fig. 7.31   The test for temperature is modeled as a decision value
             with a random variable as a child specifying the result
             of the test 226
Fig. 7.32   (a) Informational links are unnecessary in influence
             diagrams with a single decision. (b) Informational
             links only clutter up the graph 228
Fig. 7.33   (a) In some situations X is observed prior to D, while
             in others it is not. (b) By introducing an additional
             variable, we capture the situation where an observation
             on X may or may not be available 229
Fig. 7.34   The observation on Seismic is missing when the test is
             not performed 230

Fig. 7.35   This graph captures the situation where the result of
            seismic soundings may not be available 230
Fig. 7.36   Decision D selects a hypothesis of maximum probability 231
Fig. 7.37   Decision D selects a hypothesis of maximum probability 232
Fig. 7.38   Selecting a disease hypothesis with highest probability 233
Fig. 7.39   A constraint on configurations of decisions D_1 and D_2 234

Fig. 8.1   We assume the underlying process generates a
            database of cases as well as experience and knowledge
            that can be fused for learning a model of the process 238
Fig. 8.2   Three equivalent DAGs 241
Fig. 8.3   The equivalence class of the DAGs in Fig. 8.2 242
Fig. 8.4   The χ^2 density function for different degrees of freedom 244
Fig. 8.5   Skeleton representing the CDIRs $\mathcal{M}_\mathcal{D}$ of (8.2) and (8.3)............. 247
Fig. 8.6   Colliders identified from $\mathcal{M}_\mathcal{D}$ and the skeleton of Fig. 8.5 248
Fig. 8.7   Rule $\mathcal{R}_1$ for identifying derived directions 248
Fig. 8.8   Rule $\mathcal{R}_2$ for identifying derived directions 249
Fig. 8.9   Rule $\mathcal{R}_3$ for identifying derived directions 249
Fig. 8.10   Rule $\mathcal{R}_4$ for identifying derived directions. (The
            dashed line between X and V indicates that X and V
            are adjacent, i.e., connected by an edge.) 249
Fig. 8.11   Derived directions identified from the skeleton and the
            colliders identified in Example 8.7 250
Fig. 8.12   The result of structure learning 251
Fig. 8.13   The necessary path condition says that in order
            for $X \perp\!\!\!\perp Y \,|\, \{Z_1, \ldots, Z_n\}$ to be valid, there should for
            each $Z_i, i = 1, \ldots, n$ exist a path between X and Z_i
            not crossing Y and vice versa 252
Fig. 8.14   The condition graph over $\mathcal{M}_\perp$ 253
Fig. 8.15   The strongly connected component of the condition
            graph in Fig. 8.14 254
Fig. 8.16   Ambiguous edges in the skeleton due to the
            deterministic relation between Cancer, Tuberculosis,
            and Tub_or_cancer 254
Fig. 8.17   The two possible resolutions of the ambiguous region 254
Fig. 8.18   Skeleton representing the CDIRs $\mathcal{M}_\mathcal{D}$ generated
            from $\mathcal{D}$ by the PC algorithm 255
Fig. 8.19   Skeleton representing the CDIRs $\mathcal{M}_\mathcal{D}$ generated
            from $\mathcal{D}$ by the NPC algorithm after selecting minimal resolutions .. 255
Fig. 8.20   Three operations of incremental change 258
Fig. 8.21   Four data points and a linear function and a
            three-degree polynomial fitted to the data to illustrate overfitting ... 260
Fig. 8.22   Three different network structures for encoding the
            joint probability distribution in Table 8.5 262
Fig. 8.23   Five different neighbors of the DAG in Fig. 8.22b 264

Fig. 8.24   A naive Bayes model for classifying patients 266
Fig. 8.25   A maximal-weight spanning tree 268
Fig. 8.26   A tree-augmented naive Bayes model 271
Fig. 8.27   The PDAG produced by the PC algorithm 272
Fig. 8.28   The DAG $\mathcal{G}_{PC}$ produced using the PC algorithm 273
Fig. 8.29   The PDAG produced by the NPC algorithm after
            resolving ambiguous regions 274
Fig. 8.30   The DAG $\mathcal{G}_{NPC}$ produced using the NPC algorithm 274
Fig. 8.31   The DAG $\mathcal{G}_{GSS}$ produced using the greedy search and
            score-based algorithm 275
Fig. 8.32   The DAG $\mathcal{G}_{CL}$ produced using the Chow–Liu algorithm 276
Fig. 8.33   Two equivalent DAGs over X and Y 282
Fig. 8.34   Retrieval and dissemination of experience 284
Fig. 8.35   A graph specifying the independence and dependence
            relations of the Asia example 287
Fig. 8.36   The angina network 287

Fig. 9.1   A graph specifying the independence and dependence
            relations of the Asia example 292
Fig. 9.2   The impact of finding ε_S on the hypothesis $h : B =$ no 298
Fig. 9.3   The impact of finding ε_S on the hypothesis $h : B =$ yes 298
Fig. 9.4   Mr. Holmes installs a seismometer with a direct line to
            his office 300

Fig. 10.1   A graph specifying the independence and dependence
            relations of the Asia example 305
Fig. 10.2   The graph of the sensitivity function $f(t) = P(\mathsf{Sick} =$
            $\mathsf{yes}|\mathsf{Loses} = \mathsf{yes})$ as a function of $t = P(\mathsf{Dry} = \mathsf{yes})$ 314
Fig. 10.3   The graph of the sensitivity functions for $\mathsf{Sick} = \mathsf{yes}$
            and $\mathsf{Sick} = \mathsf{no}$ 317
Fig. 10.4   The graph of the sensitivity function
            $f(t_1, t_2) = P(\mathsf{Sick} = \mathsf{yes}|\mathsf{Loses} = \mathsf{yes})$
            as a function of $t_1 = P(\mathsf{Dry} = \mathsf{yes})$ and
            $t_2 = P(\mathsf{Loses} = \mathsf{yes}|\mathsf{Dry} = \mathsf{yes}, \mathsf{Sick} = \mathsf{yes})$, where
            $t_0 = (t_{1_0} = 0.1, t_{2_0} = 0.95)$ are the initial values of the
            two parameters in focus 319
Fig. 10.5   The wet grass network 324
Fig. 10.6   A graph specifying the independence and dependence
            relations of the Asia example 325

Fig. 11.1   The entropy of T 330
Fig. 11.2   The negation of the entropy of T 331
Fig. 11.3   A graph specifying the independence and dependence
            relations of the Asia example 331
Fig. 11.4   A graph representing the appendicitis example 334

Fig. 11.5   A graph specifying the independence and dependence
            relations of the Asia example 337
Fig. 11.6   An influence diagram with two decisions 338

List of Tables

Table 1.1	The four variables and their possible states for the "car would not start" problem	11
Table 1.2	Conditional probability distributions for Fuel_gauge given Fuel?, $P(\text{Fuel\_gauge} \mid \text{Fuel?})$	12
Table 1.3	Conditional probability distributions for Start? given Fuel? and Spark_plugs, $P(\text{Start?} \mid \text{Fuel?}, \text{Spark\_plugs})$	12
Table 1.4	Utility function for the "car would not start" decision problem	14
Table 2.1	The taxonomy for variables/vertices	22
Table 2.2	Vertex symbols	23
Table 2.3	Notation used for vertices, variables, and utility functions	23
Table 4.1	The conditional probability distribution $P(L \mid D, S)$	73
Table 4.2	The conditional probability distributions $P(L \mid S)$, $P(B \mid S)$, $P(T \mid A)$, and $P(X \mid E)$	75
Table 4.3	The conditional probability distribution $P(D \mid B, E)$	75
Table 4.4	Posterior distributions of the disease variables given various evidence scenarios	75
Table 4.5	The conditional probability distribution $P(\text{Seismic} \mid \text{Oil}, \text{Test} = \text{yes})$	82
Table 4.6	The conditional probability distribution $P(\text{Sick}^* \mid \text{Treat}, \text{Sick})$	87
Table 4.7	The conditional probability distribution $P(\text{Dry}^* \mid \text{Dry})$	87
Table 4.8	The conditional probability distribution $P(\text{Loses}^* \mid \text{dry}^*, \text{Sick}^*)$	87
Table 4.9	The conditional probability distribution $P(\text{Drought} \mid \text{Dry})$	99
Table 5.1	The joint expected utility function $EU(D, S, T)$	130
Table 5.2	The expected utility function $EU(T)$	130
Table 6.1	Typical causal dependence relations for different variable classes	153

U.B. Kjærulff and A.L. Madsen, *Bayesian Networks and Influence Diagrams: A Guide to Construction and Analysis*, ISS 22, DOI 10.1007/978-1-4614-5104-4, © Springer Science+Business Media New York 2013

Table 6.2   Examples of possible states of the various subtypes
                of discrete chance variables.. 181

Table 6.3   The functionality of a table generator algorithm is
                dependent on the subtype of the involved variables 181

Table 6.4   The CPT for $P(N_I \mid N_M, N_F)$ in Example 6.15
                generated from the expression $N_I = N_M + N_F$ 182

Table 6.5   The CPT for $P(\#6's \mid \#rolls, Fake\ die?)$ in the fake
                die problem of Example 6.16 generated from the
                expression Binomial(#rolls, if(Fake die?, $1/5, 1/6$)) 182

Table 6.6   The CPT for $P(C_1 \mid C_2)$ in the discretization
                problem of Example 6.17 generated from the
                expression Normal($C_2, 1$) 183

Table 7.1   The conditional probability distribution $P(Y \mid X_1, X_2, X_3)$ 193
Table 7.2   The conditional probability distribution $P(X_1 \vee X_2 \mid X_1, X_2)$ 193
Table 7.3   The conditional probability distribution $P(Y \mid X_1, \ldots, X_m)$ 194
Table 7.4   The conditional probability distribution $P(I \mid X_1, X_2)$ 195
Table 7.5   The conditional probability distribution $P(Y \mid I, X_3, \ldots, X_m)$ ... 195
Table 7.6   The conditional probability distribution $P(P \mid A, B, S)$ 198
Table 7.7   The conditional probability distribution
                $P(Sire \mid Henry, George, S)$ 199
Table 7.8   The conditional probability distribution $P(Y \mid F, X_1, X_2)$ 200
Table 7.9   The conditional probability distribution $P(C \mid S_1, \ldots, S_4)$ 203
Table 7.10   The conditional probability distribution $P(Odor \mid C)$ 208
Table 7.11   The conditional probability distribution
                $P(Obs\_Temperature \mid Quality, Temperature)$ 210
Table 7.12   The conditional probability distribution $P(Obs\_Color \mid Color)$ 210
Table 7.13   The specification of the conditional probability
                distribution $P(Bronchitis \mid Smoker, Experts)$ 212
Table 7.14   The conditional probability distribution
                $P(Bronchitis \mid Smoker)$ after absorbing Experts from
                the distribution shown in Table 7.13 214
Table 7.15   (a) The conditional probability
                distribution $P(B \mid A)$. (b) The conditional
                probability distribution $P(B \mid I, A)$ 215
Table 7.16   The conditional probability
                distribution $P(SoreThroat \mid Angina, Cold)$
                with a zero background event probability......................... 220
Table 7.17   The conditional probability
                distribution $P(SoreThroat \mid Angina, Cold)$
                with a nonzero background event probability...................... 220
Table 7.18   The conditional probability distribution $P(Seismic \mid Test, Oil)$ 225
Table 7.19   The conditional probability distribution
                $P(Seismic \mid Test, Oil)$ where Seismic has a no result state 225
Table 7.20   The utility function $U(Pr, D)$ 228

Table 7.21   The conditional probability distribution $P(\text{Obs}|\text{Seismic})$ 231
Table 7.22   The utility table $U(D_1, D_2)$ 234
Table 7.23   The utility function $U(\text{Surgery}, \text{Appendicitis})$ 236

Table 8.1   A database of cases 239
Table 8.2   The number of possible DAGs for values of n from
              one to ten as calculated using $f(n)$ 240
Table 8.3   Contingency table for testing marginal
              independence of X and Y 245
Table 8.4   A set of values (x_i, y_i) for variables X and Y 259
Table 8.5   Joint probability distributions over X, Y, and Z 262
Table 8.6   Data $\mathcal{D}$ over X, Y, and Z 262
Table 8.7   The conditional probability distributions
              $P(Y|X)$ and $P(Z|Y)$ 263
Table 8.8   The conditional probability distribution $P(Z|X, Y)$ 263
Table 8.9   BIC scores for neighbors of $\mathcal{G}$ in Fig. 8.22b 264
Table 8.10   Contingency table for L and A 266
Table 8.11   The conditional probability distribution $P(A|L)$ 266
Table 8.12   Contingency table for D and S 267
Table 8.13   Mutual information $I(X, Y)$ for each pair 268
Table 8.14   Data frequencies for D, L, and S 270
Table 8.15   Conditional mutual information $I(X, Y | L)$ for each pair 271
Table 8.16   Different model quality scores for the five DAGs
              considered in the example 277
Table 8.17   Data frequencies for estimating the distributions
              of X and Y 282
Table 8.18   Joint probability distributions over X and Y 283
Table 8.19   Experience counts for B, L, and S before and after adaptation .. 285
Table 8.20   Data frequencies for the class variable 287
Table 8.21   Data frequencies for the class variable and Population............ 288
Table 8.22   Data frequencies for the class variable and Cap_Shape........... 288
Table 8.23   Data frequencies for the class variable and Odor 288
Table 8.24   Data frequencies for A, D, S, and X 288

Table 9.1   The log-normalized likelihood of each possible
              instantiation of each variable in Example 9.1..................... 296
Table 9.2   The conditional probability distributions $P(W|A)$
              and $P(S|B, E)$ 300
Table 9.3   The conditional probability distribution $P(A|B, E, F)$ 300

Table 10.1   Sensitivity of the posterior probability distribution
              of the hypothesis variable B to findings on A 306
Table 10.2   Normalized likelihood of hypothesis h_B given all
              subsets of the evidence ε 307
Table 10.3   Discrimination between hypothesis h_B and the
              alternative hypothesis h_L 308

Table 10.4   What-if analysis on findings 309

Table 10.5   Findings impact analysis 310

Table 10.6   Cost-of-omission and distance in posterior beliefs
              of the hypothesis for each finding 311

Table 10.7   Parameter values that satisfy the constraint
              $P(\text{Sick} = \text{yes}|\varepsilon) \geq 0.35$ 322

Table 10.8   The conditional probability distribution
              $P(\text{Holmes' lawn}|\text{Rain, Sprinkler})$ 324

Table 10.9   The conditional probability distribution
              $P(\text{Gibbon's lawn}|\text{Rain})$ 324

Table 10.10   The conditional probability distribution
              $P(\text{Watson's lawn}|\text{Rain})$ 324

Table 11.1   Mutual information between Bronchitis and other
              variables given no observations 332

Table 11.2   Mutual information between Bronchitis and other
              variables given Dyspnoea = yes 332

Table 11.3   The conditional probability distribution
              $P(\text{Fever}|\text{ Appendicitis})$ 334

Table 11.4   The conditional probability distribution
              $P(\text{Pain}|\text{Appendicitis})$ 334

Table 11.5   The conditional probability distribution
              $P(\text{White\_Blood\_Cells}|\text{Appendicitis})$ 334

Table 11.6   The utility function $U(\text{Appendicitis, Operate})$ 334

Table 11.7   The joint probability distribution $P(\text{Fever, Flu})$ 339

List of Symbols

$\triangleq$	Defined as
$\equiv$	Equivalent to
$:=$	Assignment
$\propto$	Proportional to
$\\| \cdot \\|$	Domain size (i.e., $\\|X\\| = \|\mathrm{dom}(X)\|$)
$\perp\!\!\!\perp$	Independent
$\perp\!\!\!\perp_p$	Independent with respect to distribution p
$\not\perp\!\!\!\perp$	Dependent
$\not\perp\!\!\!\perp_p$	Dependent with respect to distribution p
$\perp$	d-separated
$\perp_{\mathcal{G}}$	d-separated in graph $\mathcal{G}$
$\not\perp$	d-connected
$\not\perp_{\mathcal{G}}$	d-connected in graph $\mathcal{G}$
$\mathcal{S}$	Separator set
$\mathcal{E}$	Evidence function (potential)
ε	Evidence
$\mathcal{E}_{\varepsilon}$	Evidence function for $\mathcal{X}(\varepsilon)$
η	Normalization operator
$\sim$	Connected
$\rightarrow$	Connected by directed edge
$\leftarrow$	Connected by directed edge
$-$	Connected by undirected edge
$\overset{\mathcal{G}}{\sim}$	Connected in graph $\mathcal{G}$
$\overset{\mathcal{G}}{\rightarrow}$	Connected by directed edge in graph $\mathcal{G}$
$\overset{\mathcal{G}}{-}$	Connected by undirected edge in graph $\mathcal{G}$
$\not\sim$	Not connected
$\overset{\mathcal{G}}{\not\sim}$	Not connected in graph $\mathcal{G}$
$\langle u, \ldots, v \rangle$	Path from u to v
dom	Domain (of variable or set of variables)

U.B. Kjærulff and A.L. Madsen, *Bayesian Networks and Influence Diagrams: A Guide to Construction and Analysis*, ISS 22, DOI 10.1007/978-1-4614-5104-4,
© Springer Science+Business Media New York 2013

$\mid$	"given" (e.g., "$a \mid b$" means "a given b")
pa	Parents of
fa	Family of
ch	Children of
an	Ancestors of
An	Ancestral set of
de	Descendants of
nd	Non-descendants of
true	Boolean value "true"
false	Boolean value "false"
$\mathcal{J}$	Past
EU	Expected utility
MEU	Maximum expected utility
$\mathbb{N}$	Normal (Gaussian) distribution
$\mathbb{N}_k$	k-dimensional Normal distribution
$\mathcal{L}$	Law of (e.g., $\mathcal{L}(X) = \mathbb{N}(\mu, \sigma^2)$, also denoted $X \sim \mathbb{N}(\mu, \sigma^2)$)
$\mathbb{R}$	The set of all real numbers
$\mathbb{S}$	Scope
$\mathcal{I}$	Input variables
$\mathcal{H}$	Private (hidden) variables
$\mathcal{O}$	Output variables
$\mathcal{P}$	Public variables (input + output)
$\mathcal{X}(\varepsilon)$	Evidence variables (i.e., subset of $\mathcal{X}$ for which their values are known)

References

Andersen, S. K., Olesen, K. G., Jensen, F. V. & Jensen, F. (1989). HUGIN — a Shell for Building Bayesian Belief Universes for Expert Systems, *Proceedings of the Eleventh International Joint Conference on Artificial Intelligence*, pp. 1080–1085.

Bender, E. A. (1996). *Mathematical Methods in Artificial Intelligence*, IEEE Computer Society Press.

Boyen, X. & Koller, D. (1998). Tractable inference for complex stochastic processes, *Proceedings of the Fourteenth Conference on Uncertainty in Artificial Intelligence*, pp. 33–42.

Castillo, E., Gutiérrez, J. M. & Hadi, A. S. (1997). Sensitivity analysis in discrete Bayesian Networks, *IEEE Transactions on Systems, Man and Cybernetics, Part A*, pp. 412–423.

Chan, H. & Darwiche, A. (2002a). A distance measure for bounding probabilistic belief change, *Proceedings of the Eighteenth National Conference on Artificial Intelligence*, pp. 539–545.

Chan, H. & Darwiche, A. (2002b). When do numbers really matter?, *Journal of Artificial Intelligence Research* **17**: 265–287.

Chickering, D. M. (1996). Learning Bayesian networks is NP-complete, *Learning from Data: Artificial Intelligence and Statistics V*, pp. 121–130.

Chickering, D. M. (2002). Optimal structure identification with greedy search, *Journal of Machine Learning Research* **3**: 507–554.

Chickering, D. M. & Meek, C. (2003). Monotone dag faithfulness: A bad assumption, *Technical Report MSR-TR-2003-16*, Microsoft Research.

Chow, C. K. & Liu, C. N. (1968). Approximating discrete probability distributions with dependence trees, *IEEE Transactions on Information Theory* **IT-14**(3): 462–467.

Cobb, B. R. & Shenoy, P. P. (2004). Hybrid influence diagrams using mixtures of truncated exponentials, *Proceedings of the Twentieth Conference on Uncertainty in Artificial Intelligence*, AUAI Press, Arlington, VA, pp. 85–93.

Cooper, G. F. (1990). The computational complexity of probabilistic inference using Bayesian belief networks, *Artificial Intelligence* **42**(2–3): 393–405.

Cooper, G. F. & Herskovits, E. (1992). A bayesian method for the induction of probabilistic networks from data, *Machine Learning* **9**: 309–347.

Coupé, V. M. & van der Gaag, L. C. (1998). Practicable sensitivity analysis of Bayesian belief networks, *Prague Stochastics '98 – Proceedings of the Joint Session of the 6th Prague Symposium of Asymptotic Statistics and the 13th Prague Conference on Information Theory, Statistical Decision Functions and Random Processes*, pp. 81–86.

Cover, T. M. & Thomas, J. A. (1991). *Elements of Information Theory*, John Wiley & Sons.

Cowell, R. G. (2001). Conditions Under Which Conditional Independence and Scoring Methods Lead to Identical Selection of Bayesian Network Models, *Proceedings of the Seventeenth Conference on Uncertainty in Artificial Intelligence*, pp. 91–97.

U.B. Kjærulff and A.L. Madsen, *Bayesian Networks and Influence Diagrams: A Guide to Construction and Analysis*, ISS 22, DOI 10.1007/978-1-4614-5104-4,
© Springer Science+Business Media New York 2013

Cowell, R. G., Dawid, A. P., Lauritzen, S. L. & Spiegelhalter, D. J. (1999). *Probabilistic Networks and Expert Systems*, Springer-Verlag.

Dagum, P. & Luby, M. (1993). Approximating probabilistic inference in Bayesian belief netwoks is NP-hard, *Artificial Intelligence* **60**: 141–153.

Darwiche, A. (2009). *Modeling and Reasoning with Bayesian Networks*, Cambridge University Press.

Dawid, A. P. (1992). Applications of a general propagation algorithm for probabilistic expert systems, *Statistics and Computing* **2**: 25–36.

DeGroot, M. (1986). *Probability and Statistics*, second edn, Addison-Wesley. Reprint with corrections, 1989.

Dempster, A. P. (1968). A generalization of Bayesian inference, *Journal of the Royal Statistical Society, Series B* **30**: 205–247.

Diez, F. J. (1993). Parameter adjustment in Bayes networks, the generalized noisy OR-gate, *Proceedings of the Ninth Conference on Uncertainty in Artificial Intelligence*, pp. 99–105.

Duda, R. O. & Hart, P. E. (1973). *Pattern Classification and Scene Analysis*, New York: John Wiley & Sons.

Fenton, N., Neil, M. & Lagnado, D. (2011). Modelling mutually exclusive causes in Bayesian networks, *Submitted to IEEE Transactions on Knowledge and Data Engineering, April 2011*.

Friedman, N. (1998). The Bayesian Structural EM Algorithm, *Proceedings of the Fourteenth Conference on Uncertainty in Artificial Intelligence*, pp. 129–138.

Friedman, N., Geiger, D. & Goldszmidt, M. (1997). Bayesian network classifiers, *Machine Learning* pp. 1–37.

Frydenberg, M. (1989). The chain graph Markov property, *Scandinavian Journal of Statistics* **17**: 333–353.

Geiger, D., Verma, T. & Pearl, J. (1990). Identifying independence in Bayesian networks, *Networks* **20**(5): 507–534. Special Issue on Influence Diagrams.

Grinstead, C. M. & Snell, J. L. (1997). *Introduction to Probability*, 2nd edn, American Mathematical Society.

Heckerman, D. (1986). Probabilistic interpretations for MYCIN's certainty factors, *Uncertainty in Artificial Intelligence*, Elsevier Science Publishers B. V. (North-Holland), Amsterdam, pp. 298–312.

Heckerman, D. (1993). Causal independence for knowledge acquisition and inference, *Proceedings of the Ninth Conference on Uncertainty in Artificial Intelligence*, pp. 122–127.

Heckerman, D. (1998). A tutorial on learning with bayesian networks, *Learning in Graphical Models*, MIT Press, pp. 301–354.

Heckerman, D. & Breese, J. S. (1994). A new look at causal independence, *Proceedings of the Tenth Conference on Uncertainty in Artificial Intelligence*, pp. 286–292.

Howard, R. A. & Matheson, J. E. (1981). Influence diagrams, *Readings in Decision Analysis*, Strategic Decisions Group, Menlo Park, CA, chapter 38, pp. 763–771.

HUGIN (2006). HUGIN GUI Help, HUGIN Expert A/S, www.hugin.com. Reference Manual for the HUGIN Graphical User Interface.

Jensen, F. (2010). HUGIN API Reference Manual, HUGIN EXPERT A/S, www.hugin.com. Reference Manual for the HUGIN Decision Engine APIs version 7.3.

Jensen, F., Jensen, F. V. & Dittmer, S. (1994). From influence diagrams to junction trees, *Proceedings of the Tenth Conference on Uncertainty in Artificial Intelligence*, Morgan Kaufmann Publishers, San Francisco, pp. 367–373.

Jensen, F. V. (1996). *An Introduction to Bayesian Networks*, UCL Press, London.

Jensen, F. V. (2001). *Bayesian Networks and Decision Graphs*, Springer-Verlag.

Jensen, F. V. & Jensen, F. (1994). Optimal junction trees, *Proceedings of the Tenth Conference on Uncertainty in Artificial Intelligence*, pp. 360–366.

Jensen, F. V., Lauritzen, S. L. & Olesen, K. G. (1990). Bayesian updating in causal probabilistic networks by local computations, *Computational Statistics Quarterly* **4**: 269–282.

Kenley, C. R. (1986). *Influence Diagram Models with Continuous Variables*, PhD thesis, Department of Engineering-Economic Systems, Stanford University, CA.

Kim, J. H. & Pearl, J. (1983). A computational model for causal and diagnostic reasoning in inference systems, *Proceedings of the Eighth International Joint Conference on Artificial Intelligence*, pp. 190–193.

Kim, Y.-G. & Valtorta, M. (1995). On the detection of conflicts in diagnostic Bayesian networks using abstraction, *Proceedings of the Eleventh Conference on Uncertainty in Artificial Intelligence*, pp. 362–367.

Kjærulff, U. B. (1995). dHugin: a computational system for dynamic time-sliced Bayesian networks, *International Journal of Forecasting* **11**: 89–111.

Kjærulff, U. B. & van der Gaag, L. C. (2000). Making sensitivity analysis computationally efficient, *Proceedings of the Sixteenth Conference on Uncertainty in Artificial Intelligence*, pp. 317–325.

Koller, D. & Pfeffer, A. (1997). Object-oriented Bayesian networks, *Proceedings of the Thirteenth Conference on Uncertainty in Artificial Intelligence*, pp. 302–313.

Langley, P., Iba, W. & Thompson, K. (1992). An analysis of Bayesian classifiers, *Proceedings of the Eleventh National Conference on Artificial Intelligence*, pp. 223–228.

Laskey, K. B. (1991). Conflict or surprise: Heuristics for model revision, *Proceedings of the Seventh Conference on Uncertainty in Artificial Intelligence*, pp. 197–204.

Laskey, K. B. (1993). Sensitivity analysis for probability assessments in Bayesian networks, *Proceedings of the Ninth Conference on Uncertainty in Artificial Intelligence*, pp. 136–142.

Laskey, K. B. & Mahoney, S. M. (1997). Network fragments: Representing knowledge for constructing probabilistic models, *Proceedings of the Thirteenth Conference on Uncertainty in Artificial Intelligence*, pp. 334–341.

Lauritzen, S. L. (1992a). Credit Valuation Using HUGIN. Personal communication.

Lauritzen, S. L. (1992b). Propagation of probabilities, means and variances in mixed graphical association models, *Journal of the American Statistical Association* **87**(420): 1098–1108.

Lauritzen, S. L. (1995). The EM algorithm for graphical association models with missing data, *Computational Statistics & Analysis* **19**: 191–201.

Lauritzen, S. L. (1996). *Graphical models*, Oxford Statistical Science Series, Clarendon Press, Oxford, England.

Lauritzen, S. L., Dawid, A. P., Larsen, B. N. & Leimer, H.-G. (1990a). Independence properties of directed Markov fields, *Networks* **20**(5): 491–505. Special Issue on Influence Diagrams.

Lauritzen, S. L., Dawid, A. P., Larsen, B. N. & Leimer, H. G. (1990b). Independence properties of directed Markov fields, *Networks* **20**(5): 491–505.

Lauritzen, S. L. & Jensen, F. (2001). Stable local computation with mixed Gaussian distributions, *Statistics and Computing* **11**(2): 191–203.

Lauritzen, S. L. & Nilsson, D. (2001). Representing and solving decision problems with limited information, *Management Science* **47**: 1238–1251.

Lauritzen, S. L. & Spiegelhalter, D. J. (1988). Local computations with probabilities on graphical structures and their application to expert systems, *Journal of the Royal Statistical Society, Series B* **50**(2): 157–224.

Lin, Y. & Druzdzel, M. J. (1997). Computational advantages of relevance reasoning in Bayesian networks, *Proceedings of the Thirteenth Conference on Uncertainty in Artificial Intelligence*, Morgan Kaufmann Publishers, San Francisco, pp. 342–350.

Madsen, A. L. (2005). A differential semantics of lazy propagation, *Proceedings of the 21st Conference on Uncertainty in Artificial Intelligence*, pp. 364–371.

Madsen, A. L. & Jensen, F. (2005). Solving linear-quadratic conditional Gaussian influence diagrams, *International Journal of Approximate Reasoning* **38**(3): 263–282.

Madsen, A. L., Jensen, F., Kjærulff, U. & Lang, M. (2005). The HUGIN tool for probabilistic graphical models, *International Journal of Artificial Intelligence Tools* **14**(3): 507–543.

Madsen, A. L. & Jensen, F. V. (1999). Lazy propagation: A junction tree inference algorithm based on lazy evaluation, *Artificial Intelligence* **113**(1–2): 203–245.

Madsen, A. L., Nielsen, L. M. & Jensen, F. V. (1998). Probsy — a system for the calculation of probabilities in the card game bridge, *Proceedings of the Eleventh International Florida Artificial Intelligence Research Symposium Conference*, pp. 435–439.

Madsen, A. L. & Nilsson, D. (2001). Solving influence diagrams using HUGIN, Shafer–Shenoy and lazy propagation, *Proceedings of the Seventeenth Conference on Uncertainty in Artificial Intelligence*, pp. 337–345.

Meek, C. (1995). Causal inference and causal explanation with background knowledge, *Proceedings of the Eleventh Conference on Uncertainty in Artificial Intelligence*, pp. 403–410.

Neil, M., Fenton, N. & Nielsen, L. M. (2000). Building large-scale Bayesian networks, *The Knowledge Engineering Review* 15(3): 257–284.

Nielsen, T. D. (2001). Decomposition of influence diagrams, *Proceedings of the Sixth European Conference on Symbolic and Quantitative Approaches to Reasoning and Uncertainty*, pp. 144–155.

Nielsen, T. D. & Jensen, F. V. (1999). Welldefined decision scenarios, *Proceedings of the Fifteenth Conference on Uncertainty in Artificial Intelligence*, pp. 502–511.

OEIS (2010). A003024. The On-Line Encyclopedia of Integer Sequences (OEIS). URL=http://www.oeis.org [Accessed 17 November 2010].

Olesen, K. G., Kjærulff, U., Jensen, F., Jensen, F. V., Falck, B., Andreassen, S. & Andersen, S. K. (1989). A MUNIN network for the median nerve – a case study on loops, *Applied Artificial Intelligence* 3. Special issue: Towards Causal AI Models in Practice.

Olmsted, S. (1983). *On representing and solving decision problems*, PhD thesis, Department of Engineering-Economic Systems, Stanford University, Stanford, CA.

Pearl, J. (1988). *Probabilistic Reasoning in Intelligent Systems: Networks of Plausible Inference*, Series in Representation and Reasoning, Morgan Kaufmann Publishers, San Mateo, CA.

Pearl, J. (2000). *Causality: models, reasoning, and inference*, Cambridge University Press.

Poland, W. B. (1994). *Decision Analysis with Continuous and Discrete Variables: A Mixture Distribution Approach*, PhD thesis, Engineering-Economic Systems, Stanford University, Stanford, CA.

Pradhan, M., Provan, G., Middleton, B. & Henrion, M. (1994). Optimal knowledge engineering for large belief networks, *Proceedings of the Tenth Conference on Uncertainty in Artificial Intelligence*, pp. 484–490.

Raiffa, H. (1968). *Decision Analysis*, Addison-Wesley, Reading, MA.

Renooij, S. (2001). Probability elicitation for belief networks: issues to consider, *The Knowledge Engineering Review* 16(3): 255–269.

Renooij, S. & Witteman, C. L. M. (1999). Talking probabilities: communicating probabilistic information with words and numbers, *International Journal of Approximate Reasoning* 22: 169–194.

Robinson, R. W. (1977). Counting unlabelled acyclic digraphs, *Lecture Notes in Mathematics: Combinatorial Mathematics V*, Springer-Verlag.

Shachter, R. (1998). Bayes–ball: The rational pasttime (for determining irrelevance and requisite information in belief networks and influence diagrams), *Proceedings of the Fourteenth Conference on Uncertainty in Artificial Intelligence*, pp. 480–487.

Shachter, R. D. (1986). Evaluating influence diagrams, *Operations Research* 34(6): 871–882.

Shachter, R. D. (1990). Evidence absorption and propagation through arc reversals, *Uncertainty in Artificial Intelligence*, Elsevier Science Publishers B. V. (North-Holland), Amsterdam, pp. 173–190.

Shachter, R. D. & Kenley, C. R. (1989). Gaussian influence diagrams, *Management Science* 35(5): 527–549.

Shachter, R. D. & Peot, M. A. (1992). Decision making using probabilistic inference methods, *Proceedings of the Eighth Conference on Uncertainty in Artificial Intelligence*, Morgan Kaufmann Publishers, San Mateo, CA, pp. 276–283.

Shafer, G. (1976). *A Mathematical Theory of Evidence*, Princeton University Press.

Shenoy, P. P. (2006). Inference in hybrid Bayesian networks using mixtures of Gaussians, *Proceedings of the 22nd Conference on Uncertainty in Artificial Intelligence*, pp. 428–436.

Shortliffe, E. H. & Buchanan, B. G. (1975). A model of inexact reasoning in medicine, *Mathematical Biosciences* 23: 351–379.

Spiegelhalter, D. & Lauritzen, S. L. (1990). Sequential updating of conditional probabilities on directed graphical structures, *Networks* **20**: 579–605.

Spirtes, P. & Glymour, C. (1991). An algorithm for fast recovery of sparse causal graphs, *Social Science Computing Review* **9**(1): 62–72.

Spirtes, P., Glymour, C. & Scheines, R. (2000). *Causation, Prediction, and Search*, Adaptive Computation and Machine Learning, second edn, MIT Press, Cambridge, MA.

Srinivas, S. (1993). A Generalization of the Noisy-Or Model, *Proceedings of the Ninth Conference on Uncertainty in Artificial Intelligence*, Morgan Kaufmann Publishers, pp. 208–218.

Steck, H. & Tresp, V. (1999). Bayesian belief networks for data mining, *Proceedings of the Second Workshop "Data Mining und Data Warehousing als Grundlage moderner enscheidungsunterstuetzender Systeme"*.

Suermondt, H. J. (1992). *Explanation in Bayesian Belief Networks*, PhD thesis, Department of Computer Science and Medicine, Stanford University, Stanford, CA.

Titterington, D. M., Smith, A. F. M. & Makov, U. E. (1995). *Statistical Analysis of Finite Mixture Distributions*, John Wiley & Sons, New York.

Tversky, A. & Kahneman, D. (1981). The framing of decisions and the psychology of choice, *Science* **211**.

van der Gaag, L. C., Renooij, S., Witteman, C. L. M. & Taal, B. G. (2002). Probabilities for a probabilistic network: a case study in oesophageal cancer, *Artificial Intelligence in Medicine* **25**: 123–148.

van der Gaag, L. & Renooij, S. (2001). Analyzing sensitivity data from probabilistic networks, *Proceedings of the Seventeenth Conference on Uncertainty in Artificial Intelligence*, Morgan Kaufmann Publishers, pp. 530–537.

Verma, T. & Pearl, J. (1992). An algorithm for deciding if a set of observed independencies has a causal explanation, *Proceedings of the Eighth Conference on Uncertainty in Artificial Intelligence*, pp. 323–330.

Vomlelová, M. & Jensen, F. V. (2002). An extension of lazy evaluation for influence diagrams avoiding redundant variables in the potentials, *First European Workshop on Probabilistic Graphical Models*, pp. 186–193.

Wang, H., Rish, I. & Ma, S. (2002). Using sensitivity analysis for selective parameter update in Bayesian network learning, *AAAI Symposium Series*.

Wermuth, N. & Lauritzen, S. L. (1983). Graphical and recursive models for contingency tables, *Biometrika* **70**: 537–552.

Wermuth, N. & Lauritzen, S. L. (1990). On substantive research hypotheses, conditional independence graphs and graphical chain models, *Journal of the Royal Statistical Society, Series B* **52**(1): 21–50.

Zadeh, L. A. (1965). Fuzzy sets, *Information and Control* **8**: 338–353.

Zadeh, L. A. & Kacprzyk, J. (eds) (1992). *Fuzzy Logic for the Management of Uncertainty*, John Wiley and Sons, New York.

Zhang, N. L. (2004). Hierarchical latent class models for cluster analysis, *Journal of Machine Learning Research* **5**: 697–723.

Index

Symbols

$\sum$-max-$\sum$-rule, *see* influence diagram

A

$\sim$, 19

$\|\|\|$, 21

$\rightarrow$, *see* edge, directed

$\overset{g}{\rightarrow}$, *see* edge, directed

$\overset{g}{-}$, *see* edge, undirected

— , *see* edge, undirected

abductive reasoning, *see* reasoning, abductive

acyclic directed graph, *see* graph, acyclic directed

acyclic, partially directed graph, 241

admissible deviation, 316

AIC, *see* Akaike's Information Criterion

Akaike's Information Criterion, 260, 281

almost equal, 310

alternative hypothesis, 243

ambiguous edge, 252

ambiguous region, 252

An, *see* ancestral set

an, *see* ancestor

ancestor, 19

ancestral set, 19

Appendicitis, 333, 335

Apple Jack, 71, 85

arc, *see* edge

arc reversal, 56, 116

Asia, 73, 291, 297, 304, 331

auxiliary variable, 211, 230

axioms, *see* probability axioms

B

barren variable, *see* variable, barren

Bayes' factor, 58, 308

Bayes' rule, 54

    interpretation of, 55

Bayesian Information Criterion, 260, 281

Bayesian likelihood ratio, 308

Bayesian network

    conditional LG, 75–80

    direct inference approach, 116

    discrete, 71–75

    indirect inference approach, 125

    query, 114

    query based inference, 114

belief theory, 4

BIC, *see* Bayesian Information Criterion

BIC score

    family, 261

Burglary or Earthquake, 25, 30

C

case

    complete, 278

category, *see* variable, category

causal network, 25–32

    converging connection in, 29–30

    diverging connection in, 28

    flow of information in, 25–32

    serial connection in, 26–27

    types of connections in, 26

causal reasoning, *see* reasoning, causal

causality, 6, 24–25, 72, 226

    modeling, 31

U.B. Kjærulff and A.L. Madsen, *Bayesian Networks and Influence Diagrams: A Guide to Construction and Analysis*, ISS 22, DOI 10.1007/978-1-4614-5104-4, © Springer Science+Business Media New York 2013

causally independent, 216
cause variables, 192, 196, 217
causes, 217
certainty factor, 7
ch, *see* child
chain graph, *see* graph, chain
chain rule, 8, 62
    Bayesian networks, 64, 71
    CLG Bayesian networks, 77
    CLQG influence diagrams, 90
    influence diagrams, 81
    object-oriented Bayesian network models,
        100
Chest Clinic, *see* Asia
    junction tree, 121
child, 18
chord, 121
Chow-Liu tree, 266
CLG Bayesian network, *see* Bayesian network,
        conditional LG
CLG distribution, *see* conditional linear
        Gaussian distribution
clique, 118
CLQG influence diagram, *see* influence
        diagram, conditional LQG, 90
collider, 241
combination, *see* probability potential,
        combination
conditional independence, *see* independence
conditional independence and dependence
        relations, 240
conditional linear Gaussian distribution,
        76
conditional linear–quadratic Gaussian
        influence diagram, *see* influence
        diagram, conditional LQG
conditional probability, *see* probability,
        conditional
conflict analysis, 291
    cost-of-omission measure, 292
    local conflict, 295
    partial conflict, 295
    positively correlated assumption, 293
    rare case, 295
conflict measure, 292
conflict meausre
    straw model, 292
constraint, 201
constraint variable, 151, 201
contribution variable, 217
converging connection, *see* causal network,
        converging connection in

cost of omission, 297, 305
    threshold, 298
cycle, 20
    directed, 20

D
d-connection, 33
D-map, 61
d-separation, 33, 64
DAG, *see* graph, acyclic directed
DAG faithfulness assumption, 61, 240
DAG-faithful, 238
data overfitting, 259
de, *see* descendant
decision
    full control, 233
decision future, 84
decision history, 84
decision past, 84
decision variable, 80
    informational parents, 88
decision variables
    partial order, 83
    total order, 85
deductive reasoning, *see* reasoning, deductive
default inhibitor, 218, 219
default prior probability distribution, 97
degrees of freedom, 243–245, 260, 261
dependency map, 61
descendant, 19
df, *see* degrees of freedom
dHugin, 105
diagnostic reasoning, *see* reasoning, diagnostic
directed acyclic graph, *see* graph, acyclic
        directed
directed cycle, *see* cycle, directed
directed global Markov criterion, 33, 35
directed global Markov property, 64
directed graph, *see* graph, directed
Dirichlet probability distribution, 281
dissemination of experience, 284
distance measure, 304
distinguished state, 221
distributive law, *see* probability calculus,
        distributive law of
diverging connection, *see* causal network,
        diverging connection in
division, *see* probability potential, division
        by zero, 50, 55
dom, *see* variable, domain
dynamic Bayesian network, 103

E

edge, 18
    derived direction, 248
    directed, 18
    undirected, 19
effect variable, 192, 196, 217
elicitation of numbers, *see* parameters,
        elicitation of
elicitation of structure, *see* structure, elicitation
        of
elimination order, 119
    strong, 127, 129
EM algorithm
    MAP, 279, 281
    ML, 279
    penalized, 281
entropy, 328
    conditional, 329
    conditional mutual information, 329
    cross, 329
    mutual information, 329
equivalence class, 242
equivalent sample size, 284
equivalent variables, *see* variable, equivalence
essential graph, 241
EU
    expected utility, 129
EUO, 333
event, 40
event space, 40
evidence, 24
    cost of omission, 297
    hard, 24
    impact, 297
    likelihood, *see* evidence, soft
    potential, 49
    soft, 24
    sufficient, 310
    virtual, *see* evidence, soft
expected utility, 80
experience count, 281, 284
experience table, 281
expert system, 3, 111
    normative, 5
explaining away, 8, 31
explaining-away, 18
expressions, *see* parameter, elicitation of

F

factorization
    recursive, *see* probability distribution,
        recursive factorization of

faithfulness, *see* DAG faithfulness assumption
faithfulness assumption, 238
feature variables, 265
filtering, 103
finding
    important, 310
    irrelevant, 310
    redundant, 310
flat network, 100
functional uncertainty, 199
fundamental rule, 42, 54
fuzzy logic, 5

G

gamble-based approach, *see* parameter,
        elicitation of
generalized marginalization operator, 130
global conflict, 295
goodness, 256, 259
graph
    acyclic directed, 20
    chain, 17
    condition, 253
    connected, 20
    directed, 19
    equivalence class, 241
    instances, 96
    moral, 20
    moralization, 118
    skeleton, 20, 241, 247
    strongly connected component, 253
    undirected, 19

H

head, 118
head-to-head, 19
HUGIN algorithm, 123
Hypothesis driven conflict analysis, 291
hypothesis variables, 74

I

I-map, 61
IC algorithm, 246
idioms, 156
independence, 59
    conditional, 59
    represented in DAGs, 60
independence of causal influence, 216
independency map, 61
inference engine, 111

influence diagram, 70, 79
  ∑-max-∑-rule, 128
  conditional LQG, 89–93
  discrete, 80–89
  information sets, 83
  limited memory, 93–95
    minimal reduction, 139
    soluble, 140
  linear-quadratic CG
    ∑-max-∑-rule, 133
  maximum expected utility, 83
  no-forgetting links, 88
  policy, 84
  policy function, 129
  strategy, 83
    optimal, 83
information measures, 327
informational link, *see* link, informational,
    227, 229
inheritance, 101
inhibitor probability, 218
initial assessment, 313
instance, 96
intercausal reasoning, *see* explaining away
interface, 103
interface variables, 96
intermediate variables, 192
intervening action, 80
intervening decision, 85
intervention, 214

J
joint probability distribution, *see* probability
    distribution, joint
junction tree, 120, 121
  COLLECTINFORMATION, 122
  DISTRIBUTEINFORMATION, 122
  propagation, 121
  root, 122

K
kind, *see* variable, kind
Kullback-Leibler, 266

L
law of parsimony, 184
leak probability, 218
likelihood evidence, *see* evidence, soft
LIMID, *see* influence diagram, limited memory

link, *see* edge
  informational, 81
local conflict, 295

M
marginalization, 45, 50
  generalized operator, 130
Markov blanket, 20
Markov criterion, *see* directed global Markov
    criterion
Markovian, 103
maximum expected utility principle, 80
measurement uncertainty, 209
mediating variable, 192, 193, 195, 204
minimal conditioning set, 251
minimal reduction, *see* influence diagram,
    limited memory, minimal reduction
missing data mechanism, 239
Mixture of Gaussian distributions, 222
model verification, 174
modeling technique
  parent divorcing, 169, 192
monotonicity, 294
moral graph, *see* graph, moral
multivariate Gaussian distribution, 76
mutually exclusive states, 71, 76

N
nd, *see* non-descendants
necessary path condition, 251, 252
network class, 96
  default instance, 101
  internal scope, 97
neural network, 15
no-forgetting, 81
node absorption, 213
Noisy-MAX, 220
Noisy-OR, 217
non-descendants, 19
non-intervening action, 80
normalization, *see* probability potential,
    normalization of
normalization constant, 113
normalized likelihood, 56, 295, 307
normative expert system, 70
nuisance variable, *see* variable, nuisance, 213

O
object, 96
Object-oriented probabilistic graphical model,
    70